Springer Tracts in Civil Engineering

Springer Tracts in Civil Engineering (STCE) publishes the latest developments in Civil Engineering - quickly, informally and in top quality. The series scope includes monographs, professional books, graduate textbooks and edited volumes, as well as outstanding PhD theses. Its goal is to cover all the main branches of civil engineering, both theoretical and applied, including:

- Construction and Structural Mechanics
- Building Materials
- Concrete, Steel and Timber Structures
- Geotechnical Engineering
- Earthquake Engineering
- Coastal Engineering; Ocean and Offshore Engineering
- Hydraulics, Hydrology and Water Resources Engineering
- Environmental Engineering and Sustainability
- Structural Health and Monitoring
- Surveying and Geographical Information Systems
- Heating, Ventilation and Air Conditioning (HVAC)
- Transportation and Traffic
- Risk Analysis
- Safety and Security

Indexed by Scopus

To submit a proposal or request further information, please contact:
Pierpaolo Riva at Pierpaolo.Riva@springer.com (Europe and Americas)
Wayne Hu at wayne.hu@springer.com (China)

Eduard Doujak · Sebastian Muntean ·
Daniel Patauner · Elena Vagnoni
Editors

Rethinking Hydropower: A Holistic Vision for a Sustainable Future

Springer

Editors
Eduard Doujak
Institute for Energy Systems
and Thermodynamics
TU Wien
Vienna, Austria

Sebastian Muntean
Center for Fundamental and Advanced
Technical Sciences
Romanian Academy - Timisoara Branch
Timisoara, Romania

Daniel Patauner
Institute for Energy Systems
and Thermodynamics
TU Wien
Vienna, Austria

Elena Vagnoni
PTMH
EPFL
Lausanne, Switzerland

ISSN 2366-259X ISSN 2366-2603 (electronic)
Springer Tracts in Civil Engineering
ISBN 978-3-032-33327-8 ISBN 978-3-032-33328-5 (eBook)
https://doi.org/10.1007/978-3-032-33328-5

This work was supported by European Cooperation in Science and Technology

This Springer imprint is published by the registered company Springer Nature Switzerland AG
The registered company address is: Gewerbestrasse 11, 6330 Cham, Switzerland

Foreword

Sustainability has become the defining condition for hydropower's future—not just today, not just in Europe, but everywhere.

As Europe advances its energy transition, expanding renewable capacity alone is no longer sufficient; solutions must also be environmentally sound, socially accepted, and resilient. Hydropower plays a central role in this context—yet it is also subject to increasing scrutiny.

What distinguishes hydropower from other renewable technologies is its systemic nature. Its multifunctionality and flexibility makes hydropower indispensable for integrating variable renewable energy sources such as wind and solar. But hydropower plants are not isolated assets, they are deeply embedded in river basins, ecosystems, and communities—more than any other renewable technology. Their value goes beyond electricity generation to include water management, flood protection, irrigation, and more. Therefore, it requires carefully balancing technical performance and economic viability with ecological and social considerations.

Over the past decades, the sector has undergone a profound transformation. Historically driven by engineering, technological and economic optimization, hydropower now faces broader societal expectations. Environmental protection, stakeholder engagement, and social acceptance have therefore become central. At the same time, new challenges are emerging. Climate change is reshaping hydrological conditions and increasing uncertainty. The energy transition demands flexibility and storage, while digitalization is transforming plant design and operation. Addressing these interconnected challenges requires a holistic and interdisciplinary approach, as reflected throughout this book.

Europe plays a unique role in this evolution. With a strong industrial base, leading universities and research institutions and comprehensive regulatory frameworks, it has been at the forefront of developing both technology and sustainability concepts and practices. It has helped shape global standards while acting as a laboratory for integrating policy, innovation, and implementation in complex environments. The lessons learned are therefore highly relevant for the industry beyond Europe itself.

A key consequence of this transformation, in conjunction with the mass retirement of the baby boomer generation (and its know-how?), is the growing importance of capacity building. The future of sustainable hydropower depends on a new generation of professionals who can work across disciplines—combining engineering, environmental, social, and digital expertise. This requires new educational approaches, stronger collaboration between academia and industry, and intensified knowledge exchange.

Looking ahead, the European hydropower sector benefits from clear strengths: proven and world-leading technologies, "made-in Europe" independence with a strong industrial basis across the entire value chain, including universities, consultants, civil constructors, and utilities' Operation and Maintenance know-how. However, challenges remain. Regulatory uncertainty, environmental constraints, and public acceptance issues must be addressed to unlock investment. Without decisive action, deployment, especially of pumped storage, could be slowed down, as could the modernization of existing assets.

In this context, the COST Action CA21104, PEN@Hydropower has proven both timely and impactful. By bringing together experts from academia, industry, and policy, it has established a unique platform for collaboration, knowledge exchange, and innovation, particularly in capacity building. The contributions presented in this book are the result of this collective effort. The Training School on Sustainable Hydropower, which deserves special mention, certainly seems to be a model that should be adopted and continued.

This book offers a comprehensive and forward-looking perspective on sustainable hydropower. It is both a reflection of current knowledge and a call to action. The future of hydropower will depend on our ability to embrace sustainability in all its dimensions, and to work together across disciplines and borders.

Vienna, Austria
June 2026

Peter Stettner
Andritz Hydro

The author of this foreword is: a member of the Management Committee of the referenced COST Action, a co-chair of the ETIP Hydropower Governance Board, and a member of the Hydropower Sustainability Council. He works in Market Strategy at ANDRITZ Hydro and has over 30 years of experience. While not an author of this book, he contributes as a long-standing practitioner and stakeholder.

Preface

Hydropower plays an important role in the clean energy transition and is today's backbone of energy storage. However, the field went through major transformations in the last 30 years. Being one of the oldest and most established sources of energy, it helped develop European industrial and societal wealth. However, the challenges of climate change, environmental protection, digital transformation, and societal expectations have fundamentally changed the way hydropower is perceived, designed, and operated today. An even bigger challenge is the transfer of knowledge gained over decades to enable this transformation.

This book provides a comprehensive overview. It explains how hydropower should evolve in the future, specifically addresses the subject of multi-stakeholder principles, and supports knowledge exchange between stakeholders. Starting with an overview of hydropower in the past, it then moves on to the implementation of sustainable hydropower and the role of capacity building in the future. Although knowledge transfer, research funding, multi-stakeholder thinking, and public awareness should be an integral part of any field, they were identified as essential gaps in the hydropower sector. These issues were addressed by COST Action CA21104 *Pan-European Network for Sustainable Hydropower* (PEN@Hydropower), which created a network with around 500 members. The network's strength lies in the fact that more than 50% of its members are young researchers and innovators, which makes it unique in the sector. A key activity that contributed to the large proportion of young people was the Training School on Sustainable Hydropower. Addressing the capacity-building gap and fostering knowledge exchange in the field of sustainable hydropower was of high importance. Sharing and spreading this proven and tested education concept was the main motivation for writing this book, and it is one of the core outputs of PEN@Hydropower.

Multidisciplinary teaching requires basic knowledge of the main sustainable hydropower fields. Topics include rather technical aspects that touch upon the role of flexibility, for example, but also technological evolution through innovation and digitalization. Environmental and climate change impacts, policy measures, holistic aspects, and stakeholder interaction also shape the hydropower community. Bringing all disciplines together creates a framework, where a holistic view can be developed.

Investing in training in sustainable hydropower, as verified by the European hydropower education situation at universities, should become a common interest. Diversified curricula with lower and lower in depth specialization become reality. The major concern at academia should also be the technological leadership as it is requested by the industry. Education, lifelong learning and training on the job will become more and more important, if the entire community would like to hold the leadership position in this field.

The journey should not stop here. The future is full of issues that require solving, and we need young researchers who are trained in holistic thinking and dedicated to finding solutions that cover several disciplines. The future is multidisciplinary and complex, which is precisely why we need such tools and networks.

It begins with an overview of the history of hydropower, followed by an examination of the implementation of sustainable hydropower and capacity building. It highlights the lack of educational opportunities in this field across Europe and provides an example of how sustainable, multi-stakeholder education can be promoted and realized in the future. The book links relevant hydropower topics to stakeholder groups and outlines perspectives for the future.

I would like to express my gratitude to all my colleagues who contributed to this comprehensive reflection on our past four years. This book provides an excellent overview of what dedicated and motivated individuals can achieve. I also wish to acknowledge all the members of each working group for their participation in the numerous meetings and events organized over the years. Their contributions were instrumental in compiling the material that served as the foundation for this book. Thank you all.

Vienna, Austria
June 2026

Eduard Doujak

Chair of COST Action
CA21104—Pan-European Network for
Sustainable Hydropower
(PEN@Hydropower)

Acknowledgements

This publication is based upon work from COST Action CA21104 PEN@Hydropower (Pan-European Network for Sustainable Hydropower, https://www.pen-hydropower.eu/), supported by COST (European Cooperation in Science and Technology).

COST (European Cooperation in Science and Technology) is a funding agency for research and innovation networks. Our Actions help connect research initiatives across Europe and enable scientists to grow their ideas by sharing them with their peers. This boosts their research, career and innovation. www.cost.eu.

cost
EUROPEAN COOPERATION
IN SCIENCE & TECHNOLOGY

Funded by
the European Union

PEN@HYDROPOWER·CA21104·

Contents

Contributors

Ioannis Anagnostopoulos School of Mechanical Engineering, National Technical University of Athens, Zografou, Greece;
School of Engineering, Bahrain Polytechnic, Isa Town, Kingdom of Bahrain

Maria Ubierna Aparicio Open Hydro Ltd., London, UK

Behnam Balouchi Hydropower group, AFRY Norway AS, Oslo, Norway

Mats Billstein R&D Division, Vattenfall AB, Älvkarleby, Sweden

Isabel Boavida Civil Engineering Research and Innovation for Sustainability, Instituto Superior Técnico, Universidade de Lisboa, Lisboa, Portugal

Giovanna Cavazzini Dipartimento di Ingegneria Industriale, Università degli Studi di Padova, Padova PD, Italy

Shaua Fui Chen Department of Interdisciplinary Studies of Culture, Norwegian University of Science and Technology, Trondheim, Norway

Paulo Alexandre Diogo MARE - Marine and Environmental Sciences Centre, NOVA School of Science and Technology, NOVA University Lisbon, Caparica, Portugal

Eduard Doujak Institute for Energy Systems and Thermodynamics, TU Wien, Vienna, Austria

David Christian Finger Department of Engineering, Reykjavik University, Reykjavik, Iceland

Barbara Fischer-Aupperle Global Women's Network for the Energy Transition, Vienna, Austria

Dogan Gezer TUBITAK Marmara Research Center, Ankara, Turkey

Marie-Pierre Gosselin Norwegian Institute for Nature Research (NINA), Trondheim, Norway

Marko Hočevar Faculty of Mechanical Engineering, University of Ljubljana, Ljubljana, Slovenia

Valbona Karapici Faculty of Economy, University of Tirana, Tirana, Albania

Egidijus Kasiulis Vytautas Magnus University, Kaunas, Lithuania

Alain Kilajian Hydropower Sustainability Alliance, Lisboa, Portugal

József Kádár Arava Institute for Environmental Studies, Kibbutz Ketura, Israel

Marija Lazarevikj Faculty of Mechanical Engineering, Ss. Cyril and Methodius University of Skopje, Skopje, North Macedonia

Jian-hua Meng Department of Civil and Environmental Engineering, Konstanz University of Applied Sciences HTWG, Konstanz, Germany

Sebastian Muntean Center for Fundamental and Advanced Technical Research, Romanian Academy - Timişoara Branch, Timişoara, Romania

Iraia Núñez Aguirre Zabala Innovation Consulting, Brussels, Belgium

Daniel Patauner Institute for Energy Systems and Thermodynamics, TU Wien, Vienna, Austria

Epari Ritesh Patro Water, Energy and Environmental Engineering Research Unit, University of Oulu, Oulu, Finland

Emanuele Quaranta European Commission, Joint Research Centre (JRC), Ispra, Italy

Pavel Rudolf Faculty of Mechanical Engineering, V. Kaplan Department of Fluid Engineering, Brno University of Technology, Brno, Czech Republic

Antonios Tourlidakis Department of Mechanical Engineering, University of Western Macedonia, Kozani, Greece

Elena Vagnoni Hydraulic Machinery Platform, École Polytechnique Fédérale de Lausanne, Lausanne, Switzerland

Acronyms

ABA	Administrația Bazinală de Apă
ADB	Asian Development Bank
ADRVT	Regional Development Association of the Tua Valley
aFRR	Automatic Frequency Restoration Reserve
AI	Artificial Intelligence
AMBER	Adaptive Management of Barriers in European Rivers
APA	Portuguese Environment Agency
BEP	Best-Efficiency Point
BESS	Battery Energy Storage System
BEV	Battery Electric Vehicles
BUT	Brno University of Technology
BVB	Bucharest Stock Exchange
CA	COST Action
CA21104	COST Action 21104
CBO	Capacity Building Objective
CEEP	Critical Excess Electricity Production
CEER	Council of European Energy Regulators
CEO	Chief Executive Officer
CET	Clean Energy Transition
CFD	Computational Fluid Dynamics
CG	Core Group
COST	Cooperation in Science and Technology
DDP	Dams and Development Project
DNSH	Do No Significant Harm
DWD	Drinking Water Directive
EC	European Commission
ECI	Early Career Investigators
ECTS	European Credit Transfer and Accumulation System
EDP	Energias de Portugal
EE	Eastern European Area
EEA	European Environmental Agency

EERA	European Energy Research Alliance
EIB	European Investment Bank
EPC	Engineering, Procurement and Construction
EPFL	École Polytechnique Fédérale de Lausanne
ESG	Environmental, Social and Governance
ESMAP	Energy Sector Management Assistance Program of the World Bank
ETIP	European Technology and Innovation Platforms
EU	European Union
EUPHEMIA	EU Pan-European Hybrid Electricity Market Integration Algorithm
FCR	Frequency Containment Reserve
FiPs	Feed-in Premiums
FiTs	Feed-in Tariffs
FP	Framework Program
FSI	Fluid-Structure Interaction
G1	Owners/Operators Stakeholder Group
G2	Financiers/Developers/Insurers Stakeholder Group
G3	Social Stakeholders Group
G4	Environmental Stakeholders Group
G5	Government/Policies Stakeholder Group
GE	General Electric
GHG	Greenhouse Gas
GOODD	Global Georeferenced Database of Dams
GPG	Grant Period Goals
GSM	Global System for Mobile Communication
H2O	Water
HESG	Hydropower Environmental, Social and Governance
HEU	Horizon Europe
HGIIP	Hydropower Sustainability Guidelines on Good International Industry Practice
HPP	Hydropower plant
HPU	Hydropower Unit
HS	Hydropower Sustainability
HSA	Hydropower Sustainability Alliance
HSAF	Hydropower Sustainability Assessment Forum
HSAP	Hydropower Sustainability Assessment Protocol
HSF	Hydropower Sustainability Forum
HSS	Hydropower Sustainability Standard
HVOF	High-Velocity-Oxygen-Fuel
ICT	Information and Communication Technology
IFC	International Finance Corporation
IHA	International Hydropower Association
IPCC	Intergovernmental Panel on Climate Change
IT	Information Technology

ITC	Inclusiveness Target Countries
IUCN	International Union for Conservation of Nature
JP	Joint Program
JRC	Joint Research Centre
LCA	Life Cycle Assessment
LCOE	Levelized Cost of Electricity
LIFE	L'Instrument Financier pour l'Environnement
LoRa	Long Range
LSG	Local Stakeholder Group
LSTM	Long Short-Term Memory (LSTM)
LTU	Lulea University of Technology
MACSE	Mercato a Termine degli Stoccaggi
MC	Management Committee
mFRR	Manual Frequency Restoration Reserve
MoU	Memorandum of Understanding
MS	Member State
NEA	North European Area
NECP	National Energy and Climate Plans
NGO	Non-Governmental Organization
NIB	Nordic Investment Bank
NIS	Network and Information Security
NNC	Near Neighbor Countries
NREAP	National Renewable Energy Action Plan
NTUA	National Technical University of Athens
NZIA	Net-Zero Industry Act
O&M	Operation and Maintenance
OPEX	Operational Expenditures
PEN@Hydropower	Pan-European Network for Sustainable Hydropower
PhD	Doctor of Philosophy
PLC	Programmable Logic Controller
PNEC	National Energy and Climate Plan
PPC	Public Power Corporation
PR	Public Relations
PRS	Pressure Recovery System
PSH	Pumped Storage Hydropower
PSHP	Pumped Storage Hydropower Plant
PV	Photovoltaic
R&D	Research and Development
RCO	Research Coordination Objectives
RE100	Renewable Energy 100 Initiative
REC	Renewable Energy Certificates
RED	Renewable Energy Directive
RES	Renewable Energy Sources
RoCoF	Rate of Change of Frequency
RoR	Run-of-River

SAM	Sahabat Alam Malaysia
SCADA	Supervisory Control and Data Acquisition
SCF	Social Climate Fund
SCI	Site of Community Importance
SCORE	Sarawak Corridor of Renewable Energy
SDAC	Single Day-Ahead Coupling
SDG	Sustainable Development Goals
SEA	Southern European Area
SEB	Sarawak Energy Berhad
SEEA	South East European Area
SHPP	Small Hydropower Plant
SMP	Stakeholder Management Plan
SPS	Storage Pumping Station
STEM	Science, Technology, Engineering and Mathematics
STSM	Short-Term Scientific Mission
SWEA	South West European Area
SWOT	Strengths, Weaknesses, Opportunities and Threats
TDG	Total Dissolved Gas
TS	Training School
TS1	First Training School on Sustainable Hydropower
TS2-1	Second Training School on Sustainable Hydropower
TS2-2	Third Training School on Sustainable Hydropower
TS3	Fourth Training School on Sustainable Hydropower
TS4	Fifth Training School on Sustainable Hydropower
TSC	Technical Screening Criteria
TSO	Transmission System Operator
UN	United Nations
UNCED	United Nations Conference on Environment and Development
UNESCO	United Nations Educational, Scientific and Cultural Organization
UOWM	University of Western Macedonia
UU	Uppsala University
VLH	Very Low Head
VPP	Virtual Power Plants
VRES	Variable Renewable Energy Sources
WCD	World Commission on Dams
WEA	Western European Area
WEFE	Water–Energy–Food–Ecosystems
WESE	Water-Energy-Society-Ecosystem
WFD	Water Framework Directive
WG	Working Group
WG1	Working Group 1—Hydropower Role in Flexible Energy Synergies

WG2	Working Group 2—Technological Evolution Through Innovation and Digitalisation
WG3	Working Group 3—Sustainable Hydropower and Climate Change Adaptation
WG4	Working Group 4—Clean Energy Transition and Policy Measures
WG5	Working Group 5—Holistic Assessment and Stakeholder Interaction
WiFi	Wireless Fidelity
WRS	Water Resilience Strategy
WWF	World Wildlife Fund
WWTP	Wastewater Treatment Plants
YRI	Young Researchers and Innovators

Part I
Introduction

The use of hydropower for energy generation has undergone a fundamental structural transformation over the past few decades. Whereas in the past the focus was primarily on engineering and economic optimization processes, the sector now faces a multidimensional set of requirements. In addition to addressing ongoing technical challenges, ecological and socio-economic side effects are increasingly coming into focus in scientific and practice-oriented debates. The societal discourse and the accompanying pressure to justify these projects lend the demand for ecologically sound, sustainable solutions new epistemological and pragmatic relevance. For the future viability of hydropower projects, well-founded social acceptance has thus become a conditio sine qua non.

To adequately address these complex transformation processes, both new regulatory instruments and transdisciplinary networks have emerged at the international level. A key outcome of this evolutionary development over the past 25 years is the Hydropower Sustainability Standard. This holistic framework now serves as a global reference for the certification of hydropower plants according to defined sustainability criteria. Beyond its function as a pure auditing tool, this framework finds direct application in academic and professional education, particularly within the Training School of the PEN@Hydropower network.

Ensuring sustainable development in academic teaching and training requires innovative and attractive didactic concepts. To effectively counteract the current trend of decreasing student numbers and the reduction of specific courses in the field of university hydropower education, systematic capacity building has been defined as a core strategic goal. The Training School, initiated in this context, is explicitly dedicated to the in-depth training of young researchers and innovators, grounded in the principles of the Hydropower Sustainability Standard.

Based on the educational survey, conducted within the COST action CA21104, in the following chapters, the theoretical base, methodological implementation, and practical experiences within the framework of a COST Action are comprehensively presented and discussed.

This work is divided into four complementary main chapters covering the theoretical and regulatory foundations of sustainability certification, institutional networking, practical implementation, and the promotion of young talent.

The first chapter is devoted to the historical and conceptual genesis of the Hydropower Sustainability Standard. The roots of this set of rules date back to the groundbreaking publications of the World Commission on Dams (WCD). The text reconstructs the evolutionary path that, through multilateral consultation processes involving various stakeholders, led to the establishment of today's certification standard. In addition to the substantive milestones, the analysis focuses in particular on the institutional context and the definition of stakeholder responsibilities.

The second chapter shows an in-depth exploration of COST Action CA21104, commonly referred to as PEN@Hydropower (Pan-European Network for Sustainable Hydropower). Established in 2022 in response to a clear demand for international collaboration, this network acts as a crucial forum for scientific dialogue and cross-disciplinary professional growth concerning the sustainable utilization of hydropower. The chapter delves into the network's governance framework and scrutinizes the strategic initiatives and funding mechanisms accessible to the involved stakeholders.

Chapter three focuses on a core concern inherent in this COST Action: systematic capacity building. The authors do not limit themselves here to a critical assessment of current deficits in university and vocational education. Rather, the far-reaching problem of the demographic and technical shortage of young talent within the entire hydropower community is examined empirically and theoretically to underscore the urgency of proactive countermeasures.

The fourth chapter incorporates the topics from earlier sections and determines how the Hydropower Sustainability Standard is practically applied within the framework of the COST Action. This analysis delves into the systemic connections and mutual interactions among the various participants. The main aim of this chapter is to present a sustainable and replicable model for the educational training and qualification of upcoming professionals and researchers in this scientific field.

Hydropower's Pathway to Sustainability

Barbara Fischer-Aupperle and Jian-hua Meng

Sustainability matters.

Abstract The history of hydropower is a long one and—as compared to other renewables—old, ranging back to the industrial development beyond water mills in the 19th century. The past of this technology the last decades was not always glorious, at least from a stakeholder point of view, including environmental considerations. The description of the change in doing hydropower projects therefore is including the industry development since the WCD (World Commission and Dams) Report over the last 25 years. It triggered a change in industry thinking, made proponents and critics to think and find the solution in a sustainability definition of doing hydropower planning, installation and operation, and consequentially drove the development of measuring of this sustainability—into the Hydropower Sustainability Standard. Procedures and assessments to apply the Standard an prove aspects of sustainability in a multi-stakeholder governing structure, are described in their relating parts with a view on the currently running review of the Standard.

Keywords Hydropower · Water · Mills · Measuring sustainability · Hydropower sustainability standard · Stakeholders · Environment · Industry development · World commission on dams · Governing structure · Sustainability

Hydropower—power from water, which had been generated by mills over centuries, turned into modern plant technology during industrialization. Wooden water wheels were substituted for metal wheels, to connect shafts for the process of milling wheat to

B. Fischer-Aupperle (✉)
Global Women's Network for the Energy Transition, Vienna, Austria
e-mail: barbara.fischer-aupperle@globalwomennet.org

J. Meng
Department of Civil and Environmental Engineering, Konstanz University of Applied Sciences HTWG, Konstanz, Germany
e-mail: jmeng@htwg-konstanz.de

E. Doujak et al. (eds.), *Rethinking Hydropower: A Holistic Vision for a Sustainable Future*, Springer Tracts in Civil Engineering,
https://doi.org/10.1007/978-3-032-33328-5_1

flour or to a jig-saw for cutting wood changed into transmission gears with generating equipment for electricity generation.

Maybe it was due to the "oldness" of this revolution over one hundred years ago that, hydropower was considered to be old fashioned and was not really included in the idea of "regenerative" or renewable power generation, as solar and wind generation developed from single small units to 2–4 MW plants with blades or panels from materials that could be scaled up over decades to the present time. Even though developing countries were searching for their own local sources and started planning for hydropower as the ideal solutions for their rivers, hydropower was not counted as part of new and modern renewable electricity generation. Engineers in operator companies, equipment corporation suppliers and consultancy firms did their jobs, meticulously and in an orderly way, as they had been taught. Over the last 100 years, machines became bigger, and more performant, and increased efficiencies above 90%, a unique achievement compared to any other electricity generation unit in fossil or renewable machinery. Technical issues involving cavitation, regulation, controls, erosion, or lubrication were continuously developed towards improvement and higher safety as well as better adaptability to varying water flows, discharges, and heads.

All of this happened quite "internally" and out of the public eye, but it was the result of silent, inside-oriented engineering work all over the globe. Most equipment manufacturers and consultancy services were headquartered in Europe but did business on all continents. Development banks such as the World Bank, the African Development Bank, and the Inter-American Development Bank contributed financing to support the development of energy independence in countries with sufficient water resources, thus eliminating the need to purchase costly oil and gas.

Due to growing environmental concerns (Club of Rome, 1972) and the use of resources at the expense of the environment and society, the first regulations and conditions for deliveries to operators, exporting companies, and many other players were established to obtain financing for a project or, at least, export credit insurance for their produced equipment.

The move towards considering social impacts grew alongside ever-growing projects, machines sizes, and the number of units reaching incredible figures: 10 m runner diameters per unit. Almost 1000 MW for the same units and projects combined with wide flooding areas, high dams, sluices, and ship-lifting structures collided with resettlement, land loss, compensation issues and environmental impacts, which fueled resistance. NGOs strongly supported the demand that stakeholders be meaningfully consulted in a timely manner during the preparation stage and be sufficiently informed about their stakes to be mitigated or compensated. The latter were very creative in building their public perception and pressuring financing institutions and companies in the "West", even when projects were in Asia, Africa, or South America.

This eventually led to a lengthy discussion in the world of financing, triggering the World Bank's call for a round table on "Dams" in 1998. The round-table brought together lists of projects and their implications, as well as requests to introduce a moratorium on all future new hydroelectric projects. While the industry itself did not realize the impact of this process on their future business, and mostly stayed away from the round-table, the report published without industry participation turned out to be a shock wave for the sector. The World Commission on Dams report sounded the death knell for hydropower worldwide.

1 Wake-Up Call for the Sector

So, when more than 25 years back the World Commission on Dams (WCD) published its report in November 2000, it raised serious concerns about how the environmental and social impacts of large dams should be assessed. The report proposed a framework to guide more responsible and sustainable development. The question remained: How can this be done in practice?

Even though, a quarter of a century later, with absolutely tangible progress made, the debates about hydropower's social and environmental footprint are not ended. While the WCD report is still referenced around the world, the hydropower sector has worked hard to put recommendations and measuring of footprints into practice.

Even though a quarter of a century has passed and tangible progress has been made, debates about the social and environmental footprint of hydropower are still ongoing.

The WCD report is still referenced around the world, and the hydropower sector has worked hard to implement its recommendations and measure its footprints. In that sense, the WCD process, which included a round-table of stakeholders hosted by the World Bank and culminated in a published report, could be considered a wake-up call for the sector at the time.

The dynamic dissemination came as a shock to industry stakeholders who thought they were simply doing excellent engineering work for any hydropower station planned and built around the world. Manufacturing companies, consultants, and operators realized that they had overlooked aspects of hydropower planning and installation that came with starting generation in a plant from social and environmental points of view.

They suddenly had to admit that they had never looked beyond their limited responsibility for a portion of a project portion they had contracted for. Being blamed for doing "bad projects" and accused by NGOs was a very new experience for most of the enterprises involved.

So, what should they do now? Going out of business was not an option. Neither was giving up on the unique character of hydropower as a large renewable source of energy while carbon balancing was in its beginning stages. The question was how to join forces and act collectively with such a distributed value chain that embraces responsibilities of countless players in a project.

The uncertainty of venturing into new territory and establishing practical criteria for "good" or "bad" projects placed the greatest demand on the International Hydropower Association (IHA). Recognized as the main representative of the global hydropower sector, the IHA's primary goal was to promote hydropower as the largest and lowest-carbon-emitting form of energy generation, as well as a truly renewable energy source with unmatched multi-purpose capabilities.

However, the answer to the question of hydropower's true sustainability was a long time coming from the association's first steps. Over the course of more than two decades, it led the association and its members from a newly formed round-table with key stakeholders to the development of a measuring tool that could be named a "standard." Transitioning from initial definitions to practical tools within a holistic framework was akin to the work of a think tank. This could only help assess and rate hydropower projects based on their degree of sustainability, improving their planning and implementation for the benefit of many more stakeholders than before.

Environmental protection studies were the least of the exercises compared to the request to consider social stakeholders and determine how they could be compensated fairly or how impacts could be mitigated and balanced. This included the question of who would fund all these new burdens on planners, authorities, and industry.

2 Sustainability in Hydropower

2.1 General Definition of Sustainability

Looking for a definition of sustainability will provide a number of interpretations. The most fundamental definition was delivered by the so-called Brundtland report in 1987 [1]. Here, sustainable development is defined as

> the ability [of mankind] to make development sustainable to ensure that it meets the needs of the present without compromising the ability of future generations to meet their own needs.
>
> ...
>
> Yet in the end, sustainable development is not a fixed state of harmony, but rather a process of change in which the exploitation of resources, the direction of investments, the orientation of technological development, and institutional change are made consistent with the future as well as with the present.

Later, the conference *"The United Nations Conference on Environment and Development (UNCED), also known as the 'Earth Summit', held in Rio de Janeiro, Brazil, from 3–14 June 1992"* [2], highlighted how different social, economic and environmental factors interdepend and evolve together. These three pillars should be considered equally to provide a sustainable concept that takes into account economic, ecological, and social aspects.

2.2 Why a Global Standard Is Needed

One of the reasons to develop ESG criteria and defined performance levels was the growing confusion among investors, governments, and communities about what "sustainable hydropower" actually means in practice:

- The need to address ESG risks consistently across all project types, sizes, locations and life cycle stages;
- A clear and straightforward system to certify hydropower projects that meet minimum Environmental, Social, and Governance (ESG) performance expectations. This system addresses any confusion about the sustainability of hydropower projects and, crucially, helps unblock green investment and licensing decisions. All of this is put into place by the Hydropower Sustainability Standard (HSS);
- The HSS is the only global certification scheme for hydropower—and at the same time unique across the entire renewables sector;
- The International Hydropower Association's (IHA) San José Declaration (24 September 2021 at the conclusion of the IHA World Hydropower Congress) states that "the only acceptable hydropower is sustainable hydropower". So, the Standard defines the criteria and the practical steps for good and best international practice.

2.3 Why Certification Matters—The Hydropower Sustainability (HS) Certification System

From the perspective of framework and governance: The certification against the HSS demonstrates a hydropower project's communicable compliance with international sustainability practices.

Also, a hydropower project that achieves certification gains a credible, independently verified, multi-stakeholder governed ESG performance report card.

Certified projects demonstrate alignment with many internationally relevant frameworks and rule sets, including the Climate Bonds Initiative, the EU Taxonomy, electricity sales to RE100 companies, and World Bank and IFC performance standards—the latter being a prerequisite for many development bank loans.

Beyond finance, the assessment process itself has a risk management value comparable to a health check: Project risks, strengths and weaknesses become visible and addressable before significant or irreversible investment decisions are made. This reduces the likelihood of costly delays, permit refusals, or loss of the social license to operate.

From a business perspective: Sustainability performance is becoming an increasingly importants factor in market competition. Owners and operators of hydropower assets are recognizing the HSS Certification as a quality attribute and a source of competitive advantage.

Wins for all project development and operations phases: In short, certification simultaneously reduces project risks, opens financing opportunities, and provides a powerful capacity-building and communication tool. This tool has demonstrated its value across continents, climatic zones, and project sizes ranging from very small (kW scale) to very large (>3000 MW), including storage, run-of-river, pumped storage, and single- and multi-purpose projects.

HydroSelect, the hydropower sustainability early-stage decision-making tool, extends this ecosystem upstream. By embedding ESG considerations from the outset of project identification, HydroSelect prevents costly delays, community conflicts, and environmental impacts before any significant investment commitment is made. This makes HydroSelect the natural first step before a full HSS certification process.

2.4 Governing Body: The Hydropower Sustainability Alliance

In 2023 the HS Alliance was established 2023 as an independent and multi-stakeholder that sets standards and oversees the HS Certification System.

It is structured to fulfill its task. The HS Council has four chambers—Industry, Governments, Civil Society and Financial Institutions—which elect representatives to the HS Governance Committee. The HS Secretariat manages day-to-day operations.

The Council ensures multi-stakeholder input and confidence in the content quality, relevance and assurance of the Hydropower Sustainability Certification System.

3 History of Standard Development: From the HSAF to Today

3.1 IHA and Early Sustainability Thinking (1995–2006)

The International Hydropower Association (IHA) was founded under the auspices of UNESCO in 1995 as a professional body to represent and advance the hydropower sector at a time of intense environmental and social scrutiny.

In 2000, the landmark report by the World Commission on Dams (WCD) challenged existing practices and initiated a shift in the planning and development of hydropower towards sustainability and the rights of affected communities (see introductory part).

The UN Environment Programme launched the Dams and Development Project (DDP) as a multi-stakeholder WCD follow-up while IHA began work on its Sustainability Guidelines in 2004, drawing on WCD Strategic Priorities, World Bank Safeguard Policies, IFC Performance Standards, and the Equator Principles.

Two years later, in 2006, IHA released an early assessment protocol which was developed to evaluate performance against the 2004 guidelines.

3.2 The Hydropower Sustainability Assessment Forum (HSAF) and the HSAP (2007–2010)

This led to a renewed international dialogue, and the subsequent initiation of a new, multi-stakeholder round-table: the Hydropower Sustainability Forum (HSAF); IHA provided a secretariat, with financial support from industry, development banks and governmental agencies.

Forum members and observers included representatives of environmental NGOs (WWF, The Nature Conservancy), social NGOs (Oxfam, Transparency International), development banks, commercial banks of the equator principles banks group, governmental agencies from China, Zambia, Iceland, Norway, Germany, and owners, operators, suppliers and consultants from the hydropower sector.

After extensive consultation and trialing in 16 countries across six continents; the final version was published in 2010 and formally launched at the IHA World Congress in 2011.

3.3 The HSAP in Practice and Evolution (2010–2020)

The HSAP replaced uncertainty and confusion about "sustainability in hydropower" with a globally applicable. Sustainability performance became objectively measurable and internationally comparable, and project risks, strengths and weaknesses could be systematically identified and addressed before costly and irreversible decisions were made.

The HSAP was applied in over 25 countries to projects at all life cycle stages; the World Bank assessed it as a useful tool for guiding sustainable development in developing countries; major commercial banks referenced it in their sector guidance.

The HSAP was updated in 2018 to include a topic on climate change resilience and mitigation. Simultaneously, the HESG Gap Analysis Tool and the HGIIP (Hydropower Sustainability Guidelines on Good International Industry Practice) were introduced, and both re-aligned with World Bank and IFC standards.

3.4 *From Tools to a Global Standard (2020–2026 and Onward)*

A 90-day global consultation commenced in November 2020, and aimed at elevating the Hydropower Sustainability Tools to a global standard. The outcome led to the Hydropower Sustainability Standard (HSS), formally launched by the Hydropower Sustainability Council in September 2021. The HSS is now an independent not-for-profit entity based in Lisbon, Portugal.

IHA simultaneously adopted a new Charter for Sustainable Hydropower on its 25th anniversary, signaling that all members were expected to certify new projects under the forthcoming Standard, complemented by the San José Declaration stating that "only sustainable hydropower is acceptable hydropower" in September 2021. In May 2026 (as of the writing of this text), a formal review process has been launched with a global consultation. The objective is to ensure that the HSS ensures it remains relevant, clear and aligned with global good practice as expectations continue to evolve.

4 HSS Overview, Structure, Content and Knowledge Base

4.1 *The Key Documents of the HS Certification System*

The HS Standard (HSS v1.2, Table 1) is the central document in a broader ecosystem:

Table 1 Documents, purpose and primary users

Document	Purpose	Primary users
HS standard	Theory of change, scope, performance requirements, certification process	All stakeholders
HS assurance system	All procedural elements: roles, certification steps, assessor accreditation, appeals, claims, governance, monitoring and evaluation	Organizations and practitioners
Reporting templates	Assessment report templates	Project proponents and accredited assessors
Guidance documents	Good practice guidelines (HGIIP) and topic-specific how-to guides	Project proponents and practitioners
Training manuals	Accredited assessor manuals and further training materials	Accredited assessors

4.2 Roles and Responsibilities

- **HS Secretariat**: manages day-to-day operations of the certification system; accredits assessors; publishes reports; manages the public comment period; proposes certification status to the Governance Committee; administers fees, renewals, conflicts of interest, and the appeals mechanism;
- **HS Council/Governance Committee**: the multi-stakeholder governing body; reviews and approves changes to the Standard; undertakes a formal five-yearly review; supervises appeals; ensures the integrity of the entire system;
- **Accredited Assessors**: independently trained and licensed individuals who conduct assessments against the Standard; can only operate with a valid accreditation; must avoid conflicts of interest; bear responsibility for assessment findings and results; maintain an evidence repository throughout the certification period;
- **Project Proponents**: developers, owners or operators who commission and fund the assessment; bear full responsibility for costs; must submit all required evidence and cooperate fully with the assessor.

4.3 Scope and Eligibility

- All and any hydropower types: run-of-river, storage, pumped storage, multipurpose, globally applicable regardless of size or location;
- Three life cycle stages: Preparation, Implementation, Operation (including major refurbishments)
- Harmonization pathway: a recent HSAP or HESG assessment can be carried forward into an HSS certification application.

4.4 Theory of Change

- Three intervention spheres—Policy, Promotion, Practice (the "3-Ps");
- Desired outcomes: increased uptake of the HS Standard; greater transparency and inclusivity in the certification system; increased knowledge of and trust in sustainable hydropower;
- Four long-term impact areas: Healthy Ecosystems, Resilient Infrastructure, Prosperous Communities, and Good Governance;
- Sustainability framed as a continuous journey: good practice (meeting minimum requirements) towards best practice (meeting advanced requirements). Important principle: ongoing improvement over the project life cycle.

4.5 Performance Requirements—The 12 ESG Topics

Each topic is structured around a principle statement, a scope definition, and two tiers of criteria: minimum requirements (good practice, mandatory for certification) and advanced requirements (best practice, determining Silver or Gold status).

- Labor and working conditions
- Water quality and sediments
- Community impacts and infrastructure safety
- Resettlement
- Biodiversity and invasive species
- Indigenous peoples
- Cultural heritage
- Governance and procurement
- Communications and consultation
- Hydrological resource
- Climate change mitigation and resilience.

4.6 Certification Process

Four steps:

1. **Assessment**—conducted by an Accredited Assessor against all relevant topics; any assessment is a snapshot in time and is designed for repeat application across life cycle stages; gaps against minimum requirements are flagged and advanced requirements met are scored
2. **Publication and Finalization**—the Preliminary Assessment Report is published on the HSA website for a public comment period; the assessor must respond to every comment and determine whether to amend the report; the Final Assessment Report is accepted by the HS Secretariat
3. **Application**—the Project Proponent submits the HS Certification Application Form; only possible after completing Steps 1 and 2
4. **Certification**– the HS Secretariat notifies the Governance Committee; certification is issued if no objection is raised; the Final Report and certification rating are published on the HSA website.

4.7 The HSS Knowledge Base: How-to Guides and Supporting Tools

- The How-to Guides provide practical steps to for improving sustainability of hydropower development and operation, in alignment with the Hydropower Sustainability Standard
- Guides published to date cover all 12 Standard topics, including: Environmental and Social Assessment and Management; Labor and Working Conditions; Infrastructure Safety; Indigenous Peoples; Biodiversity and Invasive Species; Downstream Flow Regimes; Resettlement; Benefit Sharing; Communications and Consultation; Erosion and Sedimentation; Climate Change Resilience; and Climate Change Mitigation (latest, November 2025)
- Complementary resources: the HGIIP (26 good-practice guidelines aligned with World Bank, IFC and Equator Principles); the Hydropower Sustainability Training Academy (delivered globally, including with World Bank ESMAP support); and the full library of published HSAP and HSS assessment reports on the HSA website
- Finance and policy linkages: the Standard is recognized by the Climate Bonds Initiative's Hydropower Criteria, supports electricity sales to RE100 companies, and is aligned with the EU Taxonomy for Sustainable Investment and World Bank/IFC performance standards.

4.8 HydroSelect—The Hydropower Sustainability Early-Stage Decision-Making Tool

- HydroSelect is the Hydropower Sustainability Early-Stage Decision-making Tool
- Developed to guide developers, governments, financial institutions, and civil society toward more sustainable hydropower project decisions by providing a clear, objective framework for evaluating potential schemes at the very outset
- It covers the same 12 ESG risk areas as the HS Standard and it sits upstream of the HSAP/HESG tools and the full HSS certification process—serving as the entry-level screening instrument that embeds sustainability considerations from the initial stages of project conception.

5 Stakeholders in Sustainable Hydropower

In examining the definition of sustainability in relation to hydropower, it becomes evident that sustainable hydropower necessitates the involvement of a diverse array of stakeholders. The hydropower sustainability standard underscores the need for a multi-stakeholder body to oversee the entire process. This body, known as the

Hydropower Sustainability Council, comprises representatives from civil society, industry, government, and the financial sector. The inclusion of these diverse stakeholder groups ensures that all perspectives are equitably considered, collectively shaping the future of this standard.

The selection of the various chambers within the hydropower sustainability standard ensures the appropriate allocation of responsibilities and fulfills several key functions. Firstly, democratic representation is essential, with chamber members elected to represent specific stakeholder groups. Secondly, the input from different stakeholders should be systematically integrated into the standard. Groups such as NGOs, governments, financial institutions, and industry must be included to represent a broad spectrum of opinions and contributions. Thirdly, the chambers serve as sources of expertise, contributing specialized knowledge and perspectives. The involvement of specific invited groups plays a crucial role in the ongoing development of the standard. Lastly, the chambers are responsible for the continuous review and discussion of standard topics and further development.

Currently, the Hydropower Sustainability Council is comprised of four chambers. The first chamber, representing civil society and NGOs, focuses on social and environmental issues, including local communities and various hydropower stakeholders. The second chamber represents the industry, aiming to enhance ESG performance within companies. The third chamber incorporates financial institutions to facilitate the implementation of the standard in funding models, thereby safeguarding investments and promoting investment security. The fourth chamber is dedicated to government, ensuring that the hydropower sustainability standard is integrated into national policies and regulations. Topics addressed here include fair livelihoods, access to clean water, and clean energy.

Competing Interests The authors have no conflicts of interest to declare that are relevant to the content of this chapter.

References

1. Brundtland GH (1987) Our common future. In: Report of the world commission on environment and development. United Nations. https://www.brundtland.co.za/wp-content/uploads/2022/08/Brundtland-Report-1987-Our-Common-Future.pdf
2. United Nations Conference on Environment and Development, Rio de Janeiro, Brazil, 3–14 June 1992. United Nations (1992). https://www.un.org/en/conferences/environment/rio1992, http://www.rsc.org/dose/titleofsubordinatedocument. Cited 15 Jan 1999

The COST Action CA21104 Network

Eduard Doujak

There is no cooperation without network.

Abstract COST Actions are primarily projects aimed at knowledge exchange and networking. They are initiated either by individuals or existing communities and proposed for funding. This process took place in 2021 and led to the current COST Action CA21104, the Pan-European Network for Sustainable Hydropower. The motivation for establishing this network was to identify future research priorities by leveraging existing expertise. Another key point was the training of Young Researchers and Innovators for future tasks in this field. To organize all these tasks, a group of leaders was established. This core group is responsible for organizing and carrying out the activities. The following chapter outlines the structures and activities of the present COST Action.

Keywords Hydropower · COST · PEN@Hydropower

COST stands for Cooperation in Science and Technology [1] and is one of the oldest funding schemes within the EU. Dating back to the 1970s of the last century this program has been running for more than 50 years now. It is open to all disciplines and themes. The proposed topic must be new and challenging for a large community. COST funds networking and knowledge exchange activities. No funding is given for research projects except Short Term Scientific missions, which is a combination of research exchange activities and capacity building. The expected outputs are a better understanding of the proposed topic and network for future activities.

Hydropower has a long history, and it is currently experiencing a renaissance. Several European associations support this by addressing different areas within the hydropower community. One of these larger associations is the European Energy

E. Doujak (✉)
Institute for Energy Systems and Thermodynamics, TU Wien, Vienna, Austria
e-mail: eduard.doujak@tuwien.ac.at

E. Doujak et al. (eds.), *Rethinking Hydropower: A Holistic Vision for a Sustainable Future*, Springer Tracts in Civil Engineering,
https://doi.org/10.1007/978-3-032-33328-5_2

Research Alliance (EERA), which established the Joint Program Hydropower (JP Hydropower) in 2019 [2]. In 2020, the management board discussed applying for a COST proposal, and some colleagues began developing it.

1 Origins

The COST Action 21104 PEN@Hydropower (Pan-European Network for Sustainable Hydropower) originated from a group of enthusiasts who recognized an urgent need for change. National and European funding for hydropower research had been declining, largely due to a 2020 trend where wind and solar began to dominate the research landscape.

There was a clear need for increased knowledge exchange, a network to address funding issues, and a focus on capacity building, as many scientists involved in initial applications were nearing the end of their careers. This created an ecosystem where a larger network was necessary to exchange ideas and best practices.

A group of 52 primary and secondary proposers joined forces to submit a proposal, which was selected for funding in 2022. The official kick-off took place on September 14, 2022, in Brussels. This group established a leadership structure for the coming years, which included electing a Chair and Vice-chair.

The network is organized into five working groups, each led by a leader and a mentor. Additionally, special coordinators were appointed for communication, grant awarding, and capacity building. Each leadership position was filled according to COST considerations, ensuring the involvement of Young Researchers and Innovators (YRI), representatives from Inclusiveness Target Countries (ITC), and a balance of gender and geographical distribution.

This core group is responsible for leading the working groups to achieve the objectives of COST Action 21104. This includes organizing events, meetings, and conferences, as well as facilitating short-term scientific missions and conference grants for young researchers.

During the project period (2022–2026), three positions in the core group changed. The leadership of working groups 2 and 5 changed, and a gender equality coordinator was appointed. Additionally, each working group appointed task leaders according to their needs (Fig. 1).

1.1 Main Objectives and Goals

The COST Action CA 21104 delineates two primary objectives: research coordination objectives (RCO) and capacity building objectives (CBO). Within the RCOs, the Pan-European network for sustainable hydropower aims to contribute to the digitalization of hydropower in the context of the clean energy transition (CET) and to establish a cohesive network of researchers, engineers, scholars, and other stakehold-

Fig. 1 Core group of CA21104

ers, including representatives from companies, governments, and civil society. The objective is to foster close collaboration among various European research groups and to support the submission of research project proposals.

The principal research coordination objectives are associated with the working groups: RCO1 of WG1 seeks to evaluate and emphasize the evolving role of hydropower and pumped hydro storage (PSH) within the energy system sector, considering perspectives from the near to long-term future, specifically from 2030 to 2050. This involves addressing the flexibility and storage requirements in the future renewable energy landscape and considering the water-energy-food nexus through interdisciplinary events to achieve the objective.

The second objective (RCO2) aims to support the scientific framework for hydropower producers and investors to enhance the efficiency and performance of both existing and new hydropower plants. This will encompass ongoing technological innovations from past design and operation stages, culminating in retrofitting strategies. Concepts such as digitalization and predictive monitoring are pivotal in achieving the research objectives of WG2.

For WG3, RCO3 involves establishing a collaborative platform for scientists and stakeholders from diverse disciplines to conduct a holistic assessment supporting the sustainable development of hydropower plants. Key topics include resilient infrastructure needs, environmental and societal conditions, and climate change adaptation and mitigation measures for the coming decades, which are crucial for knowledge exchange within this group.

Regarding WG4, the research coordination objective (RCO4) involves mapping the current European Union legislative and market frameworks. In CET scenarios, it is essential to identify policy gaps and barriers to achieving an optimal balance between hydropower production and environmental impact. Broad stakeholder interaction and discussions are vital to improving the current situation.

The research coordination objectives (RCO5) for WG5 are based on developing a comprehensive scientific hydropower community strategy to harmonize the RCOs of WG1 to WG4. A deeper understanding of production, environmental impact, and social acceptance, with all associated controversies, needs to be investigated and integrated into a unique knowledge base.

In addressing the capacity building objectives (CBO), the framework can be categorized into four primary topics.

The first objective (CBO1) aims to expand the existing network of technical sciences by incorporating additional disciplines. This expansion can be facilitated through extensive stakeholder engagement and various activities, such as thematic workshops and international conferences.

The second objective (CBO2) addresses a significant challenge within the hydropower community, which is to promote Young Researchers and Innovators (YRIs) to leadership positions while ensuring gender balance. This will be achieved by appointing selected individuals to the management committee, core group, and other activities within the COST action. A dedicated mentoring program will be established to support female YRIs.

Closely related to CBO2 is the third objective, which focuses on fostering the career development of YRIs through specialized knowledge transfer, including training schools and other initiatives. Activities such as short-term scientific missions and conference grants will be integral components of this third capacity building objective (CBO3).

All these objectives require a shared foundation to gain recognition and create awareness. Specific dissemination activities must be developed to ensure proper implementation within this COST action. Social media campaigns will aid in fostering a broader understanding and awareness of the challenges faced by YRIs in the hydropower sector and their integration into the community (CBO4).

These overarching objectives of the COST action CA 2104 must be established and integrated into the framework of such cooperative networks and the grant system. As grants are approved annually, the main objectives must be divided into grant period goals. Consequently, each working group must subdivide their overarching objective into four grant period goals, which should be evenly distributed over the four granting periods.

2 How the Network Works

The newly created network establishes a yearly work and budget plan to guide the network's structure (MC, CG, and special coordinators). Each year, actions must be defined and announced. All activities and objectives are based on the memorandum of understanding (MoU), which every member must accept. Membership growth during the project period is based on these rules, and new members must apply by declaring their research background, interests, and contributions to the COST action, as well as their objectives and actions. Each application must be approved by the Action Chair on behalf of the MC. Membership development between 2022 and 2026 is shown in Fig. 2.

In order to fulfill the objectives and goals of CA21104, several actions are defined within the annotated rules. Those can be:

- Meetings (physical or online)
- Webinars
- Training schools
- Short term scientific missions
- Conference grants
- Dissemination actions (papers, social media, books, etc.)

Those network actions have to be embedded into the WGs yearly grant period goals (GPGs) in order to support the entire action and its paramount objectives. Periodic reports monitor the successful implementation and output of those events and actions. Working group members contribute to these goals with their active participation at the events. Of course, CBOs play an integrative role, as incorporating YRIs into the hydropower field is of general interest.

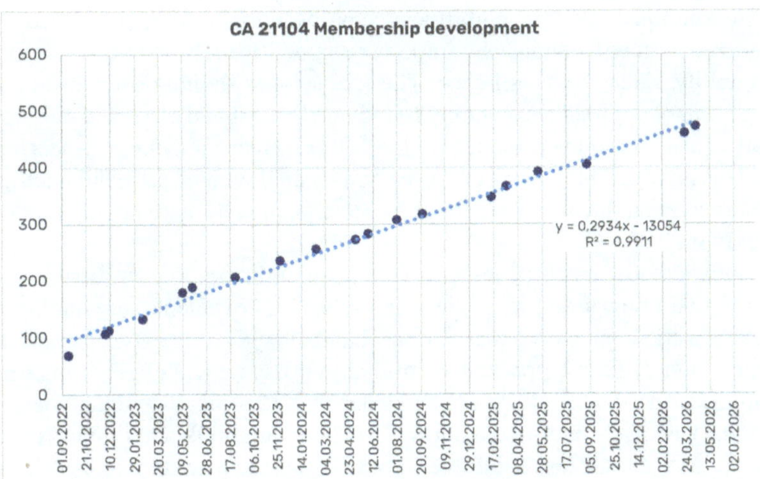

Fig. 2 CA21104 membership development from 2022 to 2026

Within this framework of possible actions, yearly grants, a growing network of experts, and common goals defined by the MoU [3], installing Working Groups (WGs) and Special Coordinators is an important task. CA21104 is organized into five WGs. All of them are related to the aforementioned RCOs and contribute to the CBOs.

2.1 WG 1–Hydropower Role in Flexible Energy Synergies

The Working Group primarily focuses on the conditions necessary for utilizing hydropower within flexible systems. The group prioritizes analyzing assumptions and trends related to the collection of electricity from various water sources and conditions, as well as investigating innovative systems and monitoring methods.

Key activities will involve: discussing the importance of hydropower use, emphasizing its adaptability, storage capabilities, and reliability during COST meetings; incorporating local stakeholder groups (LSGs) that include academic, industrial, and governmental participants in the electrical sector of hydropower use to pinpoint potential weaknesses in hydropower expansion; and engaging in dialogue with LSGs about the conditions required to establish hydropower as a crucial element of clean and secure electricity for the future.

2.2 WG 2–Technological Evolution Through Innovation and Digitalization

This working group examines the technological advancements needed to develop innovative solutions to address ambitious scenarios. Furthermore, digitalization must be considered, as it will enhance hydropower competitiveness and improve grid flexibility, stability, safety, and resilience. These objectives should be achieved through multidisciplinary approaches, encompassing mechanical and electrical engineering, information and communication technology (ICT), environmental engineering, and trading. This involves integrating extensive sensor deployment, artificial intelligence, big data processing, the development of forecasting models, and the implementation of digital twins.

Activities include: establishing pan-European collaboration on disruptive technologies in hydropower and strategies for further improvements; organizing multidisciplinary workshops with experts in mechanical engineering (turbine and equipment innovations, lifetime predictions, digital twin), electrical engineering (generator, smart grid, distributed energy sources), business (trading, blockchain, workforce management), ecology and sustainability (weather variability and climate changes), and IT (big data, clouds, artificial intelligence, cybersecurity) to address all aspects of digitalization and their interconnections.

2.3 WG 3–Sustainable Hydropower and Its Adaptation to Climate Change

Cross-disciplinary collaboration is essential to advancing sustainable development in the hydropower sector, accounting for industrial, environmental, and societal needs in a changing climate. To fully harness the potential of hydropower within a flexible, modern energy system, it is imperative to integrate new technological advancements with cutting-edge climate projections and environmental and societal considerations to develop planning and implementation tools for effective legislation and governance.

The activities include organizing workshops, summer schools, and webinars that focus on examining environmental impacts at the local level, taking into account regional conditions and their susceptibility to climate change.

2.4 WG 4–Clean Energy Transition and Policy Measures

The focus of this working group is to identify and comprehend significant trends in the restructuring of the power sector over the forthcoming decades. In this endeavor, interconnections among sectors are identified, and the evolving role of hydropower evaluated. EU frameworks and directives, CET, and prevailing trends inform this evaluation.

Activities encompass discussions on energy transition and the facilitation of cooperation among energy sectors, including universities and companies, during conferences, meetings, and summer schools, engaging all pertinent branches of science. The focus of interaction with equipment and energy producers, the EU Commission and agencies, EERA, organizations, NGOs, regulators, and others is the energy transition.

2.5 WG 5–Holistic Assessment and Stakeholder Interaction

The urgent need for a thorough assessment of hydropower, coupled with enhanced stakeholder engagement, has been acknowledged. Comprehensively, the Action refers to the consideration of all dimensions of hydropower, encompassing not only technical aspects, such as extending its operational lifespan, but also environmental considerations, such as biodiversity and local climate change, as well as societal acceptance by local communities and financial components, such as investment returns and reasonable profitability. Policy recommendations are informed by PEN@Hydropower's best practices, future scenarios, and initiatives to engage and communicate with stakeholders.

Activities include incorporating the Working Groups' outcomes into policy strategies and roadmaps. This involves collaboration among partners and stakeholders, including European, national, regional, and local hydropower operators.

2.6 Special Coordinators

In addition to the five working groups, some special coordinators handle specific tasks. These are the Science Communication Coordinator, the Grant Awarding Coordinator, the Capacity Building Coordinator, and the Gender Equality Coordinator. They support the Chair, Vice-Chair, and the five working groups by handling overarching tasks to achieve the general objectives. In particular, the capacity-building goals and raising awareness of sustainable hydropower are in their hands.

The Chair functions as the leader of the entire Core Group, with support from the Vice-Chair. Together, they oversee the network and execute the decisions of the Management Committee (MC). As specified in the overarching objectives and goals, the inclusion of young researchers and innovators is of paramount importance. Consequently, throughout the project, concerted efforts were made to encourage as many young researchers and innovators as possible to participate in this field of research.

3 Engaging Younger Generations

Hydropower is undergoing a major transformation in the areas of operation and maintenance, driven by the increasing need for flexibility within modern energy systems. This shift is largely caused by the growing share of renewable energy sources such as wind and solar power in the overall energy mix. As a result, hydropower systems must operate more dynamically to support grid stability and ensure a reliable energy supply while minimizing interruptions. These evolving operational and maintenance challenges, together with the integration of additional renewable energy sources, highlight the need for extensive future research and the development of expertise among the next generation of researchers.

To achieve these objectives, it is essential to expand the existing technical network by incorporating additional disciplines, such as computer science, environmental studies and social sciences, through a broad range of stakeholder activities. Furthermore, it is crucial to involve young researchers and innovators in leadership roles, considering gender aspects as well as national and geographical diversity. These young researchers require support for career development through participation in training schools, joint PhD programs, or short-term scientific missions.

Incorporating young researchers and innovators (YRIs) is not only a priority of the COST Action but should also be supported by raising awareness among policymakers and industry stakeholders. Additionally, an educational survey among MC

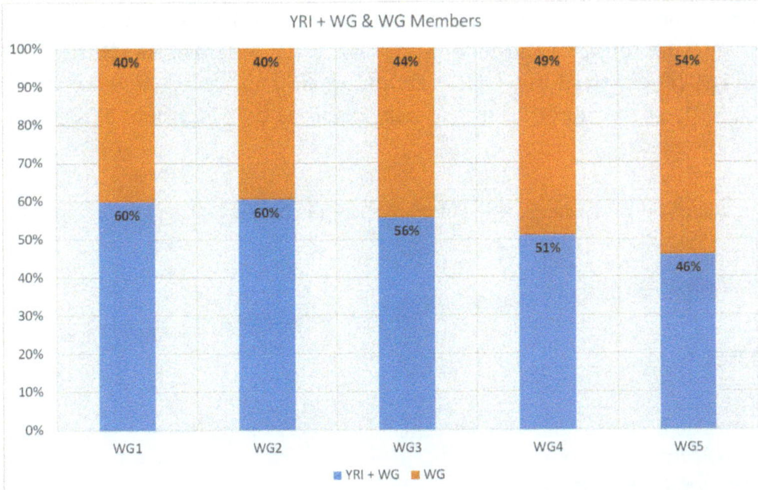

Fig. 3 CA21104 YRI distribution in 2026

member states revealed a varied distribution of ongoing hydropower education across European countries, with notable differences in the quality of courses offered by different universities across Europe. The results of this educational survey can be found in the chapter on the status of European hydropower education (see Part II Chapter "Status Quo of European Hydropower Education").

Figure 3 shows the distribution of YRIs at the five working groups. Note that in every working group the percentage of YRI is higher than 50%, which makes this action and network outstanding.

4 Supporting a Holistic Perspective

The COST Action for Sustainable Hydropower was deliberately designed to take a holistic perspective. Holism has long been recognized as the philosophy of wholeness. The word "holos" originates from ancient Greek and conveys the idea that systems are more than the sum of their parts. Accordingly, science should also move away from silo thinking and take a more holistic view of the challenges of the future. The collaboration of the five working groups, involving all relevant stakeholders, is intended to reflect this. This fundamental concept will be practiced within the framework of the Training School, where the Hydropower Sustainability Standard will be applied across various stakeholder groups. Here, stereotypical perspectives are to be broken down, and each participant's horizons are holistically expanded.

Young researchers, in particular, are to benefit from this perspective and adopt and carry forward the fundamental idea in their professional environments. This would achieve a key objective of this COST action: raising awareness that complex issues such as hydropower cannot be holistically resolved by individual experts but only by groups of experts.

Competing Interests The author has no conflicts of interest to declare that are relevant to the content of this chapter.

References

1. European Cooperation in Science and Technology (COST). https://www.cost.eu/
2. European Energy Research Alliance (EERA)—Joint Programme Hydropower (JP Hydropower). https://www.eera-hydropower.eu/
3. European Cooperation in Science and Technology (COST), Action CA21104—Pan-European Network for Sustainable Hydropower (PEN@Hydropower), Memorandum of Understanding (MoU). https://e-services.cost.eu/files/domain_files/CA/Action_CA21104/mou/CA21104-e.pdf

Capacity Building

Marko Hočevar, Barbara Fischer-Aupperle, Giovanna Cavazzini, and Eduard Doujak

Educating the future generation is a valuable investment.

Abstract Hydropower is an old technical discipline, and one might expect that capacity building measures have already been carefully developed and fully refined over many decades. However, this assumption is not correct, as the environment in which hydropower operates has significantly changed in the last decades. The sector has experienced important shifts, including increased regulatory scrutiny of projects, changes in education systems that favor a more general approach instead of detailed technical design, and a transition from providing primarily baseload power to offering flexibility to the energy system. To adapt to these continuous changes, capacity building in hydropower must also become more flexible and responsive, strengthening the technical, organizational and institutional capacities of stakeholders. Examples of such initiatives include the PEN@Hydropower training schools on sustainable hydropower, as well as targeted programs such as mentoring initiatives for women.

Keywords Hydropower · Education · Capacity building · Skills · Gender balance

M. Hočevar (✉)
Faculty of Mechanical Engineering, University of Ljubljana, Ljubljana, Slovenia
e-mail: marko.hocevar@fs.uni-lj.si

B. Fischer-Aupperle
Global Women's Network for the Energy Transition, Vienna, Austria
e-mail: barbara.fischer-aupperle@globalwomennet.org

G. Cavazzini
Dipartimento di Ingegneria Industriale, Università degli Studi di Padova, Padova PD, Italy
e-mail: giovanna.cavazzini@unipd.it

E. Doujak
Institute for Energy Systems and Thermodynamics, TU Wien, Vienna, Austria
e-mail: eduard.doujak@tuwien.ac.at

E. Doujak et al. (eds.), *Rethinking Hydropower: A Holistic Vision for a Sustainable Future*, Springer Tracts in Civil Engineering,
https://doi.org/10.1007/978-3-032-33328-5_3

1 Why Is Capacity Building for Hydropower Needed?

Hydropower is complex, capital-intensive, and increasingly challenged by environmental, social, and climate pressures [3]. Adequate human and institutional capacity is needed to help overcome pressures such as scientific and technical complexity, energy transition and grid integration and economic and financial intensity, to adapt to climate and hydrology changes, to improve environmental and social aspects of sustainability, to strengthen institutions and governance, knowledge transfer and the hiring of a young workforce, education and gender balance, etc.

Hydropower is inherently a scientifically and technically complex sector, requiring specialized knowledge across multiple engineering disciplines, including civil, mechanical, and electrical systems [4]. The ongoing energy transition, together with the increasing demand for flexibility and resilience, has pushed the entire sector towards strong technological innovation. However, the scientific publications on specific competences required to tackle this complexity are few and the authors underscore the need for skilled technical capacity to implement and manage innovations. The design of the new generation of hydropower plants (dams, spillways, turbines, generators etc. and control systems) demands multidisciplinary scientific knowledge, rigorous safety analysis, and thorough understanding of site-specific hydrological and geological conditions. Operational management adds further complexity.

Capacity building in this area focuses on educating and training skilled engineers and technicians capable of designing, operating, and maintaining hydropower infrastructure safely and effectively. Capacity building also includes training in advanced diagnostic tools, monitoring technologies, and maintenance strategies to reduce failures, extend asset lifetimes, and optimize energy generation and environmental interference.

Hydropower plays a pivotal role in supporting the energy transition by providing flexible, controllable generation that complements variable renewable sources such as wind and solar [1]. This also includes the development of operational and maintenance strategies for pumped storage and hybrid systems, enabling hydropower to act as a buffer for renewable intermittency. Effective integration into modern electricity grids hence requires specialized knowledge of power system dynamics, market operations, and ancillary services, including frequency regulation, spinning reserves, and load balancing. Capacity building in this area focuses on training engineers, grid operators, and system planners to manage these complex interactions, optimize hydropower dispatch, and participate in electricity markets.

Hydropower is characterized by high economic and financial intensity due to its substantial capital requirements, long construction periods, and extended asset lifetimes. Unlike many other renewable technologies, a large share of hydropower costs is incurred during the planning and construction phases, making projects very sensitive to cost overruns, financing conditions, and financial risks. Capacity building therefore focuses on strengthening expertise in project evaluation, financial strategies, risk management, and lifecycle cost analysis. It is, however, not limited to education and training activities, but also includes the strengthening of organizational capabilities, institutional coordination, knowledge transfer mechanisms, and managerial practices that improve the ability of organizations to prepare financially viable projects, engage with investors, and manage contracts effectively.

Another factor increasingly affecting hydropower sector is the climate change. Hydropower is highly dependent on variations in river flows, precipitation patterns, and seasonal water availability [2]. Adapting to these changes requires specialized knowledge in hydrology, climate modelling, and reservoir operation. Capacity building focuses on training hydropower planners and operators to anticipate and respond to altered water regimes, optimize storage and release strategies, and maintain reliable generation under variable conditions. To address this factor, capacity building is needed not only in terms of education and training but also in terms of activities for improving institutional and governance response. On one side, scientific papers highlights the need of improving operational resilience, which requires advanced modelling, forecasting, and hydrological expertise—indirectly supporting the need for education and training in climate adaptation. On the other side, local case studies also highlight empirical evidence of hydrological change impacting generation potential, which further justifies institutional capacity to respond. Effective hydropower development depends not only on engineering or financial capacity but also on institutional and governance frameworks such as organizational structures, decision-making processes, and regulatory capacities on project planning, construction, and operation. From this perspective capacity building is needed to target policymakers, regulators, and utility managers for efficient project management and compliance and performance assessment. There is indeed an evident need of focusing on the development of clear legal frameworks, standards, and procedures that coordinate multiple stakeholders: public authorities, private developers, and local communities.

The last factor highlighting the key role of capacity building is related to the workforce in the hydropower sector in terms of building the next generation of professionals. Hydropower is an old technical discipline, and as experienced engineers and operators retire, there is a growing risk of losing practical, tacit knowledge accumulated over decades of plant operation, maintenance, and project management [4]. Capacity building in hydropower can support the knowledge transfer and the recruitment of a younger workforce. At academic level, education plays a profound role by providing engineers with the theoretical updated knowledge, analytical skills, and practical competencies needed across the sector. Then, structured programs such as mentoring, workshops, and on-the-job training help preserve and disseminate this knowledge to new professionals. Targeted recruitment and development of young engineers, technicians, and planners must ensure continuity and put novel ideas into the sector. Capacity-building efforts must focus on bridging the gap between classroom learning and real-world hydropower operations, ensuring that trainees are prepared to contribute effectively to the sector.

But to successfully achieve the goal of preparing the next generation of professionals, capacity building is expected also to promote and implement gender balance in the hydropower sector. Historically, engineering and energy industries have been dominated by men. This has a double negative effect on the next workforce generation: on one side, it limits the diversity of perspectives and problem-solving approaches that different gender backgrounds can bring to the sector; on the other, it significantly reduces the potential pool of future professionals by discouraging or excluding female participation in hydropower careers. Addressing this imbalance requires targeted education, training, and mentorship programs, like the successful women mentoring program implemented within PEN@Hydropower to encourage women to enter and advance within hydropower careers. Capacity-building initiatives can include scholarships, internships, networking opportunities, and leadership development programs designed to remove barriers and achieve inclusive participation.

Academic programs in engineering, renewable energy, and environmental management provide essential understanding of fluid mechanics, energy conversion, and hydrological processes, although dedicated courses on hydropower are sadly slowly disappearing across Europe.

2 Limitations and Vulnerabilities in Sustainable Hydropower

Sustainable hydropower plays and will continue to play an important role in the European energy transition, but its future development is negatively affected by several limitations and vulnerabilities. The most important limitation is related to education. PEN@Hydropower [5] has shown that higher education systems are not fully aligned with the real needs of the hydropower industry. In most European university curricula, hydropower is no longer taught as a separate and specialized subject. Instead, it is often included within general courses on renewable energy or turbomachinery. Because of this, students receive limited knowledge about key hydropower topics, such as hydraulic turbine designs (Pelton, Francis, Kaplan, bulb), hydraulic systems, and plant components. As a result, many graduates understand neither the complexity nor the renewed importance of hydropower, and they are less interested in choosing hydropower field for their careers. This is also the case for international study programs. Here too educational programs often focus more on newer and more visible technologies such as solar energy, wind power, or hydrogen. International renewable energy programs dedicate only limited space to hydropower, which creates an imbalance between education priorities and real energy system needs.

Economic aspects represent the next vulnerability of hydropower. Many hydropower manufacturing companies are small or medium sized enterprises, which have limited capacity for research and development. This combines with the perception of the limited role played in the energy transition in comparison with newer technology under development. Because of this, talented students may choose other fields that promise more prosperous growth, evolving working environment and better paid jobs. Gradually and over time this leads to a shortage of qualified engineers in hydropower. Economic aspects relate also to research funding and visibility. In many European countries, hydropower topics are not included in research calls. Universities therefore have fewer research projects in this field. Consequently, fewer students write bachelor's, master's, or doctoral theses related to hydropower, which reduces transfer of knowledge and expertise from universities to industry. Finally, digitalization and artificial intelligence offer new opportunities but also new risks. Digital tools can support learning through simulations, digital twins, and advanced design methods. However, many educational institutions and teachers are not yet fully prepared to use these tools effectively. Without proper training and curriculum updates, digital solutions alone cannot solve existing problems in hydropower education.

3 Capacity Building Towards Flexibility and Hybridization with Other Sustainable Energy Sources and Untapped Hidden Hydro

The hydropower era began with the installation of large power plants and the exploitation of water resources with very high energy potential, providing for the then needed baseload operation. Most of these plants are still in operation and continue to provide a significant share of electricity in the global power system.

Hydropower has evolved from a traditional baseload energy source into a key provider of grid stability, particularly through flexible operation and hybridisation with other energy and energy storage technologies [1]. This transformation has triggered a strong drive for innovation, not only at the technological level, but also in operational strategies, market integration, system management, and institutional frameworks. A first outcome of this innovation push was the development of technologies aimed at expanding hydropower exploitation towards previously underutilized or economically unfeasible water resources. Newer technologies were designed to exploit watercourses with lower energy potential and lower water head (paving the way for what is known today as mini and small hydropower), and to some extend to enable a more flexible operation. Despite the availability of different technologies on the market, a large part of hydropower potential remains untapped. This potential is mainly located in non-traditional hydropower systems, such as water distribution networks, wastewater treatment plants, tailings dams, and low-head natural or artificial watercourses, including irrigation channels and marine or deep-water flows. Hidden hydro—a recently introduced term—now unlocks this potential and tackles currently unexploited locations.

But hydropower today faces new challenges, especially in terms of to the need for significantly increased flexibility, sustainability, and system integration [1]. The hydropower plant under such requirements becomes an active and flexible element of the power system, supporting grid stability, balancing variable renewable energy, and responding to market signals. We believe that one of the main reasons for the slowness of the refurbishment process (for improved flexibility and hybridization) and the sluggish unlocking of the untapped potential is the lack of sustainability oriented design, planning, and implementation capacity. Operators and engineers will be faced with flexibility services, such as fast ramping, frequency control, and load following, while electric energy production decisions will no longer be based only on water availability, but also on grid needs, forecasts, and optimization strategies. Another factor slowing down refurbishment processes is the uncertainty at regulatory and institutional levels, including delays in water concession renewals, unclear long-term policy directions regarding the future role of hydropower, and inconsistent policy support for research activities, technological innovation, and emerging flexibility and ancillary service markets.

The situation clearly highlights the need for improved capacity building. Achieving the transition to clean energy not only requires new technical solutions and well trained professionals, but also on strong institutional support and clarity on the key role of hydropower in future energy systems. Capacity building is therefore essential on several levels, including education, technical and sustainability training, institutional strengthening, development of regulatory and operational procedures, knowledge-transfer mechanisms, stakeholder networking, and interdisciplinary cooperation, while also addressing the human and social dimensions of the transition [3].

Flexible operation provides a clear example of the broader challenges associated with the evolving role of hydropower. More frequent starts, stops, and load changes increase mechanical stress on turbines, generators, and hydraulic structures, with direct implications for maintenance requirements, equipment lifetime, and overall economic sustainability. Addressing these impacts requires not only education and technical training on flexible operation and condition-based maintenance, but also a wider capacity-building effort aimed at improving awareness of the challenges to face at research level and to the long-term economic and sustainability tradeoffs associated with operational flexibility. This includes informing policy makers, strengthening cooperation and knowledge exchange among stakeholders and supporting the development of regulatory and market frameworks that properly account for asset degradation, maintenance costs, and long-term infrastructure sustainability.

An important starting point for this capacity building is the discussion on how Hydropower Sustainability Standard HSS is used to evaluate sustainability of hydropower projects on local, national and transnational levels (see Sect. 4 in Chapter "Hydropower Education and Professional Perspectives" of Part IV). Hydropower Sustainability Standard principles transfer to other courses in energy engineering. Since in part the negative picture of hydropower is also due to a substantially more detailed evaluation of hydropower projects, it is mandatory to include sustainability aspects within the capacity building. Practical skills, interdisciplinary knowledge, and understanding of sustainability principles must be strengthened through dedicated activities, such as the Training Schools presented later in this book. In addition, continuous training and lifelong learning will be necessary to support innovation, digitalization, and new business models in hydropower [4]. Without investment in skills, education, and knowledge transfer, the large, untapped potential of new hydropower applications will remain unused, limiting the hydropower contribution to a sustainable and flexible energy system.

Future capacity building must also address organizational and human aspects. Hybrid and flexible operation require closer cooperation between operators, maintenance staff, dispatchers, and external staff such as grid operators. Communication skills, decision-making under uncertainty and time or financial pressures will increasingly important. Capacity building should not be seen as a one-time training activity, but as a continuous process that involve stakeholders in continuous engagement activities and evolves together with technology and system requirements.

4 Competence Gaps in Sustainable Hydropower Workforce

Overview of education in Europe for sustainable hydropower shows competences mismatch rather than a lack of knowledge [5]. Universities provide sufficient fundamental knowledge (maths, physics, materials, thermodynamics, etc.), but industry needs in addition applied, operational competences and skills. Hydropower equipment is mostly prototype based, which requires experiential learning that cannot be gained only in classrooms or university laboratories and consequently important problem of most of current educational study programs is the lack of practical training.

In bachelor's and master's study programs students finish their education with strong theoretical knowledge but very limited hands-on experience [5]. Therefore, when graduates enter the industry, companies must invest significant time and resources to improve their skills [4]. New engineers need to participate in a few real projects, often overextending in time up to two years, before they become fully effective company engineers. Companies must provide extensive introduction and internal training due to previously insufficient practical preparation. Unfortunately, plant owners increasingly hire staff with less-than-optimal education, which increases the need for internal skill development.

The competences required for successful work in sustainable hydropower are non-academic skills that are often considered essential by companies [4]. They include, communication and foreign languages, use of software tools, sales and negotiation skills, conflict resolution, independent information searching, critical thinking, ethics, integrity, and teamwork.

5 Gender Balance

Hydropower plants are highly complex and engineered systems, requiring expertise across multiple disciplines, including mechanical, hydraulic, electrical, control, and digital technologies. Historically, however, engineering and energy-related sectors have been characterized by a significant underrepresentation of women, and the hydropower sector is no exception [4]. Although gender balance in hydropower had received limited specific attention in the scientific and industrial literature, the recent ESMAP–World Bank 2023 report [6] investigated this issue systematically for the first time. The results were largely consistent with the gender imbalance already observed across many engineering and energy industries.

The report clearly states that *"The hydropower sector, while expected to employ 3.7 million individuals by 2050, currently struggles with a significant gender disparity, with women representing only 25% of its workforce. This discrepancy underscores the urgent need to confront gender inequality in the industry and access the broadest talent pool. In response, the World Bank's Energy Sector Management Assistance Program (ESMAP) establishes the first global baseline for gender equality within the hydropower sector and presents a set of best practices to foster diversity and inclusivity."*

Based on 65 corporate and 900 individual survey responses, the report discusses the challenges women face in the field, such as:

- gender biases;
- a shortage of women with STEM qualifications;
- an unwelcoming work environment.

It emphasizes the importance of removing these obstacles to cultivate a more equitable and diverse hydropower sector. To achieve this goal, the report offers five key recommendations:

- enhancing STEM education opportunities for women;
- identifying and addressing gender disparities in the workplace;
- promoting hydropower careers to women;
- facilitating mentorship and role models;
- encouraging men to support gender equality initiatives actively.

Building on the findings and recommendations of the ESMAP–World Bank report [6], COST Action CA21104 actively promoted gender balance and inclusiveness within its organizational structure, encouraging broad female participation across leadership roles and working groups from the very beginning. At the launch of the Action, full gender parity (50:50) was achieved within the Chair, Working Group leadership, and mentoring structure (see also Fig. 1 in Part I, Chapter "The COST Action CA21104 Network", Sect. 1).

In addition, a dedicated mentoring initiative aimed at supporting and empowering female members within the Action was launched during Grant Period 2. The initiative was also intended to strengthen female participation and attract new women into the network through mentor and mentee roles. The program was implemented through a cohort of six mentor-mentee tandems, involving senior representatives from the Action leadership as mentors and female Task Leaders from the Working Groups as mentees. The first mentoring cohort was conducted between 2024 and 2025. As a further step towards strengthening gender equality within the Action, and in anticipation of later COST recommendations, a dedicated Gender Equality Coordinator was appointed as part of the Core Group. This Gender Equality Coordinator lead the graduation of cohort 1 in 2025 and catered for a second cohort with seven tandems to run from 2025 to 2026 until the termination of the Action. The first cohort attended a physical meeting of mentees and mentors in workshops on speaking skills and self-presentation capabilities, and virtual meetings with knowledge transfer.

Each cohort's mentors were especially trained in "good mentorship" practices to benefit for themselves and give benefit to their mentees through clear guidelines as to what is mentoring and what not.

With respect to PR, special posts were sent out around each year's International Women's days and clearly brought more visibility of women and display of role models. So did all posts and website entries on each virtual webinar and practical training.

The efforts of this gender work was much appreciated by COST generally, and the Action was able to present their mentoring activities in COST's "Gender Academy" in 2024. The introduction of a dedicated Gender Equality Coordinator role became a formal requirement for all new COST Actions following a decision by the COST Academy in November 2025. Notably, this role had already been established within our Action in September 2025.

A workshop on "gender bias" was conducted by a professional trainer at the last "multi-meetings" in April 2026 for all men and women present.

Nevertheless, the proportion of women in the working groups did not increase significantly compared to the growing number of Action members overall. It remained at roughly the same level as at the start, ranging from a high of 30% to a low of 18% (Fig. 1).

Since the issue of women in STEM is a broad and long-standing societal challenge, the Action could not directly address the underlying gender imbalance already present in the educational and professional pipeline feeding the hydropower sector. Its contribution therefore focuses on recognizing and supporting the recommendations of the ESMAP report [6], while promoting gender equality within its own activities and leading by example towards the broader hydropower community.

This was also evident in the training schools. The number of female applicants remained at the level as described in the study, 20–27%. However it demonstrated that highly capable young women could cater diversity and equity very well.

6 Approaches to Capacity Building

One important approach to future capacity building is strengthening the link between universities and industry. Closer cooperation can be achieved through joint projects, industry-based master theses, internships, and longer training periods. However, capacity building should not be limited to education alone. Stronger networking among universities, industry, research organizations, policy makers, and plant operators is equally important to facilitate knowledge exchange, dissemination of best practices, and coordinated responses to future hydropower challenges. In this respect, workshops, positioning papers, and stakeholder-oriented dissemination activities represent important tools to transfer knowledge beyond academia and support informed decision-making processes. In the future, this cooperation will become even more important, as hydropower systems will need to operate more flexibly to support variable renewable energy sources such as wind and solar.

Lifelong learning and continuous improvement of skills are critical capacity building components. Hydropower is a mature sector, but is changing due to digitalization, stricter environmental requirements, and new operating modes such as pumped storage and flexible generation. Engineers already working in the sector need regular training to update their sustainable hydropower skills [4]. Daily experience shows that there is a growing need for continuous competence development beyond a university education. At the same time, continuous capacity building is also required at institutional and regulatory levels to ensure that policy makers, authorities, and sector stakeholders remain aligned with evolving technological, environmental, and market conditions.

International mobility and exchange programs, such as Erasmus, represent another strong capacity building approach. The report highlights increasing numbers of exchange students and the potential for companies and universities to benefit more from this trend. In the future, such programs can help spread specialized hydropower knowledge across regions, especially between countries with strong hydropower traditions and those where educational capacity is weaker. Beyond student mobility, international cooperation networks and multidisciplinary working groups can further

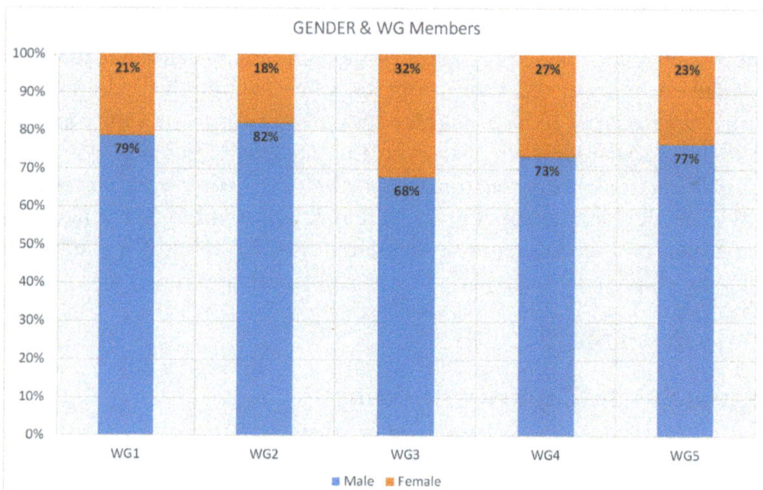

Fig. 1 Gender balance in working groups of CA21104 PEN@Hydropower

strengthen long-term knowledge transfer and foster shared approaches to sustainable hydropower development.

A forward-looking capacity building approach must also include digitalization and artificial intelligence. Despite being very traditional, education for hydropower will change through AI, for example through digital twins, simulations, and personalized learning. In the future, capacity building will not only mean learning engineering theory, but also learning how to work with AI tools, how to critically evaluate AI results, and how to combine digital methods with engineering judgement to support data-driven decision-making processes across technical, operational, and policy levels.

7 Capacity Building Implementation in CA 21104

The COST Action Pen@Hydropower aimed to identify the future requirements of hydropower and related capacity-building needs in Europe. One of the most significant contributions was the Training School on Sustainable Hydropower, which will be discussed in more detail in subsequent chapters of this book. However, Action members also carried out several other valuable activities.

Short Term Scientific Missions (STSMs) within the PEN@Hydropower COST Action contributed to capacity building in a targeted and well planned manner. These missions were closely aligned with the project proposal's objectives and the research goals of individual working groups. These short research visits were not random exchanges, but were carefully designed to support specific topics, methods, or gaps identified within the Action. Trainees had the opportunity to work with host institutions on set tasks, which helped them develop relevant technical skills and practical experience directly connected to ongoing research activities. At the same time, STSMs supported the development of soft skills, such as communication and collaboration in an international environment. Because the missions were aligned with the broader research strategy of the Action, they contributed not only to individual learning, but also to the progress of working group activities and joint outputs. In this way, STSMs strengthened both individual competencies and the overall research capacity of the hydropower community in a focused and efficient manner.

Since most members of PEN@Hydropower are from universities, their capacity building possibilities are inherently related to education. The largest impact can be achieved through courses taught at bachelor's and master's levels. Through interaction among members, it was realized that many improvements to course and study materials were implemented. In at least one case, the study materials for a course on sustainable energy sources, were updated to include more detailed technical content on machinery and its implementation.

Workshops, meetings, and other working group activities within the PEN@Hydropower COST Action also greatly contributed to capacity building. This was achieved by facilitating ways of continuous knowledge exchange and collaboration. Activities were organised around specific research topics, allowing participants to discuss hydropower related themes, share results, and align their work with the overall objectives of the Action. Through these interactions, participants improved their technical understanding and ability to work in interdisciplinary and international teams. Meetings generated joint ideas and common research approaches, strengthening cooperation between institutions and building a more connected and capable hydropower research community.

Networking, which can be viewed as a separate activity or as part of connected workshops and meetings, was related to applications for various EU and bilateral calls for projects and collaboration building. These projects increase universities' ability to attract young talent. Participation in these projects encourages students to pursue bachelor's, master's, and doctoral degrees in hydropower. It also helps prepare students for work in the hydropower sector and strengthens the competencies of teaching staff. Unfortunately, there were only a few calls dedicated specifically to hydropower. However, Action members applied to other available programs, such as Twinning and Erasmus.

Competing Interests The authors have no conflicts of interest to declare that are relevant to the content of this chapter.

References

1. Vagnoni E (2024) The new role of sustainable hydropower in flexible energy systems and its technical evolution through innovation and digitalization. Renew Energy 230
2. Soomro SEH, Soomro AR, Batool S et al (2024) How does the climate change effect on hydropower potential, freshwater fisheries, and hydrological response of snow on water availability? Appl Water Sci 14:65
3. Nazrati M (2024) Social acceptance for renewable energy technologies: the role of the energy justice framework. In: Heffron RJ, de Fontenelle L (eds) The power of energy justice & the social contract. Palgrave, Cham
4. Daw J, Stefek J, DeGeorge E, Philip RT, Cardinal A, Parkhill L (2023) U.S. Hydropower workforce: challenges and opportunities. National Renewable Energy Laboratory, Golden, CO
5. Hočevar M, Stopar RR, Patauner D, Caralis G Education–knowledge and skills for future hydropower. PEN@Hydropower COST action CA21104 event report, May 22, 2025
6. Energy Sector Management Assistance Program (ESMAP) (2023) Power with full force: getting to gender equality in the hydropower sector (English). ESMAP paper. World Bank, Washington, D.C., USA. https://www.esmap.org/Gender-and-Hydropower

Sustainable Hydropower Implementation in CA21104

Eduard Doujak

To make a hydropower plant truly sustainable, you must think holistically.

Abstract A critical component in evaluating the sustainability objectives of the Hydropower Sustainability Standard in relation to the overarching goals of the COST Action is their direct comparison. This chapter aims to illustrate that the core values of these two frameworks are not disparate but rather complementary. Initially, the respective principles are analyzed, followed by an outline of their commonalities and overlaps. Ultimately, the shared intersection of these two frameworks is elucidated.

Keywords Sustainability · Holistic · Hydropower · Stakeholders

In line with the objectives of COST Action CA21104, integrating sustainability into the hydropower sector is crucial to its future development. The long-term acceptance and viability of hydropower depend on adopting sustainable, holistic approaches. In this context, the design of this training school is informed by decades of refinement of the foundational standards. Only through the consistent application of these standards will future generations be able to reap the ecological and economic benefits of hydropower in global energy production.

1 Sustainable Versus Holistic Approaches in Hydropower

The previous chapters have provided a detailed overview of the fundamentals and definitions of sustainability. While sustainability follows a temporal concept, the holistic approach focuses on the complex systemic interrelationships at a given point

E. Doujak (✉)
Institute for Energy Systems and Thermodynamics, TU Wien, Vienna, Austria
e-mail: eduard.doujak@tuwien.ac.at

E. Doujak et al. (eds.), *Rethinking Hydropower: A Holistic Vision for a Sustainable Future*, Springer Tracts in Civil Engineering,
https://doi.org/10.1007/978-3-032-33328-5_4

in time. The following chapters will demonstrate how these two perspectives can be integrated in the field of hydropower, and specifically within the framework of COST Action CA21104.

1.1 The Sustainable Approach to Hydropower

A sustainable approach looks at the longevity, endurance, and future viability of the hydropower project. If one takes a sustainable approach to hydropower, the main focus is on time, resource depletion, and resilience:

- Sedimentation Management: Rivers carry silt. Over decades, this silt builds up behind a dam, reducing its water capacity until it can no longer generate power. A sustainable design includes advanced silt-flushing systems to prevent the dam from no longer working in 50 years.
- Climate Change Resilience: Will changing weather patterns over the next century cause severe droughts (making the dam useless) or extreme floods (overwhelming the dam)? A sustainable approach builds infrastructure that can handle the climate realities of the year 2100.
- Financial and Structural Longevity: Are there funds set aside for the continuous maintenance of the turbines? Is there a plan—and money—decommissioning the dam safely when its lifecycle finally ends, rather than leaving a dangerous, crumbling concrete wall for future generations?

The Sustainable Goal: To ensure that the clean energy produced today does not borrow from or compromise the safety, environmental health, and resource availability of tomorrow.

Within the COST Action we incorporated this approach by developing capacity building events based on the Hydropower Sustainability Standard, which was introduced above. Hydropower is the only renewable energy sources having such a guide line to create sustainable solutions for the related projects. As already shown, research and capacity building objectives focus on the future solutions in this field in order to guarantee a solid energy supply of electricity in the next decades and centuries.

1.2 The Holistic Approach to Hydropower

A holistic approach looks at the breadth of the entire ecosystem and society impacted by the hydropower station. It views the river not just as an energy source, but as a complex system where every part affects another.

When taking a holistic approach to building a power station, you consider more than just the engineering and the electricity generation. You map out and manage the ripple effects across the entire system:

- Ecological Connections: How will blocking the river affect fish migration upstream? How will the change in water temperature impact downstream aquatic life?
- Social and Human Impact: Will local communities be displaced by the new power station? How will it affect indigenous lands, local cultural sites, or downstream farmers who rely on natural flooding for their crops?
- Economic Variables: How does the power station affect local tourism (like white-water rafting) or the local commercial fishing industry?
- Geological Links: Will the massive weight of a new reservoir attached to a power station trigger micro-earthquakes, or cause bank erosion downstream?

The Holistic Goal: To understand the entire web of cause-and-effect so that maximizing electricity does not accidentally destroy the local economy, environment, or community.

Within the COST Action, this approach was adopted in the structure of the action by establishing five working groups distributed across several disciplines. Working group 5, the holistic group, addresses not only technical issues, but also environmental and climate change adaptation, legal topics, and policies. The purpose of this group is mentioned in Sect. 2.5.

1.3 Common Ground

When examining these two conflicting viewpoints, it is not immediately obvious how they can be reconciled. However, a closer look reveals that a lasting solution requires a comprehensive approach. For long-term success, it is essential to involve all stakeholders, ensuring that every need is integrated into the holistic solution and sustained over time.

Within the COST Action, both principles were embraced and applied across various initiatives. One such initiative is the training school, where sustainability guidelines were implemented alongside a holistic perspective, with the support of the working groups. It was in these training schools that efforts were made to merge these seemingly contradictory approaches into a forward-thinking program. Young researchers should understand that both principles are compatible and should be utilized in the future.

2 Stakeholders in Sustainable Hydropower

Taking this common ground into account, it becomes clear that a holistic perspective must be implemented by the various roles within a holistic solution. Incorporating the concept of diverse stakeholders into the organization of the COST Action training school requires transforming the hydropower sustainability standard chambers into a group of stakeholders. These groups should represent those affected by the planning, construction, and operation of hydropower projects and power plants. The training school offered participants five distinct stakeholder groups to encourage a shift in perspective and to step outside their daily work environment and comfort zone. These stakeholder groups are: (1) owners and operators, (2) financiers, developers, and insurers, (3) environmental experts and NGOs, (4) government and policy, and (5) social experts and NGOs.

2.1 Owners/Operators—(G1)

This group was established to evaluate the assessment of a hydropower plant from the perspective of the owner or operator. Participants are encouraged to adopt the role of this group and identify pertinent topics that are crucial for fulfilling the hydropower sustainability assessment. It is particularly important to address a variety of topics that enable operators to present their power plants in a more sustainable manner.

Additionally, considerations regarding societal acceptance could be a beneficial outcome of such an assessment. Furthermore, the potential for accessing different market remunerations may serve as a motivating factor for participating in such an assessment. These considerations and ideas should be taken into account by participants of the training school when selecting specific topics for the assessment of a power plant within this group. The same considerations and expectations apply from the owner's perspective. This rationale underpins the decision to merge owners and operators, who are not necessarily the same entity, into this initial stakeholder group within the organization of the training school on sustainable hydropower as part of this COST Action.

2.2 Financiers/Developers/Insurers—(G2)

The second group of stakeholders within the organization of the training school comprises financiers, insurers, and developers. This group employs the Hydropower Sustainability Standard to ensure sustainable investment and appropriate ethical engagement with project-affected individuals. Notably, financial institutions, such as the World Bank, had established their own regulatory frameworks to assess ethical and sustainable investments in large infrastructure projects globally, even before the introduction of the Hydropower Sustainability Standard. As hydropower increasingly assumes the role of critical infrastructure, the development of this sustainability standard represents a significant advancement. Similarly, insurance companies and project developers, overseeing project management require robust guidance and examples of best practice for developing such large-scale projects. At this stage of project development, it is logical that the Hydropower Sustainability Standard receives a specific version with special considerations. During the organization of the sustainable hydropower training school, this group was formed to provide participants engaged in technical, environmental, or social investigations with this groups perspective. It is also prudent to include this group in such training, even if the assessed power plant is already constructed and operational. For instance, insurance companies could utilize this assessment to participate in the Greenfield bond market and other markets as well.

2.3 *Social Stakeholders—(G3)*

The third group within the Hydropower Sustainability Training School is associated with the social stakeholder category. This group may include social experts and non-governmental organizations (NGOs). The focus of this group is on issues of societal relevance and impact, addressing parameters such as cultural heritage, communities affected by the project, livelihoods, labor and working conditions, and healthcare, among others. Occasionally, the expectations of society at large or a broader social group must be addressed, rather than the specific interests of an individual. Within the training school, emphasis will be placed on the broader group of individuals impacted by the operation of the hydropower plant. As this stakeholder group encompasses the whole of society, all participants are inherently involved. However, there is a distinction between addressing these topics from a professional standpoint and being personally affected by them. Participants in the training school are expected to identify key topics for assessing the use case of the power plant from a professional perspective within this stakeholder group. They should aim to determine the most beneficial outcomes for society as a whole, ensuring that the project advantages all affected individuals.

2.4 *Environmental Stakeholders—(G4)*

The fourth group within this training school organization pertains to environmental stakeholders, which may include experts in the field as well as non-governmental organizations (NGOs). These stakeholders address all aspects and topics related to both the upstream and downstream environments of a particular power plant. Their focus is on the flora and fauna both above and within the water, with the aim of ensuring minimal impact of hydropower operations on the surrounding ecosystem. Consequently, there are numerous topics to explore from the hydropower sustainability standards that assess the conditions of this critical field. In addition to experts, NGOs can also be part of this stakeholder group. Their interests are likely more society-driven, with a continued emphasis on the environmental impact of such infrastructure projects. Particularly prior to the commencement of construction, these NGOs often advocate for the preservation of nature, striving to maintain an untouched environment. This group is crucial for ensuring the responsible management of the power plant. Among the participants of the training school on sustainable hydropower, some experts may already be invited.

2.5 Government/Policies—(G5)

The fifth group of participants in the Training School on Sustainable Hydropower assumes the role of addressing governmental and policy-related issues. This group examines the legal framework articulated through policy measures to ensure the lawful operation of the power plant. There is likely an intersection with other stakeholder groups within this training school, as policies may address technical, environmental, or social aspects to ensure proper operation. Particular attention must be given to topics that are not intended to become stakeholder rights, as these require special consideration within policies to be included in any assessments or regulations. Given the strong connection between policies and governmental departments, this group should adopt this perspective and evaluate the use cases within the training school by exploring relevant topics of the hydropower sustainability standard.

2.6 Other Stakeholders

Within the framework of the COST Action, five stakeholder groups were delineated for utilization in the training schools. It is acknowledged that additional stakeholder groups exist and may be considered. Given the presence of five working groups that also offer scientific support for the topics, efforts were undertaken to identify the most significant stakeholder groups for subsequent assignment to the working groups.

3 Linkage Between Stakeholder Groups of the TSs and WGs of CA21104

Finally, the interrelationship between the stakeholder groups in the sustainability standard [1] and the COST Action [2] can be illustrated as follows (Table 1).

3.1 G1 ↔ WG1

These actors can be linked because they share a direct cause-and-effect relationship. What the grid requires from the working group (**WG1**) can only be delivered physically through the actions of the stakeholder group (**G1**). In the energy sector, this

Table 1 Interrelationship between stakeholder groups in training schools and working groups at the COST Action

HSS groups		CA21104 groups
G1—Owners/Operators	↔	WG 1—Hydropower Role in Flexible Energy Synergies
G2—Financiers/Developers/Insurers	↔	WG 2—Technological Evolution Through Innovation and Digitalization
G3—Social Stakeholders	↔	WG 5—Holistic assessment and stakeholder interaction
G4—Environmental Stakeholders	↔	WG 3—Sustainable Hydropower and its adaptation to climate change
G5—Government/Policies	↔	WG 4—Clean Energy Transition and policy measures

interconnection exists because operational capability should match grid demands. Looking to the logic behind this, the following can be explained:

1. The Power of Direct Control

"Flexible energy synergy" is a strategy that uses hydropower as a giant, responsive battery to balance the grid. When variable renewables like wind and solar suddenly drop off, hydropower must step in immediately.

- *The Link*: Owners and Operators (G1) are the ones who physically hold the keys to the control room. They are the only stakeholders with the authority and ability to open the water gates, start up turbines, or activate pumped storage. One cannot design flexibility frameworks (WG1) without the people who execute them.

2. The Reality of Mechanical Wear-and-Tear

Running a hydropower plant to provide flexibility is vastly different from running it at a steady, constant rate. Constantly ramping up and down electricity production subjects multi-ton turbines to immense hydraulic stress, vibrations, and thermal cycling.

- *The Link*: Operators understand the physical and mechanical limits of their equipment. They link with WG1 to ensure that grid expectations are realistic. They define how fast a plant can safely ramp up and how often it can do so without causing catastrophic machinery failure or shortening the plant's lifespan.

3. The Alignment of Market Economics

Grid flexibility is a specialized service, and operators expect to be compensated for the additional wear and tear on their facilities.

- *The Link*: Operators deal daily with wholesale electricity markets and grid tariffs. Linking them to WG1 ensures that when the group designs new roles for hydropower in the energy mix, those roles are structured alongside realistic financial incentives. If the market rules designed in WG1 are not profitable or viable for G1, operators will not participate.

Summary: WG1 defines the flexible grid needs from hydropower, while G1 dictates the safe and profitable operation of the machinery. They are linked because theory (WG1) cannot function without practice (G1).

3.2 G2 ↔ WG2

These two groups can be linked because they share a critical, mutually reinforcing relationship: Technology needs capital to be built, and capital needs technology to reduce risk. In the infrastructure world, this is known as the Innovation-to-Bankability pipeline. The following explains why these specific stakeholders are fundamentally tied to this working group:

1. New Technology Requires a lot of funding (The Funding Link)
Developing and deploying advanced technologies—such as AI-driven predictive analytics, digital twins, automated flow controllers, and next-generation turbine materials—is incredibly expensive.

- *The Link*: Financiers and Developers (**G2**) are the gatekeepers of the capital needed to fund these advancements. By linking them directly to **WG2**, the people inventing or proposing technological evolution can pitch to those who support it. It ensures that innovation does not stall in a lab but actually gets deployed in the field.

2. Digitalization Reduces Financial Risk (The Insurance Link)
Hydropower projects are infrastructure projects and require long-term financial involvement. Unpredicted equipment failures should be avoided through the use of machine diagnostics, as they can result in significant revenue losses and expensive repairs.

- *The Link*: Insurers and Financiers (**G2**) love data because data reduces uncertainty. **WG2**'s focus on digitalization (e.g., IoT sensors that monitor turbine health in real time) enables insurers to predict when a machine might fail accurately. By adopting these technologies, developers can demonstrate to insurers that their project is safe, thereby lowering insurance premiums and reducing interest rates on project loans.

3. Ensuring "Bankability" (The Commercial Link)
Engineers love technology for its performance; investors love technology for its return on investment (ROI). If a new technology is too experimental or complex, financiers will refuse to fund the project because it is too risky.

- *The Link*: Linking G2 to WG2 creates a vital reality check. As the working group maps out the technological evolution of hydropower, the financers and developers are eager to finance it. This ensures that the innovations emerging from WG2 are "bankable"—meaning they are commercially viable and ready for real-world market adoption.

Summary: WG2 creates the technological future of hydropower, while G2 provides the financial oxygen to make that future possible. They are linked because G2 provides the money and WG2 provides the risk-reduction that protects that money.

3.3 G3 ↔ WG5

These two groups can be linked because they share a profound, direct connection centered on human vulnerability and resilience. In the infrastructure and climate sectors, this is known as the Eco-Social Nexus. When climate change alters the water cycle, local communities are the first to feel the impacts, making them essential partners in designing adaptation strategies. These specific stakeholders are fundamentally tied to this working group for the following reasons:

1. Local Communities Feel the Direct Impacts of Climate Vulnerability
Climate change has a strong impact on the hydropower source: water. It causes erratic weather patterns, resulting in severe droughts (which shrink reservoirs) or intense flash floods (which risk catastrophic dam failures).

- *The Link*: Social Stakeholders (**G3**)—such as indigenous populations, downstream villages, and river-dependent workers—are the ones whose lives, homes, and livelihoods are directly threatened by these climate extremes. They link to **WG5** because they are the primary beneficiaries of effective climate adaptation; if a dam is retrofitted to handle a 100-year flood, it directly safeguards the human lives living below it.

2. Safeguarding the Human Right to Water (The Multi-Purpose Reservoir)
As climate change effects the water availability, a reservoir can no longer just be a tool for making electricity. It must become a shared resource.

- *The Link*: When for example the combination of **WG3** and **WG5** design adaptation strategies, social stakeholders must be involved to ensure that water management is equitable. For example, during a drought, how should the remaining water be split between generating power, irrigating local farms, and providing drinking water to nearby cities? **G3** includes an essential human perspective to ensure climate adaptation does not prioritize corporate energy profits over basic human survival.

3. Integrating Local and Indigenous Knowledge
The usage of advanced satellite modelling to predict climate change often lacks the historical, generational context of how a specific river basin behaves.

- *The Link*: Local communities have valuable knowledge about regional weather patterns. This also includes historical flood levels, and local ecology. By integrating G3 into WG5 (local stakeholders), scientists can correlate modern climate data with this traditional knowledge. Results promise a more accurate, practical, and resilient adaptation strategy.

4. Securing the "Social License to Operate"
In the past, and this was also the starting point of the HSS, many large hydropower projects failed or faced massive protests because they displaced communities or cut off water supplies without local consent.

- *The Link*: For an adaptation project to succeed—whether it is expanding a reservoir or building new flood walls—it requires a "Social License to Operate" (the ongoing acceptance of a project by local stakeholders). Involving G3 directly in the holistic planning of WG5 prevents legal blockades, costly protests, and human rights violations, ensuring that climate adaptation is done with the community, not to them.

Summary: WG5 focuses on a holistic adaptation of hydropower to survive a changing climate, while G3 represents the actual human beings living alongside that changing climate. They are linked because true sustainability cannot be achieved through engineering alone—it requires social justice and community resilience.

3.4 G4 ↔ WG3

These two groups are naturally linked because they represent the two sides of the same coin. **The ultimate goal of the clean energy transition (reducing carbon) and the ecological guardrails required to achieve it safely.** In international development, this is known as ensuring **climate-nature synergy**–making sure that solving the climate crisis doesn't accidentally accelerate the biodiversity crisis. These specific stakeholders are tied to this working group for the following reasons:

1. Defining "True" Clean Energy (The Scientific Link)
Hydropower occupies a unique space in the renewable energy world. While it produces virtually no carbon emissions, building a dam physically alters river ecosystems, blocks fish migration, and changes water quality.

- *The Link*: Environmental Stakeholders (**G4**)—such as river ecologists, conservation NGOs, and climate scientists—possess the exact data needed to define what actually constitutes a "green" project. They link with **WG3** to ensure that policy measures do not just measure a project's carbon footprint, but also account for its impact on local biodiversity, fish populations, and river health.

2. Preventing Greenwashing in Policy Measures (The Guardrail Link)
When governments and industries rush to meet aggressive net-zero carbon targets, they can sometimes look for fast, short-term infrastructure fixes that cause severe long-term ecological damage.

- *The Link*: G4 acts as the regulatory conscience of the clean energy transition. By embedding environmental advocates directly into the policy group, they ensure that international standards (like green bonds, carbon credits, or renewable energy certificates) are only awarded to hydropower projects that meet strict sustainability criteria. They prevent governments from rubber-stamping destructive projects under the generic banner of "clean energy."

3. Modernizing Environmental Legislation
Traditional environmental policies often viewed dams as purely negative. However, in a modern clean energy transition, hydropower is desperately needed to back up wind and solar power so that fossil fuel plants can be shut down permanently.

- *The Link*: Linking environmentalists to policy development creates a space for constructive compromise. Instead of just protesting projects, G4 works within WG3 to design modern policies like "eco-centric market incentives." For example, they can help draft policies that financially reward operators who install fish-friendly turbines or mandate specific "environmental flows" (releasing water to mimic natural river cycles).

4. Accelerating the Phase-Out of Fossil Fuels
Environmental groups are among the most powerful and influential advocates for moving away from coal, oil, and gas.

- *The Link*: To completely transition the energy grid, policymakers need the public campaigning power and legal expertise of environmental organizations. G4 provides the political momentum and advocacy needed to pass the aggressive transition policies being drafted in WG3, helping to shift subsidies away from fossil fuels and toward well-regulated, sustainable renewables.

Summary: WG3 designs the legal and economic blueprints for a zero-carbon future, while G4 ensures that this future does not come at the expense of our planet's rivers and wildlife. They are linked because **a transition is only truly "clean" when it protects both the atmosphere and the ecosystem.**

3.5 G5 ↔ WG4

Governments (**G5**) are the only entities with the neutral mandate, legal authority, and systemic viewpoint required to conduct a Holistic Assessment (**WG5**) and manage complex stakeholder dynamics.

In public administration, this is known as governance and consensus-building. When a major infrastructure sector like hydropower involves competing economic, social, and environmental demands, the government must step in as the ultimate coordinator.

These specific stakeholders are inextricably tied to this working group for the following reasons:

1. Only Governments Can Enforce a "Holistic" View

Every other stakeholder group has a specific, narrow bias: operators want efficiency (**G1**), financers want profit (**G2**), communities want safety (**G3**), and environmentalists want conservation (**G4**).

- *The Link*: A "holistic assessment" requires considering all these competing priorities simultaneously and making difficult trade-offs. Because Governments (G5) do not represent just one single interest group, they are uniquely positioned to oversee WG4. They use multi-criteria decision frameworks to ensure that a project benefits the nation's energy grid without destroying its rivers or bankrupting its citizens.

2. Resolving Multi-Stakeholder Conflicts (The Arbiter Link)

Hydropower projects frequently stall because of intense friction between groups—for instance, when a developer's plan (**G2**) directly threatens an indigenous community's land (**G3**) or a fish habitat (**G4**).

- *The Link*: Governments have the mandate to act as the institutional mediators within society. By coupling **G5** with **WG4's** "Stakeholder Interaction" workstream, policymakers deal with formal and legal frameworks for public consultation, mediation, and dispute resolution. This integration tries to convert potentially disorderly public protests into organized, democratic dialogues, ensuring that every voice is legally recognized.

3. Translating Assessments into Enforceable Law

A holistic assessment tool is useless if it just sits on a shelf as a theoretical report. To have any real-world impact, the criteria established by the working group must be enforced.

- *The Link*: Governments hold the exclusive power to legislate. When WG4 determines the best practices for balancing social, environmental, and economic needs, G5 policymakers can instantly translate those findings into binding regulations, building codes, and permitting processes. If a developer wants a license to build a dam, they must pass the holistic criteria codified into law by the government.

4. Balancing National Security with Public Interest

To balance this contradiction, governments are responsible for safeguarding national security, maintaining economic stability, and meeting international climate mandates, like the IPCC report. Subsequently, interactions with geopolitics, finance, and ecology are to be managed. Ensuring that a nation's energy infrastructure not only drives economic growth and guarantees power grid security, but also aligns perfectly with global carbon-reduction targets.

- *The Link*: WG4 serves as the laboratory where governments can test how individual hydropower decisions impact the entire nation. For example, a government can use WG4's data to determine if building a new pumped-storage plant will help to achieve the country's net-zero goals while keeping electricity prices affordable for low-income citizens.

Summary: WG4 provides the comprehensive data, assessment tools, and communication channels needed to look at the big picture, while **G5** provides the legal authority to turn that big picture into a national strategy. They are linked because, without government authority, holistic assessments have no power, and without holistic assessments, government policies have no direction.

Competing Interests The author has no conflicts of interest to declare that are relevant to the content of this chapter.

References

1. Hydropower Sustainability Alliance. Access online: https://www.hs-alliance.org/
2. COST Action Ca21104, Pan-European Network for Sustainable Hydropower - PEN@Hydropower. Access online: https://www.pen-hydropower.eu/

Part II
Educational Approach and Outcomes for Sustainable Hydropower

The clean energy transition (CET) is strongly dependent on the integration of renewable energy resources, among which hydropower continues to play a central role. Although hydropower is the most firmly established renewable energy technology, the sector is currently undergoing significant transformation in order to meet modern sustainability, environmental, and social requirements. In this context, the development of human capital equipped with updated knowledge and adapted skills is essential for addressing the challenges associated with the CET.

Recent analyses of the European hydropower sector have highlighted the important role of research and higher education in supporting this transition. Investigations of undergraduate and postgraduate curricula across European universities, combined with the analysis of research activities and participation in European projects, revealed both strengths and vulnerabilities in hydropower education. Based on a weighted European Credit Transfer and Accumulation System (ECTS) methodology, approximately 300 courses from 34 countries within the PEN@Hydropower network were evaluated. The results demonstrated that hydropower-related education is predominantly offered at the master's level, while hydropower itself represents the primary focus in only a limited number of courses. Furthermore, strong geographical disparities were identified, with educational and research activities concentrated mainly in Western and Northern European countries, where hydropower contributes significantly to the energy mix.

The PEN@Hydropower Education Survey further identified a major gap in capacity building related to sustainable hydropower. While sustainability requirements are increasingly shaping the future of the hydropower sector, the implementation of sustainability standards remains constrained by the limited availability of professionals trained in this field. Today, sustainable hydropower is increasingly associated with the practical application of the Hydropower Sustainability Standard (HSS), which provides a structured framework for evaluating environmental, social, technical, and governance aspects of hydropower projects and assets.

To address this educational gap, PEN@Hydropower developed, implemented, and tested a dedicated framework for capacity building in sustainable hydropower. The framework was integrated into a series of Training Schools designed to equip the next generation of professionals with both theoretical knowledge and practical competences. The PEN@Hydropower Training School on Sustainable Hydropower structure combined: (i) clearly defined learning objectives related to sustainable hydropower; (ii) a curriculum integrating theoretical foundations of the Hydropower Sustainability Standard with practical assessment activities at existing hydropower facilities; (iii) modern teaching approaches, including teamwork and interactive exercises; and (iv) individual and group-based assessment methods.

Each Training School was organised as a four-day course. The first day introduced the PEN@Hydropower initiative, presented the activities of the Working Groups (WGs) and organised participants into interdisciplinary teams. The second day focused on the theoretical background of the Hydropower Sustainability Standard and included practical exercises. During the third day, participants visited hydropower facilities and prepared assessment reports based on the HSS methodology, thereby strengthening their practical skills. The fourth day concluded with group presentations of the assessment results and the awarding of participation certificates.

In total, PEN@Hydropower organised five Training Schools across Europe, covering a broad geographical range: TS1 in Timis, oara (Romania), TS2-1 in Thessaloniki (Greece), TS2-2 in Porto (Portugal), TS3 in Uppsala (Sweden) and TS4 in Salzburg (Austria). Although all events followed the same general structure, each Training School reflected the specific characteristics of the visited hydropower facilities and the diverse backgrounds of the participants. Altogether, 111 Early Career Investigators (ECIs) participated in the Training Schools. Notably, approximately one quarter of the participants were employees of hydropower companies, which strengthened the interaction between academia and industry. The Training Schools therefore created a collaborative European network in which participants expanded their expertise in sustainable hydropower while establishing long-term professional connections.

The materials, methodologies, and experiences developed within PEN@Hydropower have subsequently contributed to the establishment of sustainable hydropower education initiatives at several European universities. Starting from the academic year 2025–2026, courses dedicated to sustainable hydropower were introduced at the École Polytechnique Fédérale de Lausanne (EPFL) and the Brno University of Technology (BUT). More than forty master's students benefited from educational materials and knowledge developed within PEN@Hydropower. In parallel, TU Wien (Vienna, Austria) initiated the integration of both theoretical and practical components of sustainable hydropower education into its master's programs, including field-based assessment activities at hydropower facilities similar to the PEN@Hydropower Training Schools.

These developments demonstrate the growing educational capacity in sustainable hydropower across Europe and highlight the importance of structured training initiatives in supporting the clean energy transition. The participants trained through these activities represent a new generation of professionals who will contribute to the implementation of sustainable practices and concepts within the hydropower sector and the broader renewable energy transition.

Status Quo of European Hydropower Education

Daniel Patauner and Eduard Doujak

Abstract This chapter examines the current state of hydropower in Europe, with particular focus on research and the role of higher education. Based on electricity data, as well as the results of the PEN@Hydropower education survey, it provides a comprehensive overview of hydropower-related teaching across European universities. The analysis covers 300 courses across 34 countries, introducing a weighted ECTS (European Credit Transfer and Accumulation System) methodology to quantify the actual share of hydropower content in academic curricula. Results show a pronounced geographical imbalance, with most courses and research activities concentrated in Western and Northern Europe, reflecting similar patterns in installed hydropower capacity and its contribution to the electricity production. Although the majority of courses are offered at Master's level, hydropower is the primary focus in only one third of lectures, which limits visibility and student engagement. The study further highlights a strong link between research funding, project participation and educational opportunities. The findings suggest that maintaining Europe's competitiveness in the hydropower sector requires increased investment in research and innovation, along with a broader and more visible integration of hydropower topics into university education. Strengthening interdisciplinary approaches and expanding educational offerings will be essential to meet future workforce demands.

Keywords Hydropower · Education · Capacity building · Research · Sustainable

1 Hydropower in Europe

Hydropower is the fourth-biggest energy producer in the European Union in 2025, trailing only nuclear, wind and natural gas [1]. Due to the expansion of renewable energy sources such as wind and solar, it is gradually becoming a complementary energy source, while its dispatch ability is constantly gaining importance. So far,

D. Patauner (✉) · E. Doujak
Institute for Energy Systems and Thermodynamics, TU Wien, Vienna, Austria
e-mail: daniel.patauner@tuwien.ac.at

E. Doujak
e-mail: eduard.doujak@tuwien.ac.at

E. Doujak et al. (eds.), *Rethinking Hydropower: A Holistic Vision for a Sustainable Future*, Springer Tracts in Civil Engineering,
https://doi.org/10.1007/978-3-032-33328-5_5

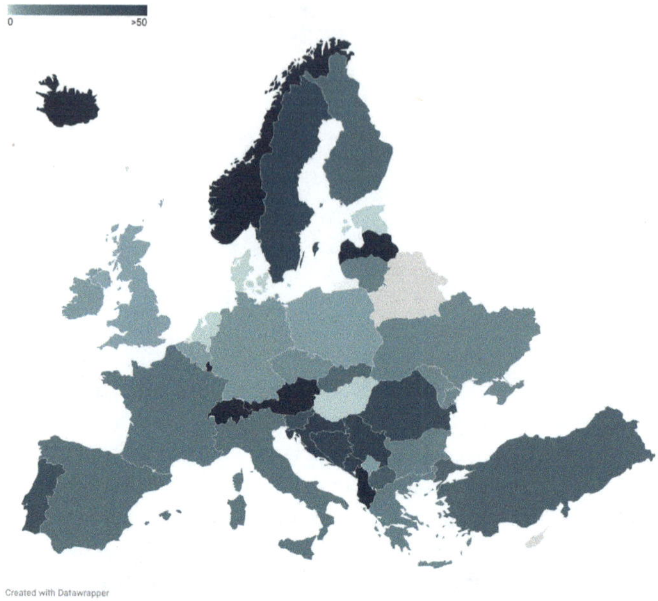

Fig. 1 Share of hydropower in % out of total electricity production [1]

hydropower continues to be the backbone of European energy storage, providing grid regulation, ancillary services, etc. Bearing in mind the importance of its role for the European economy [2], it is surprising to see that the visibility of the industry has been decreasing for years. This trend can also be observed in universities. Against this background, PEN@Hydropower assessed the educational situation at European universities and compared it with electricity statistics, offered lectures and research projects in the field.

European Electricity Statistics

The comparison starts with an overview of the role of hydropower in the electricity system, showing the relative energy production and installed capacity of the European continent. Figure 1 shows the relative hydropower production in % across Europe. Northern and central European countries e.g. Norway, Switzerland, Iceland and Austria, are at the forefront of regions that rely heavily on hydropower to cover their energy demand [1].

Since energy production by itself is only one relevant indicator, installed capacity (Fig. 2) was also assessed for completeness' sake. A high installed capacity does not automatically result in high energy production, but provides high power flexibility to stabilize the grid. Analyzing both statistics together is a good method to find out in which countries hydropower plays an important role.

Installed capacity data is not available [1] for all countries.

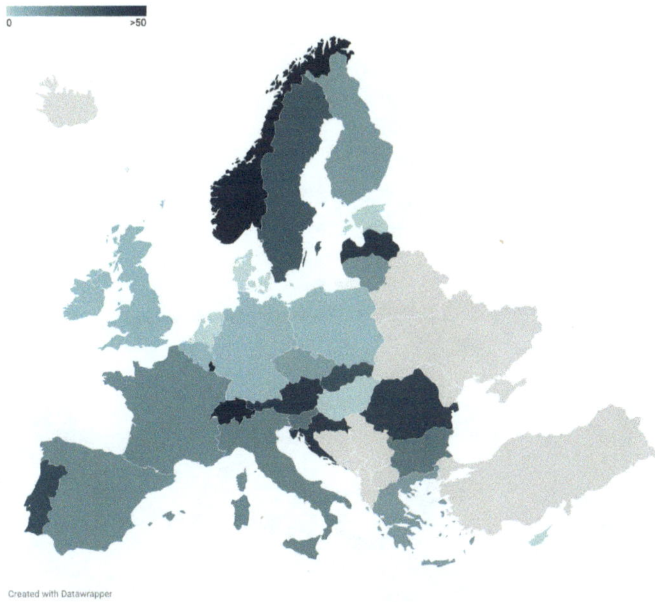

Fig. 2 Share of hydropower in % of total installed capacity [1]

2 PEN@Hydropower Education Survey

2.1 Description

To evaluate the current state of hydropower education in Europe, PEN@Hydropower conducted a survey among members and researchers that are active in the hydropower sector. The questionnaire looked at both national-level data such as general electricity production and hydropower statistics, as well as institution-specific information including lectures and courses offered at respondents' universities. Furthermore, the survey differentiated between various academic levels. In total, 66 participants from 29 countries responded to the survey (see Fig. 3). By extending the analysis beyond the initial responses, it was possible in a comprehensive review to identify 118 universities that offer approximately 300 lectures in the field of hydropower. For every lecture the following was noted: the faculty, the lecture name and, if available the description and ECTS (European Credit Transfer and Accumulation System) value. More details and all questions of the survey are shown in [3]. This study covers only university courses and not additional education possibilities e.g. summer schools etc. These additional education offers are rarely continuous and therefore less frequent as university courses and projects parts.

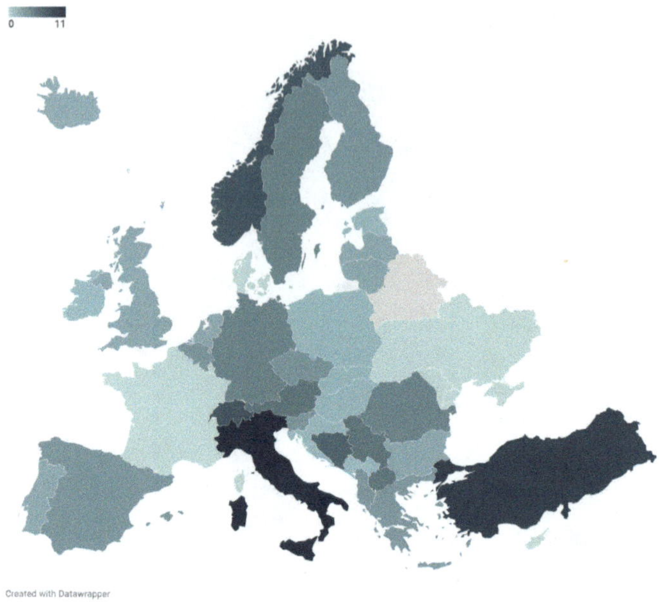

Fig. 3 Questionnaire participants (2024 and 2025)

2.2 *Results*

The data collected through the survey was analyzed using several processing approaches. A first evaluation was carried out in [3]. Courses were grouped according to education levels (Fig. 4). Most of the 295 lectures are offered at Master's level, namely over 65%. Bachelor and Phd courses make up the other remanding third. For eight courses, it was not possible to determine the level exactly.

Examining the European distribution of the identified courses (Fig. 5) reveals that Western and Northern European countries offer a greater number of courses in the field of hydropower. Comparing this figure with total hydropower capacity (Fig. 2) and production (Fig. 1) reveals a correlation between installed hydropower usage and educational opportunities within a country. To classify all lectures even more effectively, a weighting principle based on keywords was introduced. If any of the keywords appeared in the course title, the lecture was considered to be entirely hydropower-related. For other courses, e.g. Renewable Energy, a more detailed examination of the lecture description (if available) was conducted for a more accurate assessment, and a weighting of 25%, 50%, or 75% was awarded. The highest value was assigned to courses containing specific keywords in their titles.

This approach assumes that title generally reflects the course content, therefore, if particular keywords appear, a significant portion of the course is likely to address this topic in depth. The keywords used for this screening represent core technical fields in university education: hydropower, hydraulic machines, generators or large electrical

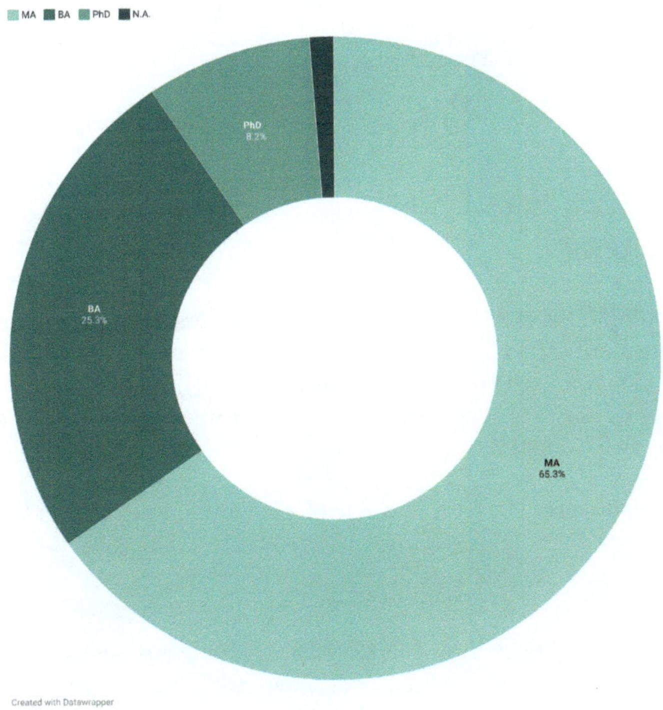

Fig. 4 Level of education

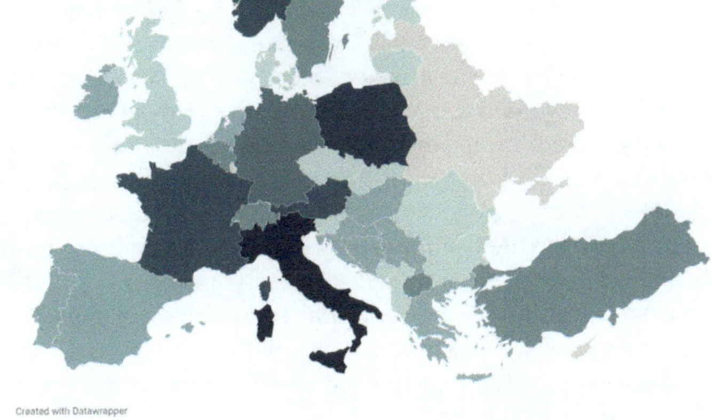

Fig. 5 Number of courses relevant to hydropower in the database

Table 1 Average weighting of course levels

Course level	Number of courses	Average weighting in %
Bachelor	90	59
Master	178	60
Phd	19	71
Total	287	61

machines, hydraulic steel construction, dams and barriers and river hydraulics. From the weighting and ECTS points, it is possible to infer the so-called weighted ECTS value. This value was used to characterize the effective hydropower education offered at a certain university, in a given city or country. Its calculation is fairly simple and illustrated by two examples:

$$\begin{aligned} \text{weighted ECTS} &= \text{ECTS value of a lecture} \times \text{weighting in \%}, \\ 4 &= 4 \times 100\% \quad \text{or} \\ 4 &= 8 \times 50\%. \end{aligned} \tag{1}$$

The average weighting of all courses is 61%. Table 1 presents the distribution across education levels. It shows an almost equal weighting between Bachelor's and Master's courses. Only dedicated Phd lectures are more targeted.

A comparison of Figs. 5 and 6 shows that hydropower-specific education does not necessarily correlate with the number of lectures. Clearly, the content of a lecture and thus its weighting play a crucial role. It was observed that more than a third of all lectures were classified as either 25% or 100%. This suggests that either the course topics fully cover hydropower or only touch upon it. The underrepresentation of the subject in two-thirds of the courses gives it much less visibility to students and decreases awareness of it. Without clear hydropower keywords in the lecture title, students will not recognize that a lecture is connected to or targets that topic.

At university level, almost all courses are offered by engineering faculties, e.g. mechanical, civil or electrical engineering. Only a small number of courses are offered in subjects such as chemical engineering or forestry. This further demonstrates that hydropower is widely regarded as an engineering task. However, when we look at the building process, operation and recommissioning more closely, we see that many other professions are needed. Large infrastructure projects such as hydropower plants also include legal, economic, environmental, political and social issues. This broadens the future need for a skilled and trained workforce, not only in the engineering sector. Analyzing lecture titles confirmed PEN@Hydropower's initial assumption that education in sustainable hydropower is limited, which strengthened the decision to create a framework for the Training School on Sustainable Hydropower. More information will be provided in Part II, Chapter "Training the Next Generation on Sustainable Hydropower in Europe".

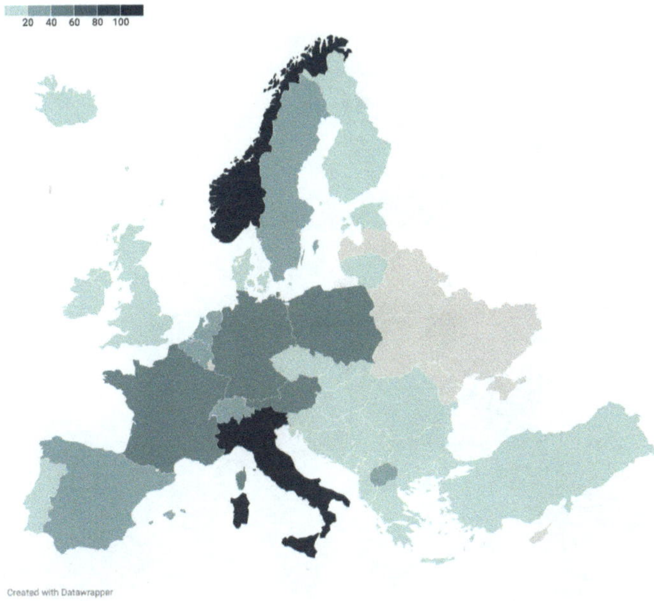

Fig. 6 Sum of weighted ECTS

3 Comparison of Research and Education

Survey participants also indicated that most of them offer Master's and PhD theses. It stands to reason that it is necessary to have research projects or contractual research with companies in order to offer these theses. For this reason, the survey also evaluated European research projects from the European Commission's Framework Programmes 8 (FP8 – Horizon 2020) and 9 (FP9 – Horizon Europe), comparing them with the lectures offered, weighted ECTS and electricity statistics [4]. The Horizon 2020 programme funded projects between 2014 and 2020, while the Horizon Europe FP9 programme will fund projects from 2021 to 2027. Within FP8 and FP9, 16 hydropower research projects were funded, resulting in a total of 231 partners and funds totalling more than 94 million euros [5]. For a certain period, project partners from Switzerland and the United Kingdom were not funded by the European Commission, but rather received national funding for projects that had been awarded. Due to their complexity, national hydropower research projects and contractual research could not be collected, evaluated and included in the graphics.

Figure 7 illustrates the distribution of project partners across European countries within all Horizon 2020 and Horizon Europe hydropower projects. It is clear that most partners are located in Central, Western and Northern Europe.

Figure 8 shows the same pattern as Fig. 7. Projects with a larger number of partners in a given country also received higher total funding. Both distributions demonstrate similar tendencies to those in Figs. 5 and 6, and reinforce the importance of research

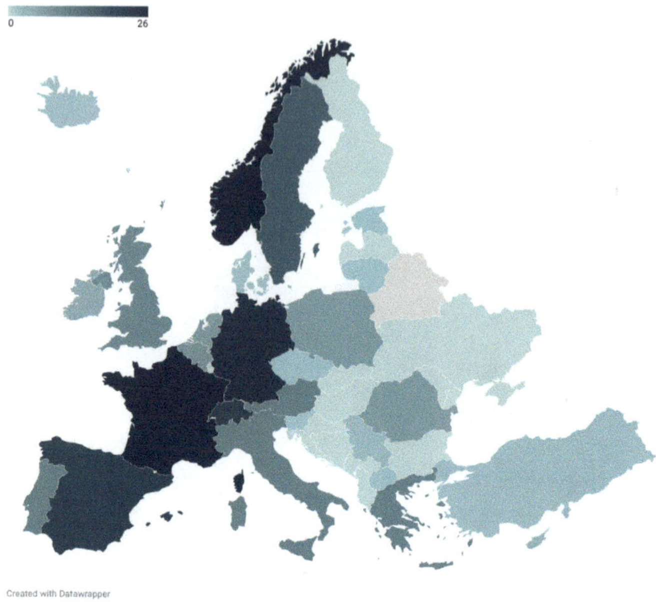

Fig. 7 No. of project partners of hydropower projects within the H2020 and HEU funding scheme [5]

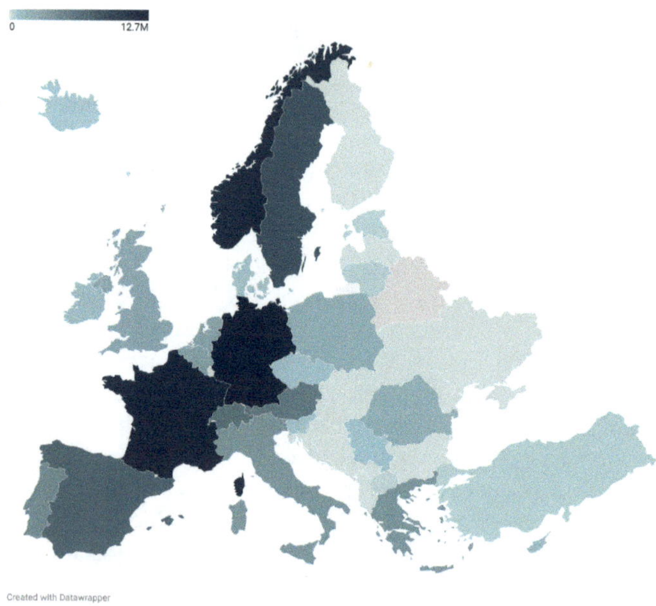

Fig. 8 H2020 and HEU funding distribution of hydropower projects in € [5]

projects for the future education and availability of workforce for the hydropower sector. In countries with a strong hydropower sector, the number of research activities and lectures correlates. It is hypothesized that fewer projects lead to fewer thesis opportunities, thus fewer students and less qualified staff for the sector. Since the EU hydropower industry still plays a leading global role, long-term competitiveness increases [2]. Maintaining that competitiveness is only possible with well-trained personnel, i.e. through research.

4 Conclusion

To examine the status quo of European hydropower education, PEN@Hydropower conducted a survey of 300 courses across 34 countries. The results were compared with electricity statistics, research projects and funding, and demonstrate the need for greater visibility and more educational opportunities in the sector. As stated in the European Commission's Competitiveness Report [2] and the Draghi Report [6], the EU needs to invest more in research and innovation to maintain its leading role in the hydropower sector.

Increased investment in research and education would result in more hydropower-related projects, creating more opportunities for professionals in this field. These projects would create more opportunities for university students, thereby increasing visibility and contributing to a better-educated workforce. Currently, project partners in European hydropower research initiatives predominantly come from Western and Northern countries. A similar pattern is observed in the analysis of weighted ECTS credits and the number of university courses across Europe. Notably, in only one-third of all lectures, hydropower is the primary focus. In another third, it plays only a minor role. The underrepresentation of hydropower in course titles reduces students' awareness of and identification with the subject. Without clear keywords in course titles, recognition of the topic becomes more difficult. This suggests that the role of hydropower in a country's energy system correlates with the availability of educational opportunities and research projects in this area.

Acknowledgements The authors would like to thank all contributors who completed the PEN@Hydropower questionnaire. Special thanks goes to the following TU Wien colleagues for their help in evaluating the survey: Magdalena Kaliba, Attila Nadaskay, Jakob Olsacher, Simon Geni and Constantin Königswieser.

Competing Interests The authors have no conflicts of interest to declare that are relevant to the content of this chapter.

References

1. Eurelectric—Elda: your electricity data assistant (2026). https://electricity-data.eurelectric.org/. Accessed 02 May 2026
2. European Commission (2025) Progress on competitiveness of clean energy technologies
3. Doujak E, Cavazzini G, Kaliba M, Righettini MS, Muntean S, Hocevar M, Anagnostopoulos I (2025) From knowledge to impact: the socio-economic relevance of hydropower education and its status. In: The 38th international conference on efficiency, cost, optimization, simulation and environmental impact of energy systems
4. European Commission—Framework programme facts and figures. https://research-and-innovation.ec.europa.eu/statistics/framework-programme-facts-and-figures. Accessed 02 May 2026
5. European Commission-CORDIS—EU research results. https://cordis.europa.eu/ . Accessed 02 May 2026
6. European Commission (2024) The Draghi report: a competitiveness strategy for Europe

Training the Next Generation on Sustainable Hydropower in Europe

Sebastian Muntean, Ioannis Anagnostopoulos, Antonios Tourlidakis, Isabel Boavida, Paulo Alexandre Diogo, Mats Billstein, and Daniel Patauner

Abstract Hydropower plays a key role in the clean energy transition. The demand for sustainable hydropower in the energy market is growing. Today, sustainable hydropower is an achievable goal if the Hydropower Sustainability Standard is applied to hydropower assets. The results of the PEN@Hydropower Education Survey, which was analyzed in the previous chapter, provide a comprehensive overview of the hydropower education system across Europe. The study identified limited capacity building in sustainable hydropower as the main gap in educational activities.

S. Muntean (✉)
Center for Fundamental and Advanced Technical Research, Romanian Academy - Timişoara Branch, Timişoara, Romania
e-mail: sebastian.muntean@upt.ro

I. Anagnostopoulos
School of Mechanical Engineering, National Technical University of Athens, Zografou, Greece
e-mail: anagno@fluid.mech.ntua.gr; ioannis.a@polytechnic.bh

School of Engineering, Bahrain Polytechnic, Isa Town, Kingdom of Bahrain

A. Tourlidakis
Department of Mechanical Engineering, University of Western Macedonia, Kozani, Greece
e-mail: atourlidakis@uowm.gr

I. Boavida
CERIS - Civil Engineering Research and Innovation for Sustainability, Instituto Superior Técnico, Universidade de Lisboa, Lisboa, Portugal
e-mail: isabelboavida@tecnico.ulisboa.pt

P. A. Diogo
MARE - Marine and Environmental Sciences Centre, NOVA School of Science and Technology, NOVA University Lisbon, Caparica, Portugal
e-mail: pad@fct.unl.pt

M. Billstein
R&D Division, Vattenfall AB, Älvkarleby, Sweden
e-mail: mats.billstein@vattenfall.com

D. Patauner
Institute for Energy Systems and Thermodynamics, TU Wien, Vienna, Austria
e-mail: daniel.patauner@tuwien.ac.at

E. Doujak et al. (eds.), *Rethinking Hydropower: A Holistic Vision for a Sustainable Future*, Springer Tracts in Civil Engineering,
https://doi.org/10.1007/978-3-032-33328-5_6

Consequently, the implementation of sustainable hydropower in the energy market is limited by a lack of human resources with the requisite application of know-how and skills. A framework for capacity-building in sustainable hydropower has been developed, applied and tested by PEN@Hydropower in order to train the new generation inmanner that allows it to fill the identified educational gap. The framework for capacity-building is put into practice in a face-to-face training school on sustainable hydropower. PEN@Hydropower organized five training schools across Europe, spanning from east to west and from north to south. A total of 111 early career investigators (ECIs) participated in all five training schools. It should be noted that around a quarter of the participants were from industry. The Training Schools on Sustainable Hydropower organized by PEN@Hydropower contribute to the transfer of theoretical knowledge and practical skills to the new generation, preparing them for teamwork at a European level. The capacity-building materials and experience gained by PEN@Hydropower members in sustainable hydropower led to the initiation of theoretical and practical knowledge transfer nuclei in European universities. Investing in the training of the next generation of hydropower experts paves the way towards a sustainable society.

Keywords Training schools · Education · Sustainable hydropower · Next generation · Methodology development · Capacity building

1 Training School Structure and Materials

Hydropower plays a key role in the clean energy transition (CET). Hydropower projects and other water—related infrastructure provide an interface between society and the environment. Therefore, sustainable hydropower is the target of more and more projects around the world. The implementation of sustainable hydropower in projects requires the participation of a number of specialists and a framework in which their knowledge and expertise in the field will be passed on to the next generation. However, the opportunities available for developing the skills for Early Career Investigators (ECIs), Master's and Ph.D. students in the Hydropower Sustainability field, as shown in the previous chapter, are fairly limited.

PEN@Hydropower aimed to bridge this gap by developing a framework for a multi-day interactive on-site training school. As highlighted in the introduction, sustainability is inherently multidisciplinary and requires expertise from various fields. Addressing these complex challenges, therefore, demands continuous expansion of knowledge and competencies. Against this background, the evolving challenges in hydropower sustainability call for strengthening the expertise of both young professionals from diverse educational backgrounds, and experienced practitioners. Knowledge transfer and skills development are crucial for enhancing economic competitiveness, fostering social inclusion, and enabling adaptation to technological change. In the long term, one of the greatest challenges remains ensuring the effective transfer of knowledge to future generations.

The chapter highlights this challenging task, addressed by COST Action CA21104 "Pan European Network for Sustainable Hydropower" (PEN@Hydropower) from 2022 to 2026 through Training Schools on Sustainable Hydropower with participation of ECIs, master and PhD students across Europe from COST Member Countries and COST Near Neighbour Countries (NNC). The training schools focus on developing the capacities of participants in the field of sustainable hydropower, fostering teamwork and bringing together different disciplinary perspectives, thereby filling a gap in the educational system. Therefore, the training schools developed and implemented by PEN@Hydropower are a key component of lifelong learning, providing an opportunity to develop and update knowledge and skills across a broad range of areas and supporting a holistic approach in the field of hydropower.

This approach aligns with the policies and strategies of the European Union, which aim to ensure that at least 47% of employees (aged 25–64) have participated in training within the last 12 months by 2025, with this figure set to rise to 60% by 2030 [19]. From the outset, the primary objective of the training schools has been to bolster the development of a European community of experts with theoretical knowledge and practical skills in sustainable hydropower, incorporating additional disciplines such as social, environmental and computer sciences through stakeholder outreach activities. Therefore, PEN@Hydropower has developed, applied and tested a framework for capacity building in sustainable hydropower to train the new generation and fill the educational gaps identified in previous chapter. This approach complements the training, assessment and certification activities carried out by the Hydropower Sustainability Alliance (HSA) [29]. To assess and certify sustainability performance, the International Hydropower Association (IHA) [34], together with multiple other stakeholders, has developed a sustainability standard for the hydropower industry. As outlined in detail in the introduction (see Chapter "Hydropower's Pathway to Sustainability" of Part I, Sect. 4), the Hydropower Sustainability Standard (HSS) covers 12 Environmental, Social and Governance (ESG) topics. It is a global certification scheme that outlines sustainability expectations for hydropower projects worldwide. The capacity-building framework takes the form of a face-to-face training school on sustainable hydropower. The training school's curriculum includes (i) learning objectives on sustainable hydropower, (ii) a theoretical knowledge base from the Hydropower Sustainability Standard [29] and practical skills developed when assessing existing hydropower assets, (iii) modern teaching methods including teamwork activities, and (iv) individual learning and group assessment strategies.

The Training Schools on Sustainable Hydropower were organized in all European areas (eastern, western, southern, and northern). All Training Schools follow the same format and include face-to-face activities over the course of four days, during which theoretical and practical knowledge is developed. The schedule is as follows:

- **Day 1**: Opening session, self-presentation of each participant, introduction into PEN@Hydropower, working group (WG) presentations, participant grouping;
- **Day 2**: Hydropower Sustainability Standard (HSS): Introduction and hands-on exercises with theoretical knowledge;

- **Day 3**: Hydropower plant visit and preparation of the assessment based on HSS to enhance practical skills;
- **Day 4**: Presentation of the assessment by each group, other hydropower initiatives, closing session.

As aforementioned (see listing above), the training school is a multi-day event covering a broad range of topics related to the field of sustainable hydropower. Each day focuses on different subjects. Creating a capacity-building framework for sustainable hydropower and preparing all the needed materials is itself a challenge.

After the introduction to PEN@Hydropower, participants will be paired and asked to introduce their colleague to the entire group. To encourage fresh interactions and unbiased first impressions, the pairs should consist of participants who do not know or have not spoken to each other before. This exercise promotes reflection, observation and interpersonal exchange among the participants.

Since participants come from diverse academic backgrounds (e.g. mechanical engineering, environmental engineering, law, etc.) and professional environments (universities, research institutes, NGO's and industry), it is first necessary to establish a common basis across the various disciplines. To this end, there will be presentations concerning the stakeholders in sustainable hydropower (see Chapter "Sustainable Hydropower Implementation in CA21104" of Part I). These presentations are designed to be interactive and address state-of-the-art approaches while also offering an outlook on future topics and challenges in the field.

On the second day, participants will receive an introduction to the concept of sustainability and the Hydropower Sustainability Standard (HSS). Within their assigned stakeholder groups (G1 to G5), participants develop their own understanding of sustainability, then they compare it by means of demanding practical examples. Each group receives guidance from on-site representatives of PEN@Hydropower, including experts from the different stakeholder areas, who provide support and share their professional perspectives throughout the discussions and exercises.

This is followed by an introduction to the HSS, including its history, implementation stages, roles, perspectives, responsibilities and assessment procedures. The theoretical knowledge gained is then applied in the assessment of an existing hydropower plant. Since participants are intentionally assigned to stakeholder groups outside their own field of expertise, the assessment is intended to broaden perspectives and encourage participants to step outside their comfort zones. Although the HSS can be applied across all phases of a hydropower project–including early-stage planning, preparation, implementation and operational phases, the assessments carried out within the training school were limited to the operation phase.

To carry out an HSS assessment, specific steps and rules must be followed. Depending on the project stage, the HSS includes 12 topics, while its predecessor, the Hydropower Sustainability Assessment Protocol, covered up to 24 topics. Each topic contains up to five criteria: assessment, management, stakeholder engagement, stakeholder support and outcomes. A scoring methodology and rating system is used to evaluate each topic. To earn the Hydropower Sustainability Certified label, a project must meet all minimum performance requirements across all relevant topics.

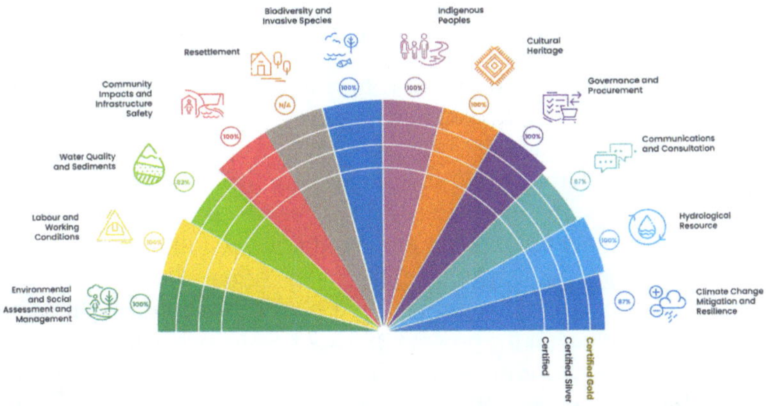

Fig. 1 Exemplary scoring of a hydropower plant assessment using the HSS [28]

The HSS distinguishes between two levels of performance requirements: minimum requirements, which represent basic good practice and are necessary for certification, and advanced requirements, which reflect international best practice and earn additional recognition. Higher levels of performance are recognised through Silver and Gold certification tiers, depending on the number of advanced requirements met. An example of a Gold certification is shown in Fig. 1. The scoring process starts with the highest possible grade (a score of five). Each significant gap identified during the assessment reduces the score by one point. A minimum score of three is required to achieve basic-good-practice certification.

Each group assesses the selected hydropower plant from its respective perspective (owners/operators, financiers/developers/insurers, social stakeholders, environmental stakeholders and government/policies) by selecting between two and four HSS topics. Preparation for the assessment includes collecting publicly available data of the hydropower plant and reviewing the corresponding HSS topics prior to the visit. The standard itself can be downloaded free of charge from the HSA website [30]. As the training school aims to increase participants' knowledge and does not constitute an official assessment contracted by the plant operator, participants rely exclusively on publicly available information, thus have no access to internal plant documentation. This limitation influences both the selection of assessment topics and the scope of the evaluation.

Since a site visit is an essential component of every assessment to enhance practical skills, it is also part of the training schools. Such visits provide several important benefits for the participants:

- for some participants it is the first-time visit to a HPP;
- better understanding of the scale and complexity of a hydropower facility;
- more comprehensive insights into the HPP and its surrounding environment;
- better preparation for discussions and questions;

- possibility of direct interaction with HPP personnel and local stakeholders;
- connection of theoretical knowledge with practical, real-world applications.

Depending on available infrastructure, time and access possibilities, more than one hydropower plant was assessed during each training school organized by PEN@Hydropower. Following the site visit, the groups process and analyse the information gathered during the visit together with their prior research. To foster collaborative project work, stakeholder groups prepare and present their work on the last day to all training school attendees. Feedback and follow-up discussions are provided by the trainers and other participants.

An import part of the training school is the networking aspect. In addition to hydropower related content, the program provides numerous opportunities for participants to interact in an informal and collaborative environment. Participants are encouraged to get to know each other beyond their professional backgrounds through group activities, common meals, visits and social events. These interactions help to create a positive and open atmosphere, strengthen intercultural exchange and foster personal and professional relationships among young researchers, industry representatives and experts in the field of sustainable hydropower.

The first PEN@Hydropower Training School on Sustainable Hydropower (TS1) was held from 9–12 May, 2023, in Timişoara and primarily targeted participants from Eastern European Area (EE) countries, including Armenia, Bulgaria, Cyprus, the Czech Republic, Georgia, Hungary, Moldova, Poland, Romania, Slovakia, Türkiye and Ukraine.

14 out of the 26 participants were from EE countries, while ten participants came from across Europe (see Fig. 2 and Table 1). Two participants represented the hydropower industry sector. The selected participants reflected a broad range of professional and academic backgrounds, including various engineering disciplines, forestry and business administration, environmental studies and sustainability, international law, energy management, environmental science and mining and energy. The hydropower facilities visited and assessed during TS1 are described in Sect. 2.2. Particular attention was given to achieving gender balance during the participant selection process. The success rate for applications submitted by young women was 100%, reflecting both their strong qualifications and high level of engagement in the hydropower field.

The activities carried out during TS1 combined the transfer of theoretical knowledge (see Fig. 3) with the development of practical assessment skills. Participants evaluated hydropower assets during field visits. Figure 4 shows the participants visiting the Zerveşti SHPP.

To attract a larger number of Early Career Investigators (ECIs), including master's and PhD students as well as young researchers within the network two Training Schools on Sustainable Hydropower were organized for participants from the Southern European Area (SEA) countries: Albania, Bosnia and Herzegovina, Croatia, Greece, Italy, Malta, Montenegro, the Republic of North Macedonia, Portugal, the Republic of Serbia, Slovenia, Spain, Andorra, San Marino and the Vatican. This decision was also influenced by the geographical configuration of the Southern European

Table 1 Distribution of participants by country at the TSs organized by PEN@Hydropower: y means yes—attended; n means no—not attended

Country	TS1	TS2-1	TS2-2	TS3	TS4
Albania (AL)	y	n	n	n	n
Austria (AT)	y	y	y	y	y
Belgium (BE)	n	n	y	n	y
Bosnia and Herzegovina (B&H)	n	y	n	n	n
Croatia (HV)	y	n	n	y	n
Czech Republic (CZ)	y	n	n	y	n
Finland (FI)	n	n	n	y	n
France (FR)	n	n	n	n	y
Georgia (GE)	y	n	n	n	n
Germany (DE)	y	n	y	y	y
Greece (GR)	n	y	n	n	y
Hungary (HU)	n	y	n	n	n
Iceland (IS)	n	y	n	n	n
Italy (IT)	y	y	y	y	y
Lithuania (LT)	n	n	n	y	n
Netherlands (NL)	n	n	n	n	y
North Macedonia (MK)	y	y	y	n	n
Norway (NO)	n	n	n	y	y
Poland (PL)	y	n	n	n	n
Portugal (PT)	y	n	y	n	n
Romania (RO)	y	y	n	n	y
Republic of Serbia (RS)	n	y	n	n	n
Slovakia (SK)	n	y	n	n	n
Slovenia (SI)	y	y	n	n	n
Spain (ES)	n	n	y	n	y
Sweden (SE)	n	n	n	y	n
Switzerland (CH)	y	n	y	y	y
Tunisia (TN)	n	y	n	n	n
Türkiye (TR)	y	y	n	n	n
Ukraine (UA)	y	n	n	n	n
United Kingdom (UK)	n	n	y	n	n
No. of countries: 31	15	13	9	10	11
No. of participants: 111	26	19	20	23	23
No. of women: 37	7	7	6	9	8
No. selected by industry: 25	2	0	8	8	7

Fig. 2 Participants in the first Training School on Sustainable Hydropower in Eastern Europe (TS1), held from 9–12 May, 2023 in Timisoara, Romania

Area, which involves long distances—for example over 2500 km between Greece and Portugal. The main objective was to maximize the resources allocated to ECI training while minimizing travel-related costs and burdens. The two Training Schools on Sustainable Hydropower were organized as follows: TS2-1 was held in Thessaloniki, Greece, from 7–10 May 2024, while TS2-2 took place in Porto, Portugal, from 3–6 September 2024.

The second Sustainable Hydropower Training School (TS2-1) took place in Thessaloniki, Greece, and targeted participants from the South East European Area (SEEA), including Albania, Bosnia and Herzegovina, Croatia, Greece, Italy, Malta, Montenegro, the Republic of North Macedonia, the Republic of Serbia, and Slovenia.

Ten out of 19 participants were from the targeted SEEA countries, while the remaining nine came from COST Near Neighbour Countries (NNC) (see Fig. 5 and Table 1). Particular attention was given to achieving gender balance during the selection process. Participants represented a broad range of engineering disciplines and professional backgrounds. The hydropower infrastructures visited during TS2-1 are detailed in Sect. 3.4.

Participants in TS2-1 received training on hydropower sustainability standard (see Fig. 6). In addition, visits to the Ilarion HPP/SHPP and Polyphyto HPP provided practical experience in the assessment of hydropower assets (see Fig. 7).

Fig. 3 Activities during TS1 in Timisoara, Romania

The third Training School on Sustainable Hydropower (TS2-2) took place in Porto, Portugal, from 3–6 September 2024. TS2-2 targeted participants from the South West European Area (SWEA), including Italy, Malta, Portugal, Spain, Andorra, San Marino, the Vatican and the United Kingdom. Out of 20 participants, eight were both from industry and the targeted SWEA countries, while the remaining four came from other COST countries (see Fig. 8 and Table 1). In total six women attended TS2-2.

During TS2-2, participants enhanced their theoretical understanding of hydropower sustainability (see Fig. 9) and further developed their practical skills through the technical visit to the Foz Tua PSHP facility (see Fig. 10) and the subsequent preparation of sustainability assessment plans.

The fourth Sustainable Hydropower Training School (TS3) was organized in Uppsala, Sweden, from 24–27 June 2025. TS3 targeted participants from the North European Area (NEA) countries, namely Denmark, Estonia, Finland, Iceland, Ire-

Fig. 4 Participants in TS1 during their visit to Zervesti SHPP

Fig. 5 Participants in the second training school on sustainable hydropower in South East Europe (TS2-1), held from 7–10 May 2024 in Thessaloniki, Greece

Fig. 6 Activities during TS2-1 in Thessaloniki, Greece

land, Latvia, Lithuania, Norway, Sweden, and the United Kingdom. In total, 23 participants attended TS3. Ten were from the targeted NEA countries and eight from the hydropower industry. Nine participants were women. The professional backgrounds of the participants covered a broad range of disciplines, including various engineering specializations, fish and wildlife conservation and management, water management and governance and land and water resources (see Fig. 11).

TS3 also strengthened the theoretical and practical skills related to hydropower sustainability by means of a technical visit to the Älvkarleby HPP facility (see Fig. 12). The case study assessed in TS3 is presented in Sect. 5.2.

Fig. 7 Participants in TS2-1 during their visit to Polyphyto HPP

Fig. 8 Participants in the third Training School on Sustainable Hydropower in South West Europe (TS2-2), held from 3–6 September 2024 at Porto, Portugal

Fig. 9 Activities during TS2-2 in Porto, Portugal

Fig. 10 Participants in TS2-2 during their visit to Foz Tua PSHP

The fifth Sustainable Hydropower Training School (TS4) was organized from 19–22 May 2026 in Salzburg, Austria. TS4 targeted participants from the following Western European Area (WEA) countries as follows: Austria, Belgium, Germany, France, Luxembourg, Monaco, The Netherlands and Switzerland. Twelve participants were from WEA target countries, seven from the hydropower industry and four from other European countries. Eight out of 23 attendees in TS4 are women. Participants have a variety of professional backgrounds, specializing in engineering, international studies, hydropower asset management, and geography.

Once again site visits and theoretical knowledge served to improve practical hydropower sustainability skills (see Figs. 13 and 14). The hydropower infrastructures visited and selected for the HSS-based assessment during TS4 are presented in Sect. 6.2.

Ultimately, more than 110 participants attended the training schools on sustainable hydropower organized by PEN@Hydropower in Europe. An overview of the countries that sent participants to the PEN@Hydropower Training Schools on Sustainable Hydropower is shown in Fig. 15 and summarized in Tab. 1. PEN@Hydropower provides opportunities for ECIs to further develop their theoretical knowledge and practical skills after participating in Training Schools, thereby strengthening the hydropower community. This community is formed by ECIs from different areas of expertise who have had the opportunity of getting to know each other, consolidating their knowledge in sustainable hydropower, and collaborating. It is this community that will develop and implement the future of hydropower.

Fig. 11 Participants in the fourth Training School on Sustainable Hydropower in North Europe (TS3), held from 24–27 June 2025 at Uppsala, Sweden

2 The First Training School on Sustainable Hydropower (TS1)

The first Training School on Sustainable Hydropower (TS1) intended to assess the hydropower assets specific to the Eastern European area. TS1 was organized from 9 to 12 May 2023 in Timisoara and hosted by the Romanian Academy – Timisoara Branch and University Politehnica Timisoara. This was no random choice: Timişoara was the first European city where large-scale electric street lighting (731 coal filament lamps) was put into operation on 12 November 1884 [11].

2.1 Hydropower in Romania

Hydropower plays a central role in the Romanian energy system as shown in Fig. 16. The maximum annual electricity production, around 65 TWh, was reached in Romania in the period from 2014–2016. Then followed a decrease by 15% to 55 TWh in 2022. As can be seen in Fig. 16, the mix of electricity production in Romania changed significantly between 2009 and 2025. The first wind turbines in Romania started generating electricity in 2010, then photovoltaic power plants started to be commissioned in 2013. In 2024, electricity generation in Romania reached a total of 52.5 TWh, of which 6.4 TWh thereof (12.1%) was attributable to wind, 3.4 TWh

Fig. 12 Activities during TS3 in Uppsala, Sweden

Fig. 13 Activities during TS4 in Salzburg, Austria

(6.5%) to photovoltaic, 14.3 TWh (27.3%) to hydropower, 10.9 TWh (20.8%) to nuclear, and the remaining 17.5 TWh (33.3%) to thermal. Electricity production from renewable sources has obviously replaced fossil fuel production in accordance with EU policies. Due to the increase in electricity production from non-dispatchable renewable sources (wind and solar), the contribution of hydropower in Romania is crucial to enable their integration into the power grid.

Figure 17 presents an overview of the trend in installed electricity production capacity in Romania from 2009 to 2025. It should be noted that Romania's total installed electricity generation capacity is approximately 18.5 GW. A reconfiguration of installed capacities in Romania is observed following its accession to the EU in 2007. Some coal- and gas-fired power plants have been replaced with renewable

Fig. 14 Participants in TS4 during their visit at Hintermuhr PSHP, Austria

energy production capacities (especially wind and photovoltaic power plants, which together reached a capacity of 6.4 GW by the beginning of 2026).

Installed hydropower capacity in Romania increased by 13% between 2009 and 2025, reaching over 6.6 GW—approximately 35% of the total installed capacity. Hydropower plants in operation and under construction on the main rivers are marked on the Romanian map in Fig. 18. The Hidroelectrica company manages 6.37 GW installed in 189 hydropower plants (HPPs), comprising: 2.9 GW in run-of-river (RoR) HPPs, 0.38 GW in storage HPPs, and 91.5 MW in storage pumping stations (SPSs). Of these HPPs, 108 have a power output greater than 10 MW, 22 have a power output between 4 MW and 10 MW, and 53 are small HPPs with a power output less than 4 MW. Hidroelectrica is a leader in electricity production and the main provider of necessary technological services in the national power system. The Hidroelectrica shareholder structure is dominated by the Romanian state. Through the Ministry of Energy, the Romanian state holds more than 80% of the share capital. The remaining shares are held by private shareholders. Hidroelectrica shares are listed on the Bucharest Stock Exchange (BVB).

The hydropower assets (HPPs and SPSs) managed by Hidroelectrica are grouped in hydropower systems on the main rivers of Romania, see Fig. 18. The source of the rivers start is in the Carpathian Mountains, where the amount of precipitation is the highest (Fig. 19), then they flow into the large rivers (Danube River, Tisa River, Siret River, etc.) in the plains.

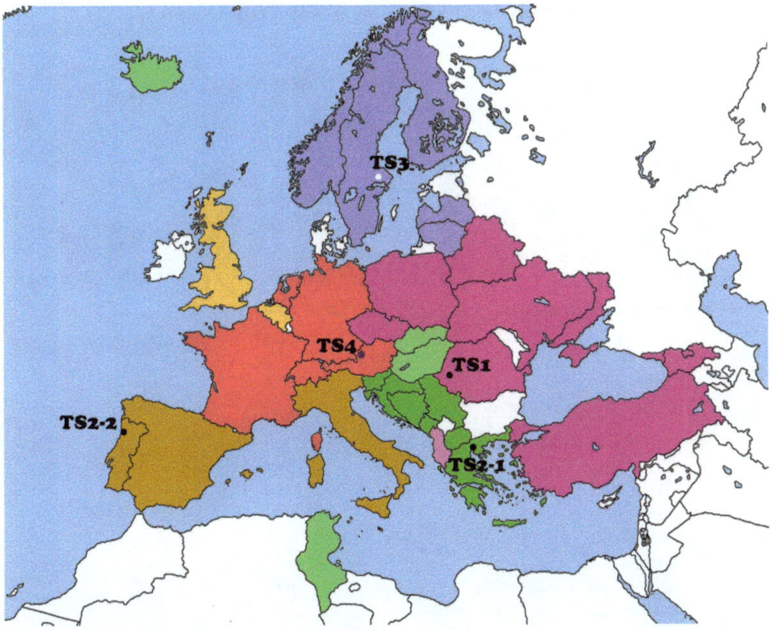

Fig. 15 Countries that participated in TSs organized by PEN@Hydropower: first Training School on Sustainable Hydropower in Eastern Europe (TS1) from 9–12 May 2023 in Timisoara, Romania (magenta); second Training School on Sustainable Hydropower in South-Eastern Europe (TS2-1) from 7–10 May 2024 in Thessaloniki, Greece (green); third Training School on Sustainable Hydropower in South-Western Europe (TS2-2) from 3–6 September 2024 in Porto, Portugal (brown); fourth Training School on Sustainable Hydropower in Northern Europe (TS3) from 24–27 June, 2025 in Uppsala, Sweden (blue) and fifth Training School on Sustainable Hydropower in Central Europe (TS4) from 19–22 May 2026 in Salzburg, Austria (red)

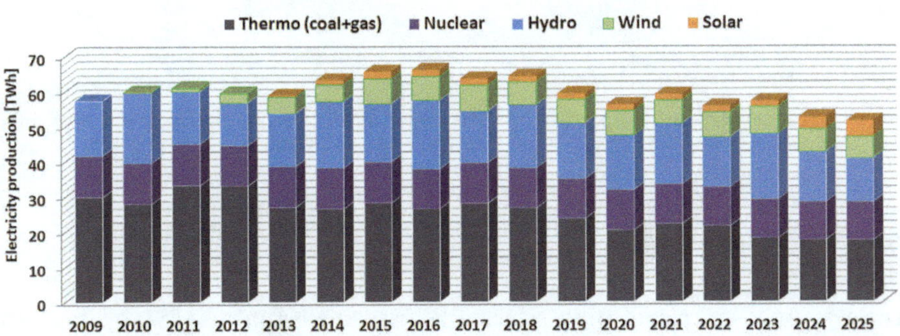

Fig. 16 Electricity production in Romania from 2009 to 2025 [31]

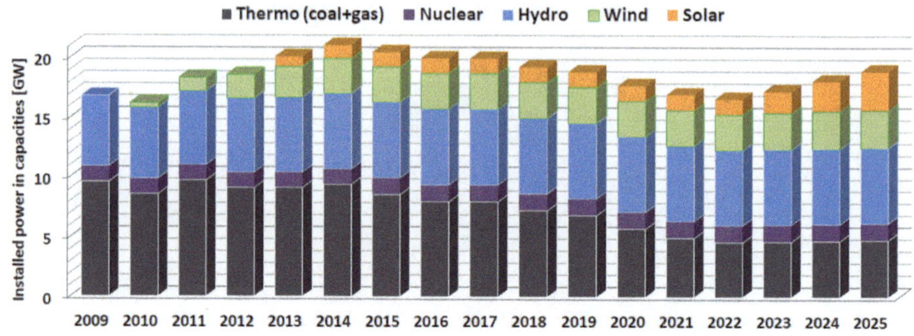

Fig. 17 Installed power in capacities in Romania from 2009 to 2025 [31]

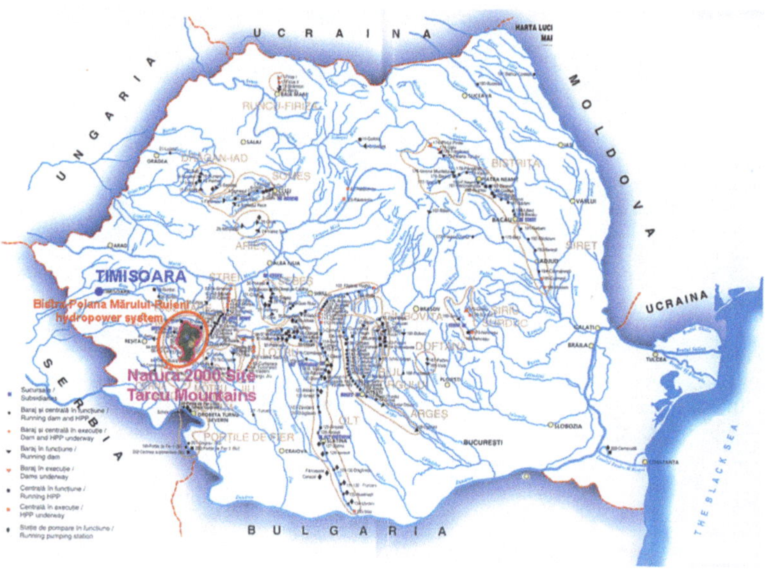

Fig. 18 Hydropower plants in operation (black dots) and under construction (red dots) in Romania. The Bistra - Poiana Marului - Ruieni hydropower system is marked with a red circle [12]

2.2 TS1 Case Studies

Three hydropower plants were selected as case studies in the TS1 hydropower assessment exercise based on the Hydropower Sustainability Standard [29]. These are: Ruieni HPP and Zerveşti SHPP from the Bistra—Poiana Mărului—Ruieni hydropower system (Fig. 18) 120 km from Timişoara and Timişoara SHPP on the Bega River installed in the urban area.

The Bistra-Poiana Mărului-Ruieni hydropower system is depicted in Fig. 20. This hydropower system was designed as a set of hydraulic assets that collect and process the flows of both main and secondary headraces. Construction of the Bistra—Poiana

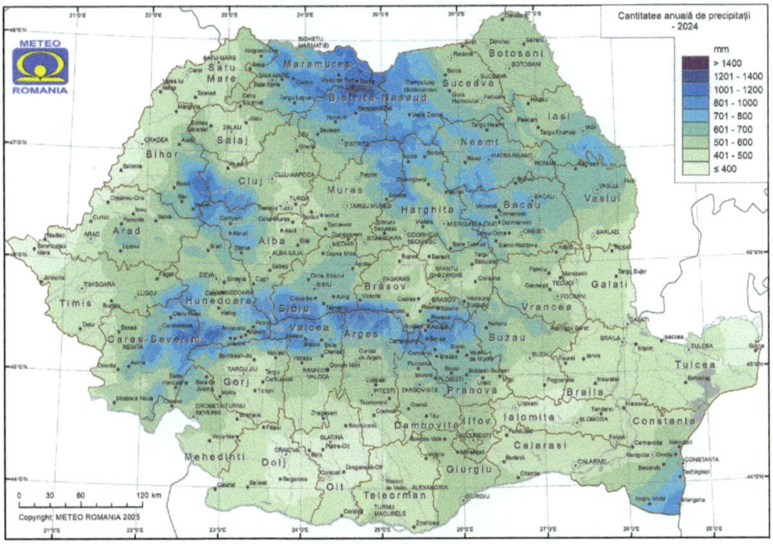

Fig. 19 The annual quantity of precipitation in Romania in 2024 [2]

Mărului—Ruieni hydropower system began in 1981. Deposits in the Măru concrete faced rock-fill dam began in 1984 and ended in 1992, when the height reached a hight of 125 m, Fig. 21.

The Bistra-Poiana Marului-Ruieni hydropower system has an annual energy production of 264 GWh in an average hydrological year. The Poiana Marului reservoir is fed by the Bistra Marului and Sucu rivers, which originate in the Tarcu Mountains. The Bistra Marului River contributes an average flow of 5.8 m^3/s, collected over a catchment area of 204 km^2. The main headrace tunnel between Poiana Marului and Ruieni has an inner concrete diameter of 4.9 m and a total length of 9.842 m.

The Poiana Mărului reservoir, which was formed by the construction of the Maru rock-fill dam, supplies water to the Turnu-Ruieni underground hydroelectric power plant (HPP). The reservoir has a surface area of 2.73 km^2 and an active volume of 89.3 million m^3. It is fed by the upper basin of the Bistra Mărului and a few neighboring catchment areas. The water streams in these catchment areas are managed by the National Administration 'Apele Romane', Banat Subsidiary (Administraţia Bazinală de Apă—ABA Banat) [1]. These water streams are also on the agenda of the Administration of the Ţarcu Mountains Natura 2000 site, as they are located in a protected area, as shown in Fig. 22 [20]. In 2007, the Tarcu Mountains area was declared a Site of Community Importance (SCI) by Order of the Minister of Environment and Sustainable Development, based on the establishment of the protected natural area regime for SCI, as an integral part of the European ecological network Natura 2000 in Romania. The Site of Community Importance Tarcu Mountains largely overlaps the Tarcu and Muntele Mic mountain areas.

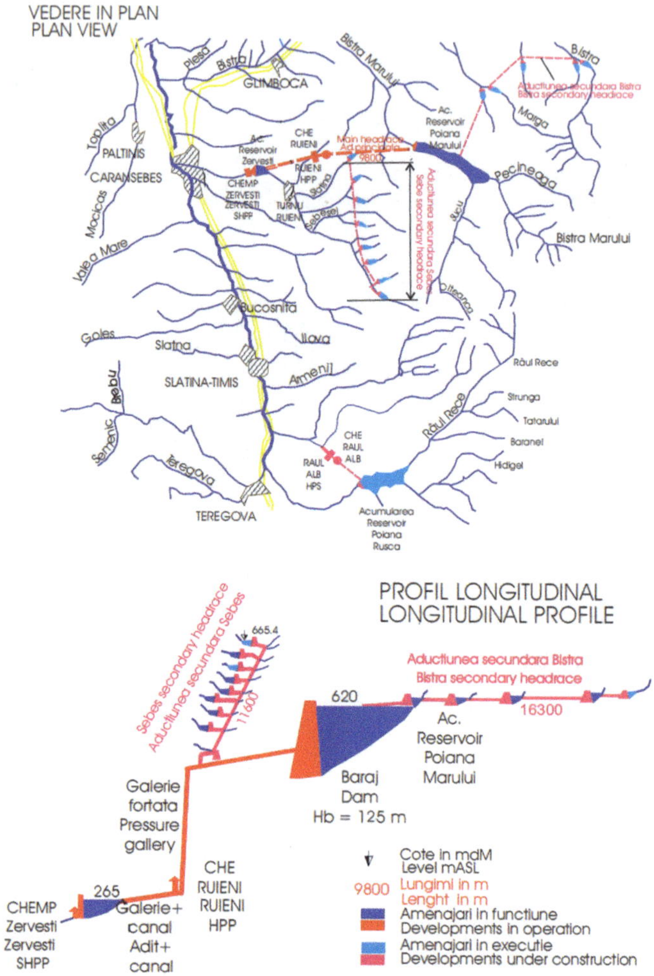

Fig. 20 Bistra—Poiana Mărului—Ruieni hydropower system [12, 43]

The Tarcu Mountains Natura 2000 site covers an area of 586.4 km^2, comprising 21 catchments, some of which are full and some partial [20]. The largest catchment is Bistra Marului, located in the north of the site, covering approximately 127 km^2, representing around 20% of the total area. The upper catchment of the Sebes River is also included in the Tarcu Mountains Natura 2000 site.

The Poiana Marului area has the following climatic conditions. There is usually abundant precipitation (more than 1000 mm per year), which is typical of a mountain climate, with high values in spring and autumn. The average annual temperature in the Poiana Marului area (at an altitude of around 500 m) is typically in the range of 7–9°C, which is typical of mid-mountain regions. The climate is sub-mountainous,

Fig. 21 Poiana Mărului reservoir view and Măru concrete faced rock-fill dam with clay core [43]

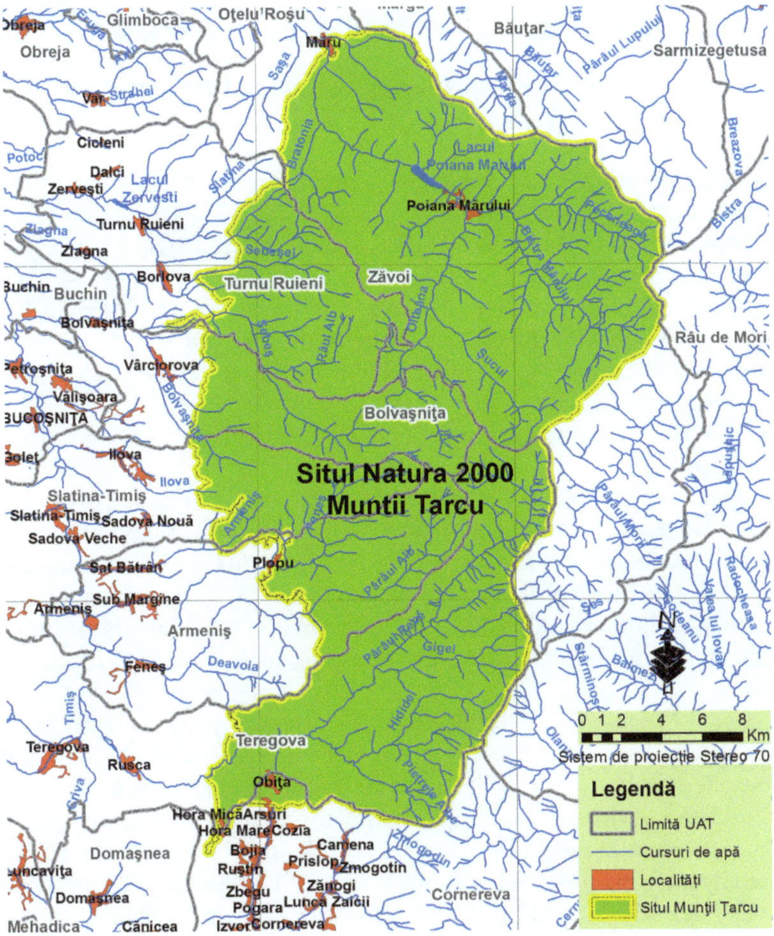

Fig. 22 Țarcu Mountains Natura 2000 Site [20]

Fig. 23 Turnu-Ruieni HPP: main access gallery, machine hall and turbine shaft of the HPU

with cool summers and winters, when frosts frequently occur. The area is renowned for its clean air. The climate is influenced by the Tarcu Mountains.

The Turnu-Ruieni underground HPP is located approximately 2 km northeast of Ruieni village, in the Turnu stream valley. The underground cavern is 64.85 m long, 38.20 m high and 19.75 m wide at the level of the machine hall (see Fig. 23), and consists of two Francis machines commissioned in 1993 and 1995 [39].

A cross-sectional view of the Ruieni HPP is shown in Fig. 24. Both units have a maximum power output of 76.5 MW at a maximum flow rate of 27.5 m^3/s and a rated net head of 326 m. The hydropower plant is equipped with two 90 MVA synchronous generators [37].

The amount of precipitation in the Zervesti area is influenced by the sub-Carpathian landform, with annual averages similar to those in the county's depressional areas. Precipitation levels range from 700 mm to 900 mm. The average annual temperature in the Zervesti area is approximately 10–11 °C, despite its location in a depressional and hilly area. The climate is continental with Mediterranean influences, characterized by mild winters and early springs, which are typical of the region's lowlands.

The main objective of the Turnu-Ruieni hydropower plant (HPP) is to provide ancillary services and regulation to the Romanian power grid. During periods of high flow rates, such as in spring, it mainly produces base load. It is also strategically located in an area with no major injection point in the national power grid, which achieves significant savings and reduces electricity transmission losses. The water processed by the HPUs installed at Turnu-Ruieni HPP is discharged into the Zerveşti buffer reservoir. The Zerveşti reservoir also benefits from the natural flow of the Sebeş River. It has a surface area of 0.218 km^2, an active volume of 1.031 million m^3 and serves as the headwater for the Zervesti small hydropower plant (SHPP). Additionally, the reservoir serves as a drinking water supply for the city of Caransebeş. An aerial view of the Zervesti hydropower system is shown in Fig. 25.

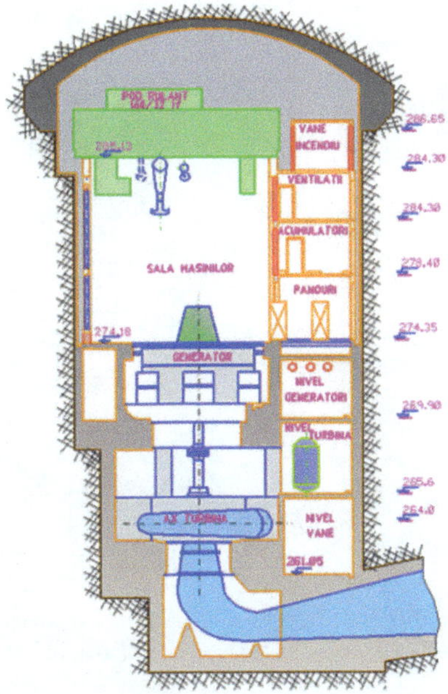

Fig. 24 Turnu-Ruieni HPP cross section view

Fig. 25 Aerial view of the Zerveşti hydropower system

Fig. 26 Zerveşti SHPP machine hall equipped with two HPUs

The SHPP consists of two HPUs equipped with cross-flow turbines (also known as Banki, Mitchell or Ossberger turbines), with a total power output of 1.52 MW (see Fig. 26). In the event of a blackout, the Zervesti SHPP ensures the start-up of the more powerful Turnu-Ruieni HPP in order to stabilize the Romanian national grid. To meet environmental regulations and objectives, the SHPP operates to deliver the minimum environmental flow rate to the downstream river. The power of each HPU is automatically adjusted depending on the net head value and the minimum flow rate delivered downstream (see Fig. 27). The protection of the Zerveşti daffodil meadow was of great importance during the construction of the reservoir. It was also designated a protected area in 2000.

Timişoara SHPP is the second test case selected in TS1. This test case is a run-of-river (RoR) type hydropower plant installed on the Bega Canal, Fig. 29. The Timişoara SHPP was commissioned in 1910 and is still in operation today [11].

Large variations in the Bega River's flow rate caused navigation problems and flooding, limiting community development and causing health problems for the population living in the marshy area around Timişoara [4]. In 1728, the Dutch engineer Maximilian Fremant proposed a hydraulic engineering solution involving a double connection between the Timis and Bega rivers to balance their flow rates [26]. Hydraulic engineering works started in 1758 to supplement the Bega River's flow rate with water from the Timiş River through the supply channel during dry periods, and to discharge some of the Bega River's flood flow into the Timiş River through the discharge channel during flood periods (Fig. 28). The flow rate of the supply channel is controlled by the Costeiu hydraulic knot, located 50 km upstream of

Fig. 27 Zerveşti SHPP surface power house and tailrace channel (view from downstream)

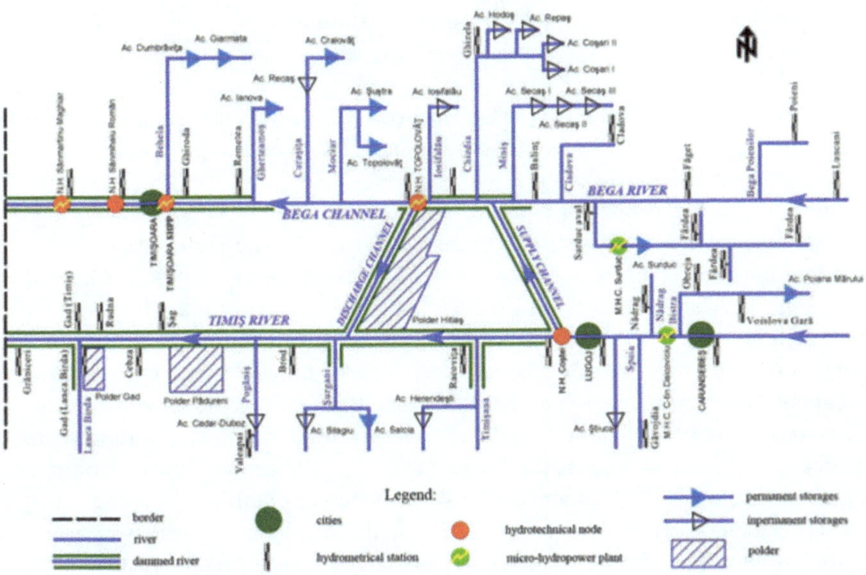

Fig. 28 Hydraulic connections between the Timiş-Bega rivers [1]

Timişoara [42]. The flow rate of the discharge channel is controlled by the Topolovăţu Mare hydraulic knot, located 30 km upstream of Timişoara. The Bega Canal begins downstream of Topolovăţu Mare village, and its navigable section is 114 km long, starting just downstream of the Timişoara SHPP and ending in the city of Titel in the Republic of Serbia. The Bega River flows into the Tisa River.

In 1902, Emil Szilárd, the city's chief engineer at the time, drew up a plan for the systematisation of Timişoara, which involved the abolition of the old water channels

Fig. 29 Aerial view of the Timişoara SHPP (from downstream)

in the Fabric neighbourhood [41]. The Timişoara SHPP was designed as part of the urban planning scheme to solve flooding issues and meet a large proportion of the city's energy needs.

The Timişoara hydropower complex was designed by Székely László, a famous chief architect of the early twentieth century [52], Fig. 30. The Timişoara hydropower plant building looks very much like a small fortress from the 1900s surrounded by water, Fig. 29. The construction of the Timişoara hydropower plant began in 1907, it was commissioned in 1910 with an annual energy production of over 4 GWh. The production of the hydropower plant covered 89% of the electricity consumption of the capital of the Banat region at that time. One kWh costed the citizens Timişoara 0.725 crowns in 1910. The hydropower plant installed in Timişoara represented a major technological advance at the beginning of the 20th century, almost completely replacing the electricity production of coal-fired power plants [11].

The cost recorded for the construction of the Timişoara SHPP was over two million crowns. The expense register shows that the costs of hydro-mechanical and electrical equipment accounted for approximately 20%, the costs of civil work (e.g. powerhouse, dam, spillways and gatehouse) accounted for 60%, and the costs of road infrastructure (bridges, access roads and site improvements) accounted for 20%. The city of Timişoara took out a loan to finance the construction of the hydropower plant [11] (Fig. 31).

The Timisoara SHPP was equipped with three hydropower units, each with a capacity of 660 CP (485 kW), manufactured by Ganz Villamossagi Rt in Budapest, which was the capital of the Austro-Hungarian Monarchy at that time (Fig. 32).

The Timisoara SHPP has a gross head of 5.6 m and the chosen technical solution for the hydraulic turbine includes three Francis runners on one shaft, with guide vanes

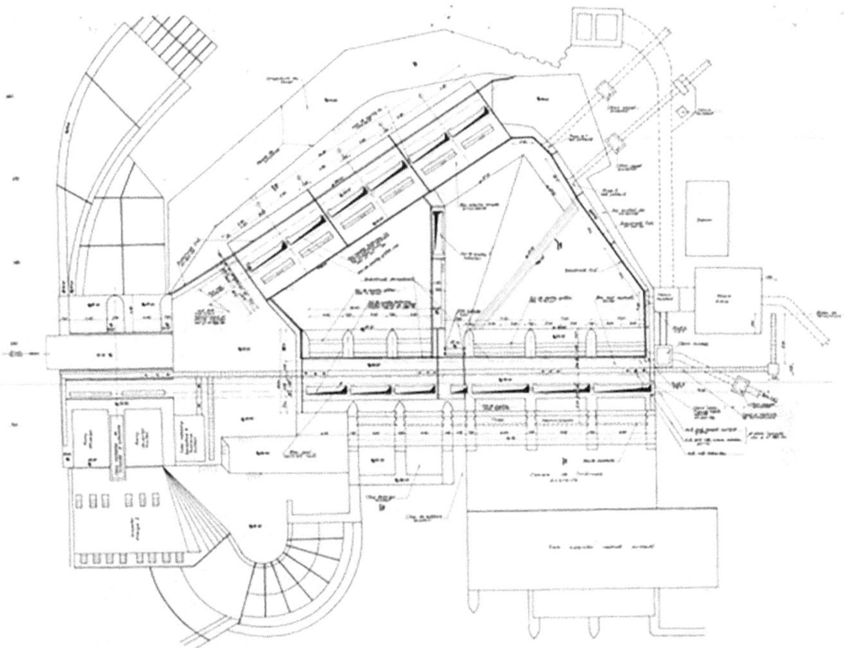

Fig. 30 Drawing of the Timişoara SHPP (top view).

for each runner and without spiral casing (see Fig. 33). This solution with Francis runners was selected for the very low head hydropower site, as the axial turbine (e.g. the Kaplan turbine) had not yet been invented. The minimum water level in the Bega canal downstream of the Timisoara SHPP is 2 m, whereas the flood protection level is 4.3 m. The average annual temperature in Timisoara is approximately 10.5°C, while the average annual precipitation generally ranges between 600 and 650 mm. Precipitation is moderate, with higher amounts in the summer months (June–July). This reflects a moderate temperate-continental climate with Mediterranean influences. This temperature makes Timişoara one of the warmest areas in Romania.

The multiannual average flow rate of the Bega canal recorded at the Remetea Mare hydrometric station located upstream of the Timisoara SHPP is 17 m^3/s. A water intake is installed on the left bank of the upstream channel of Timişoara SHPP for the water treatment plant that supplies Timişoara city with drinking water [16]. The Bega Canal is a complex multifunctional infrastructure for sports, fishing and recreation, Fig. 34.

Timişoara SHPP is one of the most challenging test cases for assessing hydropower assets according to the sustainability standard criteria, taking into account that it was commissioned over a century ago and that the operating requirements are constrained by the urban environment.

Fig. 31 Timişoara SHPP (view from upstream and downstream)

Fig. 32 Powerhouse and hydropower units installed at Timişoara SHPP

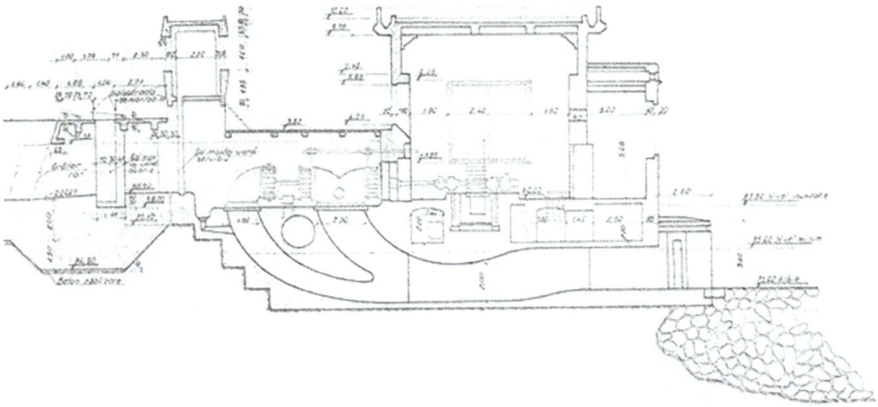

Fig. 33 Cross section view of the Timişoara SHPP

Fig. 34 View of the leisure activities on the Bega Canal in Timisoara

3 The Second Training School on Sustainable Hydropower (TS2-1)

The second Training School on Sustainable Hydropower (TS2-1) intended to assess hydropower assets specific to Southeastern European area, especially in Greece. TS2-1 was held from 7–10 May 2024 in Thessaloniki. The local organizers were the University of Western Macedonia (UOWM) and the National Technical University of Athens (NTUA).

3.1 Hydropower in Greece

Greece features more than 80% mountainous or semi-mountainous regions. This topography, combined with several major river basins (Fig. 35), provides favorable conditions for the development of hydropower infrastructure. The climate in Greece is typical of the Mediterranean region (mild, rainy winters and hot, dry summers), but the weather varies due to mountainous profile. The climate in eastern Greece is drier,

Fig. 35 Greece: Topography and major river systems

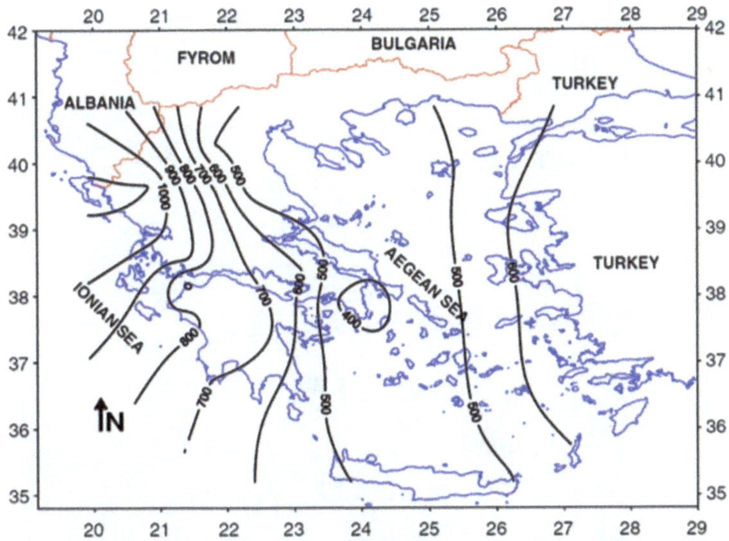

Fig. 36 Annual mean precipitation values (in mm H2O) for the 1950–2001 period [36]

whereas the climate in the northern and western regions is wet. For this reason, most large and small hydropower plants are located in the western and northern regions, where precipitation levels and river flows are relatively higher (Figs. 35 and 36).

The construction of large hydropower plants began immediately after the establishment of the Public Power Corporation (PPC) utility in 1950, when major infrastructure projects were implemented to support the rapid electrification and economic development of the country. By 1975, seven large hydropower stations had been built (Agras, Ladonas, Kremasta, Kastraki, Edassios and Polyphytos), with a total installed capacity of about 1400 MW. Since then, another seven large HPPs have been built (Pournari I and II, Sfikia, Asomata, Stratos I, Piges Aoou, Thesavros, Platanovrisi, Ilarion), adding almost 1800 MW of capacity. All large hydropower plants are owned and operated by PPC. Most of them are grouped in cascade systems along the main river basins of the country (Figs. 35 and 37), as follows:

- Acheloos River: Kremasta (437 MW), Kastraki (320 MW), and Stratos I (75 MW) (total: 907 MW);
- Aliakmon River: Ilarion (153 MW), Polyphyto (375 MW), Sfikia (315 MW), Assomata (108 MW), Agras (50 MW), and Edessaios (19 MW) (total: 1,020 MW);
- Arachthos River: Pighai Aoos (230 MW), Pournari I (300 MW), Pournari II (33.6 MW), and Louros (10.3 MW) (total: 574 MW);
- Nestos River (originating in Bulgaria): Thissavros (378 MW) and Platanovryssi (116 MW) (total: 494 MW);
- Tavropos River: N. Plastiras (130 MW);
- Ladon River: Ladon (70 MW).

Fig. 37 Locations of large hydropower plants (HPPs) and pumped storage hydropower plants (PSHPs) in Greece

The total installed power of the large HPPs operating today is almost 3.2 GW. Two of these plants, Sfikia and Thissavros are pumped storage hydropower plants (PSHPs) with a total power of around 700 MW. Also, approximately 150 small hydropower plants (SHPPs) operate in Greece (Fig. 38). Their installed capacity is roughly 300 MW, and their average annual production around 700 GWh.

Hydropower accounts for approximately 13% of the total installed generation capacity in Greece, and its share in the electricity production is about 7–10%, depending on annual precipitation (Fig. 38). The average annual production during the last, drier decade was approximately 4.1 TWh, whereas the average production since 2000 was about 4.4 TWh.

3.2 The Greek Electricity System

The electricity production mix in Greece has changed dramatically in the past two decades, as part of the country's energy transition and decarbonization policy (Fig. 39). In 2019, Greece announced plans to phase out all existing lignite-fired plants by 2023, with the exception of a newly built station which would remain operational until 2028. This would make Greece one of the fastest countries to transition away from coal. The natural gas-fired power sector is currently the main

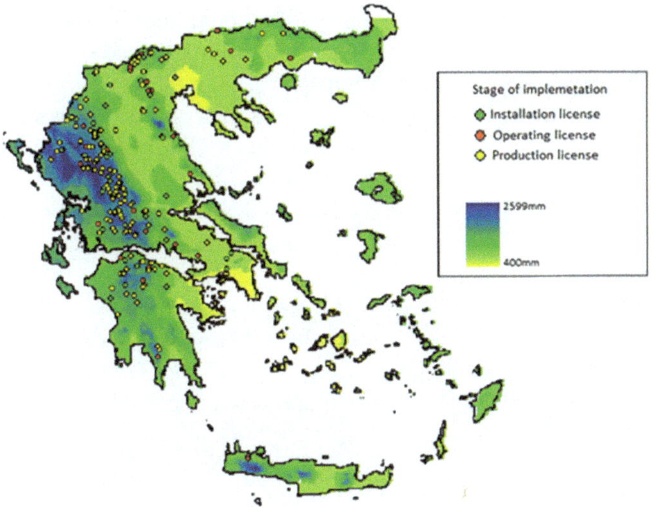

Fig. 38 Distribution of small hydropower plants (SHPPs) in Greece

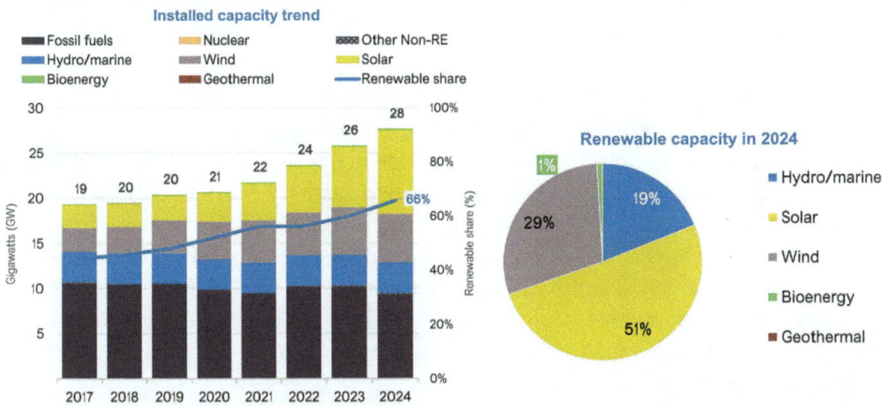

Fig. 39 Electricity production technologies installed in Greece since 2017 (left) and renewable capacity (right) [35]

source of electricity production. As of 2026, the country has approximately 6 GW of installed natural gas plants. New high-efficiency units are under construction or near completion.

Wind and solar power installations grew rapidly, reaching almost 18 GW at the end of 2025. Most of this capacity, approximately 12.2 GW, is Solar Photovoltaics (PV), which is already getting close to the target of the National Energy and Climate Plan [38], i.e. of 13.5 GW by 2030. On the other hand, wind installations are growing more slowly, primarily due to the more complicated licensing process and the environmental concerns and reactions from local communities. However, there are serious plans to develop offshore wind farms in the next decade.

Hydropower has been playing an important role in the Greek electricity system for several decades. Today, it continues to play a vital role in balancing and storing electricity, as it is a highly flexible and controllable renewable energy source that can support the integration of intermittent wind and photovoltaic production. In recent years, the two large pumped storage hydropower plants in Sfikia (on the Aliakmon River) and Thissavros (on the Nestos River) have largely been used to store electricity during periods of surplus renewable energy production (mainly from photovoltaic sources) and/or low demand, releasing it through the turbines during peak demand hours. Currently, the share of hydropower in both renewable energy source (RES) installed capacity and electricity generation is around 19% (Fig. 39). This share may decrease slightly due to the faster deployment of wind and solar RES.

3.3 Future Development in Greece

Future energy planning in Greece considers the expansion of hydropower and pumped storage capacity as an important tool to support the integration of renewable energy sources. Several projects are currently under development or under consideration. These include the completion of the 160 MW Mesochora hydropower plant on the Acheloos River and the 40 MW Metsovitikos hydropower plant on the Metsovitikos River. Another important project under construction is the pumped storage scheme near Amfilochia, with an installed capacity of about 680 MW, which is expected to become one of the largest energy storage facilities in southeastern Europe. Additional hydropower projects are being examined in the Acheloos River basin, including the Sykia and Avlaki HPPs (Acheloos river, 120 MW and 84 MW), as well as new developments in other river basins. Aditionally, PPC plans to develop around 1 GW of new pumped storage hydropower plants (PSHPs) over various locations in the coming years.

According to the national energy strategy [38], the total installed hydroelectric capacity in Greece is expected to increase gradually after 2030, potentially reaching around 4.5 GW by 2050. Also, the pumped storage installed capacity is planned to exceed 5 GW by 2050. At the same time, it is envisaged to deploy a huge number of batteries on the main grid or combined with existing and new solar and wind units (hybrid plants), to reach 12 GW of installed capacity by 2050 [27] (Fig. 40). However, this ratio between the two technologies could shift in the future in favour of pumped storage, due to the countries favourable topography [22] and the very high national added value and parallel benefits of this technology.

However, the future development of hydroelectric power in Greece is subject to environmental, social and climate constraints, which have delayed or suspended several large hydroelectric projects in recent decades. In addition, climate change is expected to affect the hydrological cycle throughout the Mediterranean region. Climate projections indicate that average annual rainfall in Greece could decrease

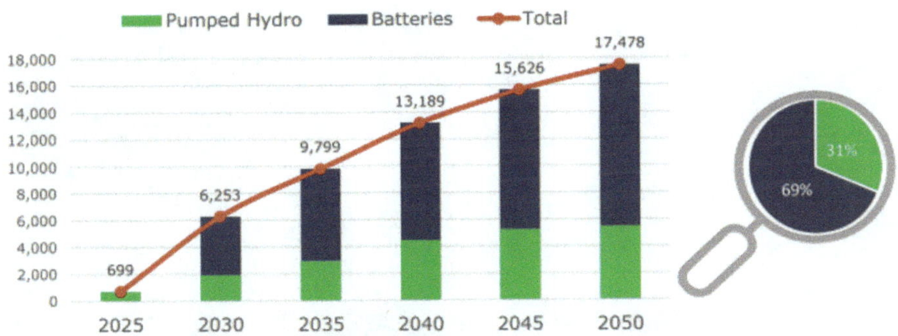

Fig. 40 Prospects of future development of electricity storage in Greece [27]

by up to 15% by the middle of the century, while droughts will be longer and more frequent. These changes may reduce the country's annual hydroelectric power production accordingly [23].

Despite these challenges, however, hydroelectricity will remain an important pillar of the energy transition in Greece, and the modernization of existing units, the development of multipurpose reservoirs, and the expansion of pumped storage capacity are expected to play a critical role in ensuring the reliability and stability of the future Greek electricity system.

3.4 TS2-1 Case Studies

Two hydropower plants were selected as case studies in the TS2-1 hydropower assessment exercise based on the Hydropower Sustainability Standard [29]. These are: Ilarion HPP and Polyphyto HPP from the Aliakmon hydropower system in Western Macedonia.

The **Aliakmon Hydropower System** (also known as the Aliakmon Cascade) is the largest and most complex hydroelectric cluster in Greece. It is managed primarily by the Public Power Corporation (PPC) and consists of a series of five main dams and several small hydroelectric plants (SHPPs) which are arranged in a "step-down" sequence along the Aliakmon River. The system is multipurpose, since it provides not only a combined capacity of approximately 1000 MW but also serves as the primary drinking water source for Thessaloniki, and is a critical irrigation supply for the areas around lake Polyphyto and the plains of Central Macedonia. Additional water is required for cooling the lignite steam power plants in the Kozani-Ptolemaida basin, and to maintain the ecological balance in the river delta, a continuous environmental flow is needed downstream of the Agia Varvara control dam (Fig. 41).

The **Aliakmon River** originates in the mountainous regions of northwestern Greece. Etymologically, the river's name goes back to the Hellenic deity Aliakmon, the mythological son of Oceanus and Tethys. It serves as a vital hydrological

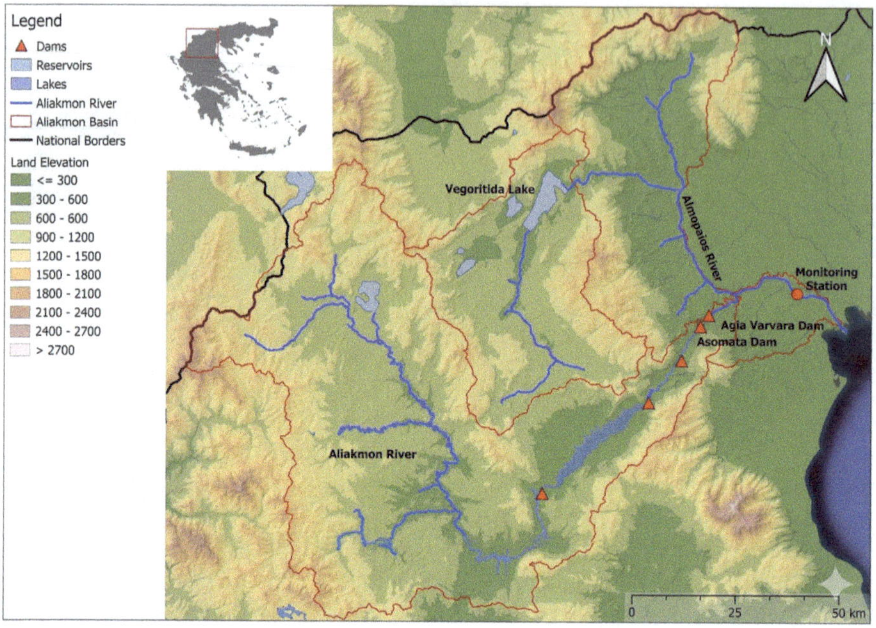

Fig. 41 The Aliakmon river basin

artery for the regions of Western and Central Macedonia, eventually discharging into the Thermaikos Gulf. The Aliakmon is the longest river flowing entirely through Greek sovereign territory. It is characterized by its transition from high-altitude, mountainous terrain to extensive lowland plains. The length of the main branch is approximately 297 km and the mean annual runoff is 2 billion m^3.

Flood protection has been an important goal of dam construction and other hydraulic works. The flow of the Aliakmon River is managed by five hydropower plants as it moves from the mountains towards the sea (Fig. 42):

1. **Ilarion** (Upstream Buffer) is the "newest" major addition; its primary technical role is to regulate the natural, unpredictable inflow of the Aliakmon. It allows for a more "smoothed" release into the downstream Polyphyto Lake by capturing heavy mountain runoff;
2. **Polyphyto** (Main Reservoir) is the largest reservoir of the system; it holds most of the water volume (1.22 billion m^3) and acts as the primary seasonal storage facility;
3. **Sfikia** (Reversible Engine) is a pumped storage hydropower plant (PSHP). It has three reversible units with a total installed capacity of 315 MW;
4. **Asomata** (Lower Buffer) is a smaller plant (108 MW) that acts as the lower reservoir for the Sfikia PSHP;
5. **Agia Varvara** (Final Regulation): A small dam that ensures a constant flow for the massive A0 irrigation canal.

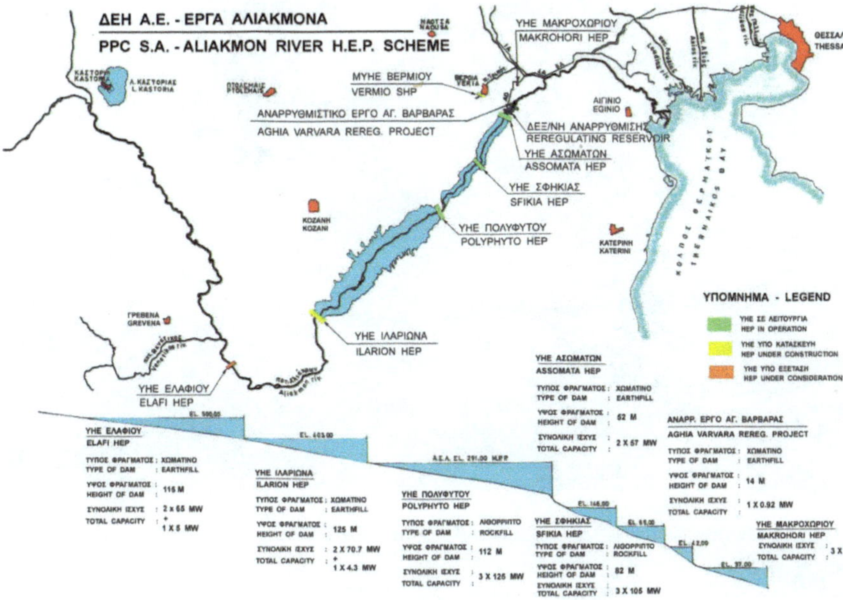

Fig. 42 General Layout of the Aliakmon river hydropower plants [6]

Figure 43 shows the catchment area. The surface of the sub-river basins is given in Table 2.

The technical specifications of the reservoirs and hydropower stations of the Aliakmon River are given in Table 3.

The Aliakmon River serves as a vital artery for biodiversity in Northern Greece and is surrounded by several Natura 2000-protected zones. Several entertainment, sports, tourist and environmental protection facilities have been developed around the Polyphyto reservoir. A graphical illustration of the various resources and demands of the Aliakmon river basin is presented in Fig. 44.

The Polyphyto hydroelectric power plant is a large hydroelectric facility on the Aliakmon river. It was commissioned in 1974 as part of the broader Polyphyto dam and reservoir project. The reservoir formed by the dam is called the Lake Polyphyto. It is one of Greece's largest artificial lakes and serves multiple purposes including power generation, flood control, water management for irrigation and urban supply, and environmental protection around the lake. The Polyphyto HPP is owned and operated by the Public Power Corporation (PPC), it has an installed capacity of 375 MW from three conventional Francis turbines (125 MW each) and generates approximately 400–600 GWh of electricity annually. More details are listed in Table 3.

Lake Polyphyto (Fig. 45) has a surface of roughly 74 km^2; its storage capacity is 1220 million m^3. The dam regulates the flow of the Aliakmon River, protecting downstream areas from seasonal flooding. The reservoir provides essential water

Fig. 43 The catchment area of the Aliakmon River upstream of the Agia Varvara dam [8]

Table 2 Aliakmon River catchment areas [6]

	Sub-River Basin	Surface (km^2)
1	Dam of Ilarion	5005
2	Dam of Polyphyto	830
3	Dam of Sfikia	175
4	Dam of Asomata	70
5	Dam of Agia Varvara	20
	Total	6100

for the agricultural plains of Imathia and Thessaloniki, as well as cooling water for nearby thermal power plants in the Ptolemaida region.

Apart from its technical function, the Polyphyto project has also shaped the geography and economy of the region: The lake is crossed by the Servia High Bridge (one of the longest in Greece). This bridge is a vital transport link for northern Greece.

The lake has developed into an important wetland, supporting numerous fish species and providing a habitat for migrating birds. Polyphyto HPP is currently undergoing a significant transition as it evolves from a traditional "workhorse" into a high-tech, digitally-integrated component of Greece's power grid. The most recent environmental and operational upgrades focus on digitalization, sustainability, and reservoir management (Figs. 46 and 47).

The Ilarion hydropower plant is a major power facility, and is the first and most upstream reservoir in a cascade of five plants on the river. Its commercial operation started in 2014. The hydropower plant has a total installed capacity of 153 MW and is equipped with two Francis turbines, each rated at 76.5 MW. There is a separate small

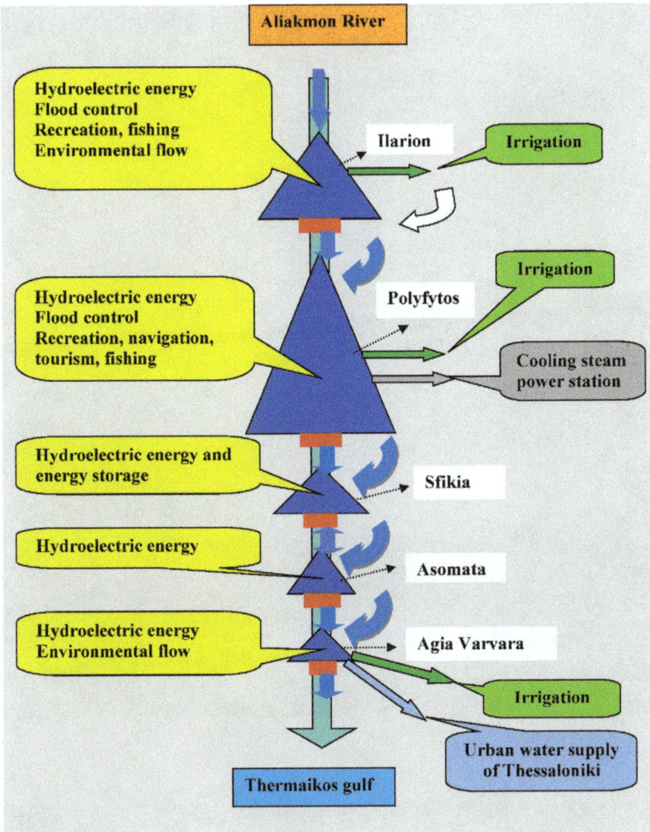

Fig. 44 Schematic view of the Aliakmon River basin, resources and demands [8]

Fig. 45 View of Lake Polyphyto

Fig. 46 The Polyphyto dam

Fig. 47 Polyphyto hydropower plant

Table 3 Technical characteristics of reservoirs and hydropower stations of the Aliakmon River [6]

Dam	Ilarion		Polyphyto	Sfikia	Asomata	Agia Varvara
Type and size of hydropower facilities	HPP	SHPP	HPP	PSHP	HPP	SHPP
Dam height (m)	130		112	82	52	
Flood storage level (m)	403.5		293	147	89	42.5
Maximum operating level (m)	398.5		291	146	85.5	42
Minimum operating level (m)	366		270	141.8	81	38.75
Gross storage (hm^3)	379.64		1939	99	53	4.5
Net storage (hm^3)	270.14		1220	18	10	3
Water spread (km^2)	21.9		74	4.3	2.6	0.9
Generator capacity (MW)	155.32 (2×77.66)	4.2	375 (3×125)	315 (3×105)	110 (2×55)	0.92
Water discharge (m^3/s)	160	6	311	635	303	8
Net head (m)	104		145.6	62	42	15
Specific consumption for production (m^3/kWh)	4.1	5	3.2	7.2	10	32
Adjustment ability	Annually		Annually	Daily	Daily	Daily
1st year of operation	2012	2015	1974	1985-86	1985	2008

hydropower unit (Ilarion SHPP) at the site with a capacity of 4.2 MW for utilizing environmental flow. Depending on hydrological conditions, the annual energy production is estimated at 320 GWh. The gross head is approximately 100 m. Regarding the dam and reservoir specifications, the dam is an earth-fill (gravel) structure with an impervious clay core. The dam is 130 m high from the foundation, with a crest level of around 403 m. Active storage is approximately 270 million m^3, which allows for significant seasonal water regulation. The catchment area is around 5005 km^2 (Fig. 48).

The plant is a multipurpose project, which means that it serves several functions beyond just generating electricity. It provides flexible, renewable energy and ancillary services (e.g. frequency regulation) to the Greek national grid. It is a critical source of drinking water for the city of Thessaloniki and provides water for agricultural needs in the surrounding regions. The SHPP maintains a constant environmental flow of approximately 4.5 m^3/s downstream to protect the local ecosystem. An important contribution of Ilarion is the stabilisation of the water level in Polyphyto Lake (Fig. 49).

Fig. 48 The Ilarion dam (downstream and upstream view)

Fig. 49 The Ilarion Small HPP (SHPP)

The Ilarion project serves as a prime example of modern multipurpose dam management, which must achieve a delicate balance between hydropower demands and the protection of a vital river ecosystem. As the Aliakmon is Greece's longest river and a critical water source for Thessaloniki, the project was subject to some of the nation's most stringent environmental regulations.

In order to mitigate the cultural impact of flooding the reservoir, a substantial archaeological rescue mission was conducted (peaking between 2012 and 2014), one of the largest in Greek history. This effort uncovered over 100 acres of history, ranging from Bronze Age settlements to Hellenistic-era homes.

A notable "environmental-cultural" success was the protection of the Zavorda and Tornikiou monasteries. By stabilizing the reservoir's slopes, engineers prevented the erosion that would otherwise have threatened these Byzantine-era structures, effectively incorporating them into the 'Cultural Landscape' protection of the site. Most notably, the 12th-century Monastery of Panagia Tornikiou was saved from the rising waters through an impressive engineering feat: the entire church was relocated to higher ground (see Fig. 50).

Fig. 50 Relocation of the church of Panagia Tornikiou to a higher ground

4 The Third Training School on Sustainable Hydropower (TS2-2)

The third Training School on Sustainable Hydropower (TS2-2) held in Porto, Portugal, aimed to strengthen interdisciplinary capacity-building on hydropower sustainability by combining engineering, ecological, and socio-environmental perspectives of hydropower assets in the Southwest European Area. TS2-2 took place between 3 and 6 September 2024 at the Faculdade de Engenharia do Porto, University of Porto, and was organized by Instituto Superior Técnico, University of Lisbon, and NOVA School of Science and Technology. The training, in which 20 participants from academic and private-sector institutions across Europe took part, offered a platform to deepen the participants' understanding of sustainable hydropower practices, and focused particularly on southwestern Europe.

4.1 Hydropower in Portugal

Hydropower has been the main source of renewable electricity generation on the Iberian Peninsula for several decades. However, its relative importance varies significantly between countries. In Portugal, hydropower remains a dominant renewable energy source, typically contributing between 10% and 30% of electricity generation depending on hydrological conditions [53]. In contrast, hydropower tends to play more of a complementary role in Spain, where wind and solar energy contribute more strongly to the electricity mix [50]. As of 2024, the installed hydropower capacity in Portugal was 8.352 GW. Pumped storage hydropower (PSH) accounts for 3.867 GW of this capacity, highlighting its importance for energy storage and system flexibility. Overall, hydropower generation amounted to around 15 TWh in 2024 [34], corresponding to roughly 30% of Portugal's total electricity production. Between 1 January and 31 December 2025, a total of 48.9 TWh of electricity was generated in mainland Portugal, of which 75.6% came from renewable sources. In contrast, Spain's hydropower system is larger but structurally different. As of 2024, hydropower installed capacity in Spain reached 22.747 GW, including 5.65 GW of pumped storage capacity. The annual hydropower generation reached approximately 35 TWh. While hydropower serves primarily as a balancing and flexibility resource in Spain's power system, recent policy and investment developments highlight a growing strategic emphasis on pumped storage hydropower (PSH) to support system flexibility. Under its National Energy and Climate Plan, Spain is considering the introduction of a capacity market to ensure long-term grid reliability and facilitate the integration of variable renewables. Several large pumped-storage projects are already making progress, including Iberdrola's Alcántara project (440 MW, 15 GWh) and the Valdecañas scheme (275 MW, 210 GWh), alongside new initiatives with more than 800 MW of additional capacity [34]. (see Fig. 51).

Recent developments in hydropower in Portugal reflect a strategic shift toward system flexibility and large-scale energy storage within an increasingly renewable-based electricity mix. While traditional large-scale reservoirs remain foundational to the national electricity system, there has been renewed emphasis on modernizing existing infrastructure in order to improve efficiency and environmental performance, including upgrades to turbines and fish passage solutions. Pumped storage facilities are playing an increasingly strategic role in integrating variable renewable generation, particularly wind and solar, by providing energy storage and grid stability.

By the end of 2025, total renewable installed capacity reached 21.9 GW [14], while the National Energy and Climate Plan (PNEC 2030) set a target of 42 GW to be met by 2030, indicating that approximately 20 GW of additional renewable capacity is yet to be deployed to meet national objectives. Within this context, hydropower continues to play a pivotal role, particularly through pumped storage schemes. The most significant recent addition is the commissioning of the 160 MW Alto Tâmega hydropower plant, marking the final stage of the Tâmega Hydropower Complex,

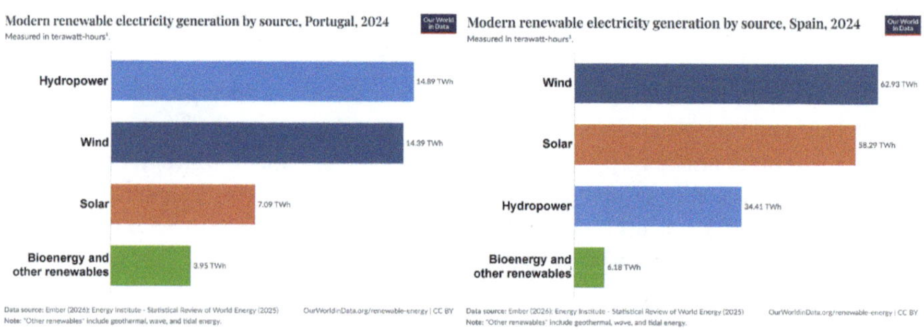

Fig. 51 Renewable electricity generation by source for (a) Portugal, and (b) Spain, in 2024

one of the largest hydroelectric developments in Europe in recent decades. The Alto Tâmega plant is located at the base of a 105 m-high double-curved arch dam, constructed with approximately 220,000 m^3 of concrete and spanning 335 m at the crest. Upon full commissioning of the complex, total installed capacity will have reacehd 1158 MW, comprising the 880 MW Gouvães pumped storage facility, the 118 MW Daivões plant, and Alto Tâmega [34].

According to the Portuguese Environment Agency (APA), Portugal has approximately 260 large dams distributed across its territory [3]. The country's total reservoir storage capacity is estimated to be 13.29 km^3, with the largest reservoirs located in the Guadiana and Tagus basins, both of which originate in Spain (Fig. 52) and play a strategic role in cross-border water and energy management [51]. Altogether, this development substantially strengthens Portugal's electricity storage capacity and reinforces hydropower's strategic role in enabling a power system with a high share of renewables.

Hydropower generation is particularly vulnerable to climate change, as it significantly impacts global water resources through shifts in precipitation timing, magnitude and spatial distribution, as well as changes in runoff regimes [44]. These impacts are highly region-specific and modulated by local climatic and physiographic conditions [45]. The Iberian Peninsula is not generally classified as a water-scarce region. This is demonstrated by various comparative studies that position it favourably in terms of average water availability in comparison to other European countries or those in the Mediterranean basin [33]. However, the climatic, geomorphological,

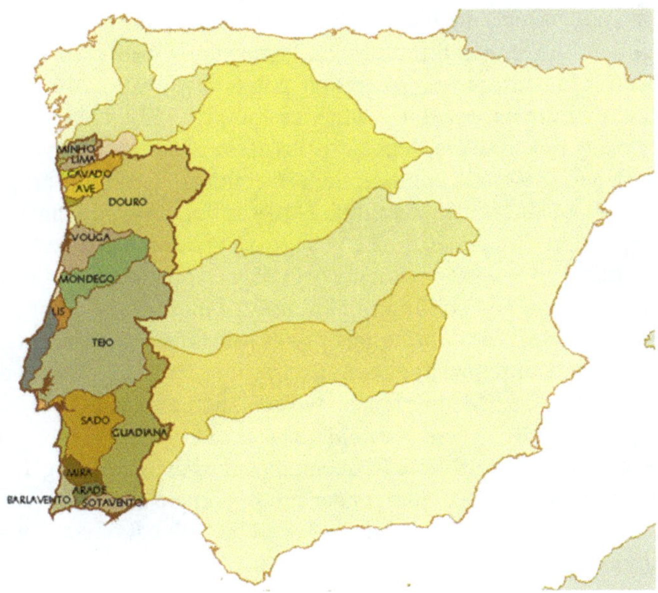

Fig. 52 Transboundary river basins in Portugal [15]

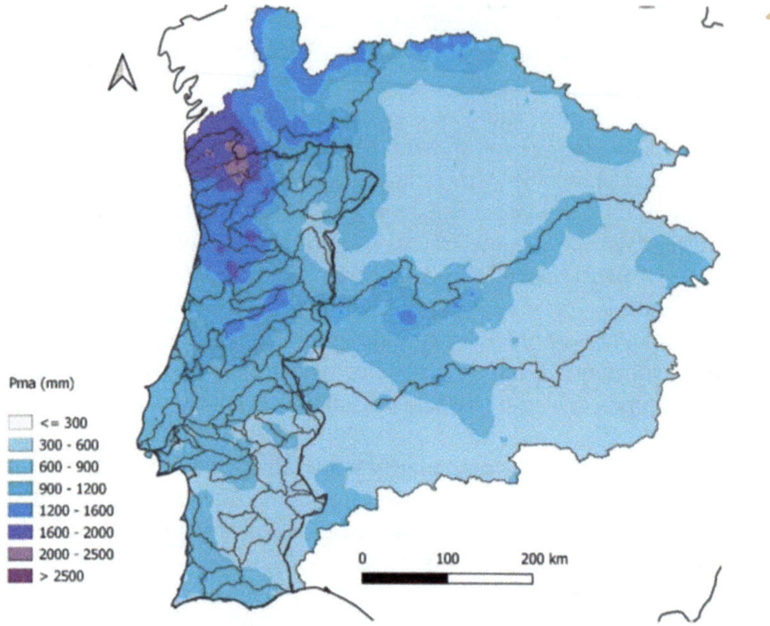

Fig. 53 Mean annual precipitation in Portugal and upstream river basin areas in Spain, 1930–2016 [3]

geological, and socioeconomic characteristics of the Iberian Peninsula present several challenges. In this context, the Iberian Peninsula is projected to experience a substantial decline in hydropower generation, potentially by up to 40% by 2070. The strongest reductions are expected in southern regions [5, 53]. This pronounced vulnerability is closely linked to the characteristics of the Mediterranean climate, which is marked by strong seasonal contrasts, high interannual variability in precipitation and a natural predominance of winter rainfall [13], as well as significant regional variations in precipitation (Fig. 53). Climate change is expected to exacerbate these features by increasing temperatures, reducing overall precipitation, intensifying drought frequency and duration, and shifting rainfall toward more intense but less frequent events. As a result, reduced runoff, altered reservoir inflow seasonality, and higher evaporative losses are anticipated, directly constraining hydropower production and operational reliability. The Mediterranean basin is therefore widely recognised as a climate change "hot spot", where warming rates exceed global averages and hydrological extremes are becoming more frequent, thus further amplifying the challenges for sustainable hydropower operation under future climate scenarios [21].

In recent years the Iberian Peninsula has been the subject of several studies that exlored the impact of climate change on the region. According to IPCC assessments, Southern Europe and the Mediterranean region are expected to see particularly detrimental effects. This is especially true for the Iberian Peninsula, in particular the region south of River Tagus, where temperature substantially increases and diminished pre-

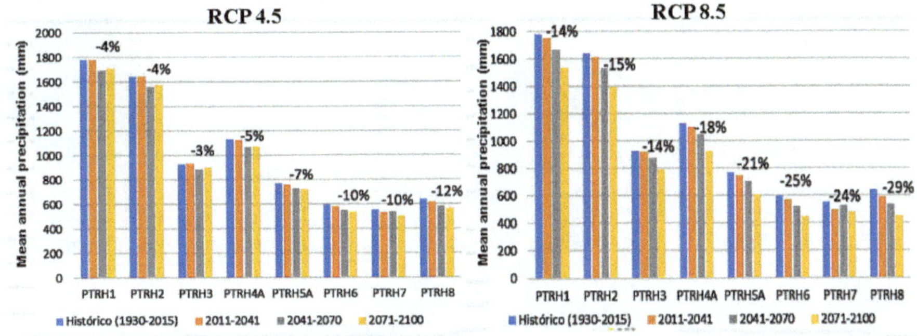

Fig. 54 Mean annual precipitation under climate change scenarios in Portugal [3] River Basins identification: PTRH1—Minho, Lima; PTRH2—Cávado, Ave, Leça; PTRH3—Douro; PTRH4—Vouga, Mondego e Lis; PTRH5—Tejo; PTRH6—Sado e Mira; PTRH7—Guadiana; PTRH8—Ribeiras do Algarve

cipitation (Fig. 54) and runoff are forecasted by 2100 [10]. Thus, the hydropower potential in Spain and Portugal will likely decrease due to reduced runoff, while higher temperatures will likely increase summer electricity demand [25]. This variability requires water storage solutions to handle frequent instances of water scarcity and the promotion of water management policies to balance water demand with availability.

In Portugal, hydropower assets are managed by a relatively small number of large operators, resulting in a highly concentrated ownership and operational structure. The national hydropower system is dominated by EDP, which operates most large storage, run-of-river (RoR) and pumped storage hydropower plants (PSHPs) across the main river basins, including the Douro, Cávado, Lima, Mondego, Tagus, and Guadiana. Other major international players with an important role include Iberdrola and Movhera, the latter being a joint venture between ENGIE, Crédit Agricole Assurances (through the Predica fund) and Natixis Investment Managers (through the Mirova fund).

Hydropower schemes are mainly located in mountainous areas and upstream river reaches, which are often transboundary basins shared with Spain, where the hydrological potential is highest before the water flows into more regulated downstream sections. This concentration of large operators shapes not only electricity generation but also river regulation, water management practices and the implementation of environmental mitigation measures. Although photovoltaic production has registered a significant increase in recent years, hydropower remains the main renewable energy resource in Portugal (Fig. 55).

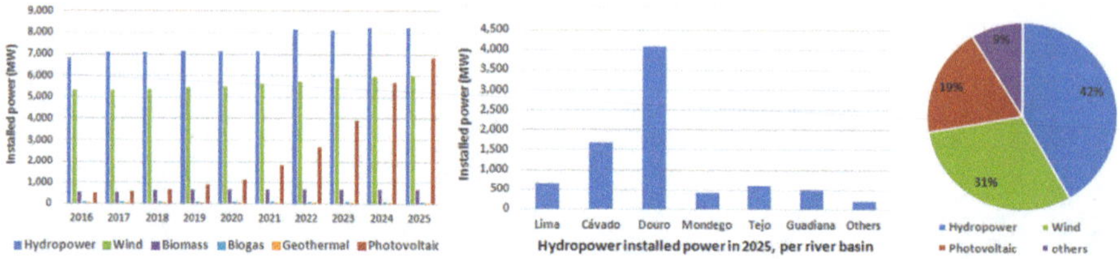

Fig. 55 Installed capacity of renewable energy sources (**a**, **b**) and production percentage (**c**) in Portugal in 2025 [14]

4.2 *TS2-2 Case Study*

The hydropower plant selected for the TS2-2 was the Foz Tua pumped storage hydropower plant (PSHP), located near the confluence of the Tua River with the Douro River, in northern Portugal, approximately 120 km from Porto (Fig. 56). The Tua River drainage basin includes 700 km^2 in Spain and totals an area of 3809 km^2. The estimated mean annual flow is around 1.2 million m^3, of which around 80% is attributable to the period between November and April. Commissioned in 2017, the plant is a multipurpose hydropower facility consisting of a double-curvature arch dam and an underground powerhouse (Fig. 57). The facility is equipped with two reversible Francis turbines, providing a total installed capacity of 274 MW and an

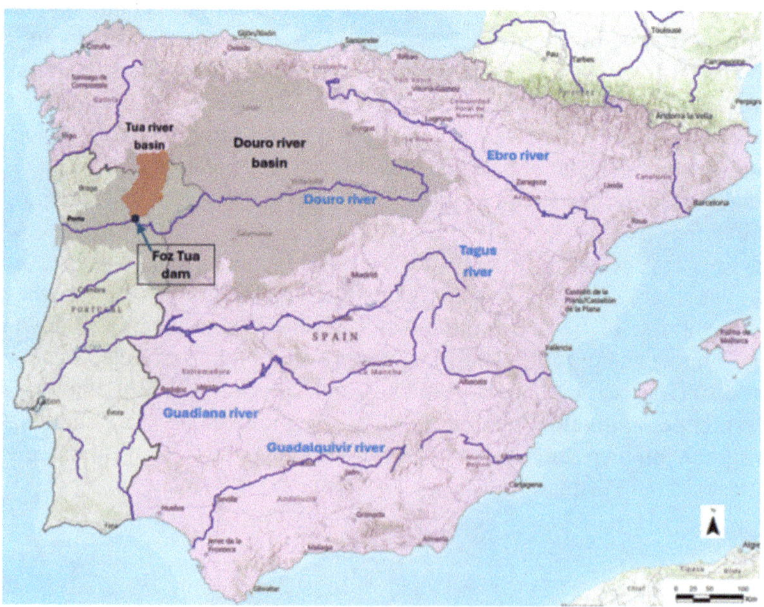

Fig. 56 Foz Tua dam location (adapted from ArcGIS Online)

Fig. 57 Foz Tua pumped storage hydropower plant (PSHP); **a** Downstream view, and **b** right bank view [14]

average annual electricity generation of 305 GWh. In addition, the plant has a pumping capacity of 246 MW and is operated by Movhera. The double arch dam has a height of 108 m and a crest length of 275 m (Figs. 57 and 58). The reservoir has an effective storage capacity of 10.8 million m^3 of water.

The project was developed in a highly sensitive environmental and cultural context, as the Tua Valley is characterised by high ecological value and landscape significance and lies within the Alto Douro Wine Region, a UNESCO World Heritage landscape of exceptional cultural and historical importance. Owing to this setting, strong emphasis was placed on architectural integration and landscape compatibility. The dam and its associated infrastructure were designed by the renowned Portuguese architect Eduardo Souto de Moura, whose guiding principle was to suppress the structure's conventional architectural character and instead express it as a functional machine embedded in the landscape. This design philosophy translated into a series of concrete solutions aimed at minimising visual intrusion, including a compact substation integrated into the powerhouse building, the elimination of spillway towers, the use of steep excavation slopes to reduce apparent height and the strategic concealment of the dam body by the powerhouse itself. Additional measures, such as avoiding access routes along the dam crest and defining non-intervention zones to preserve riparian vegetation, further contributed to reducing landscape fragmentation and visual impact. Altogether, this approach reflects a deliberate effort to reconcile large-scale energy infrastructure with cultural heritage preservation, making Foz Tua a notable example of how architectural design can enhance the social acceptance of hydropower developments in protected landscapes.

In addition, the Foz Tua project incorporates a comprehensive set of mitigation and compensation measures, including an environmental flow regime and ecological monitoring programmes. Today, the Foz Tua hydropower plant represents a relevant case study for examining trade-offs between renewable energy production, river ecosystem protection and socio-territorial impacts within Mediterranean river systems.

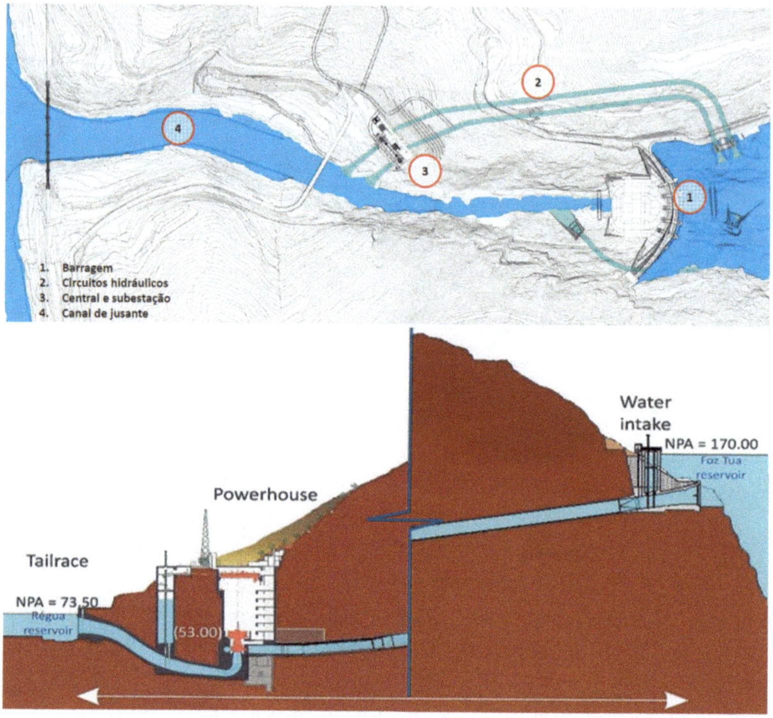

Fig. 58 Layout and hydraulic scheme of the Foz Tua pumped storage hydropower plant (PSHP). Top: General plan view showing the main components of the scheme: (1) dam, (2) water tunnels, (3) powerhouse and substation, and (4) downstream channel. Bottom: Longitudinal cross-section illustrating the water conveyance system from the Foz Tua reservoir to the powerhouse, including the intake, pressure tunnel, reversible Francis turbines, and tailrace discharge into the downstream reservoir.

5 The Fourth Training School on Sustainable Hydropower (TS3)

The fourth Training School on Sustainable Hydropower (TS3) aimed to assess hydropower assets specific to Northern European area and especially in Sweden. TS3 was held from 24-27 June 2025 in Uppsala. Luleå University of Technology (LTU) and Uppsala University (UU) acted as local organizers.

5.1 Hydropower in Sweden

Electricity production in Sweden between 2012 and 2023 is shown in Fig. 59. Annual electricity production in Sweden increased by around 12%, from 152 TWh in 2013

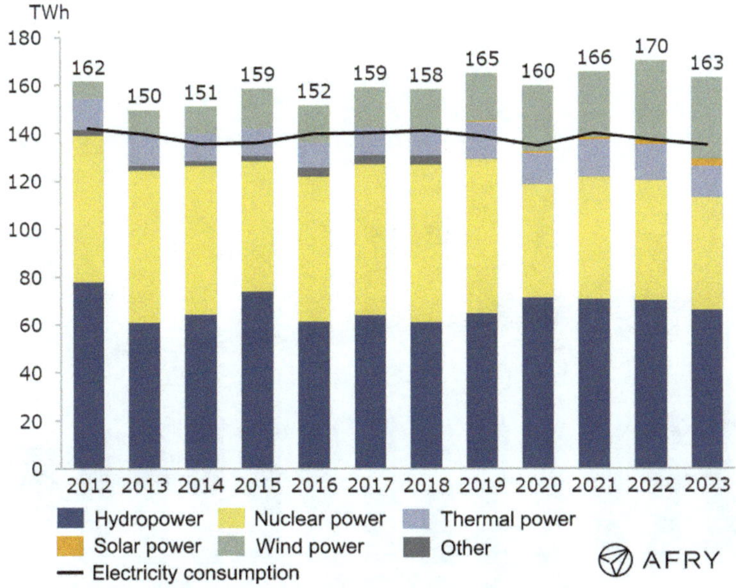

Fig. 59 Electricity production in Sweden from 2012 to 2023 [18]

to around 170 TWh in 2022. Hydroelectric and nuclear power production are the main sources of electricity in Sweden. It should be noted that there has been an approximately 5-fold increase in wind energy production between 2012 and 2023, while the contribution of solar energy production is very limited in the electricity mix.

An overview of the trend of installed capacity for electricity production in Sweden from 2012 to 2023 is presented in Fig. 60. It is noted that the total installed capacity for electricity generation in Sweden increased from 36 GW to 50 GW, up almost 40% from 2010 to 2023. The power production capacities of hydro and thermal power plants were maintained, while some nuclear power plants have been put on hold. It is important to note that the installed power of wind turbines increased approximately 10 times between 2010 and 2023, while the installed power of PVs started in 2015 and reached 4 GW (25% of the installed wind power) at the end of 2023.

More than 2100 hydropower plants with a total capacity of approximately 16 GW are installed in Sweden, see Fig. 61. It can be seen on the map in Fig. 61 that large hydropower plants (marked with large red circles) are installed on most rivers that cross Sweden from west to east and flow into the Gulf of Bothnia, the northernmost arm of the Baltic Sea. Hydroelectric power production was 66 TWh in 2023. The distribution of hydropower assets by category and their contribution to hydropower production in Sweden are presented in Table 4.

The national plan for modern environmental measures aims to revise and update the environmental permits for Swedish hydropower plants to meet modern environ-

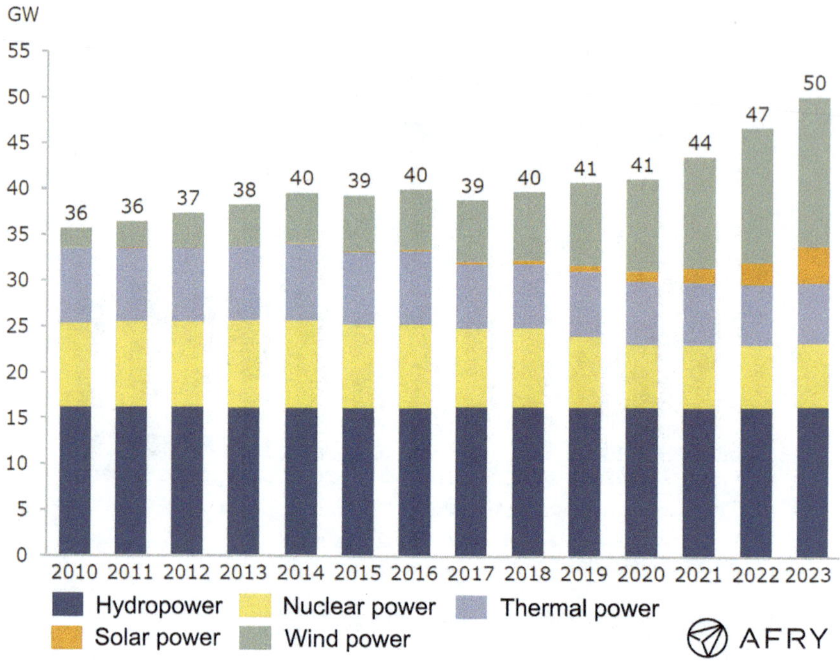

Fig. 60 Installed capacity in Sweden from 2010 to 2023 [18]

mental standards. The hydropower plants (about 200 marked with different colors in Fig. 61) were divided in different"application groups". Under this plan, almost all hydropower plants and their associated regulatory dams have prepared an application for revision of their permits to Land and Environmental Courts. The National Plan covers an operational period of 20 years, starting in February 2022.

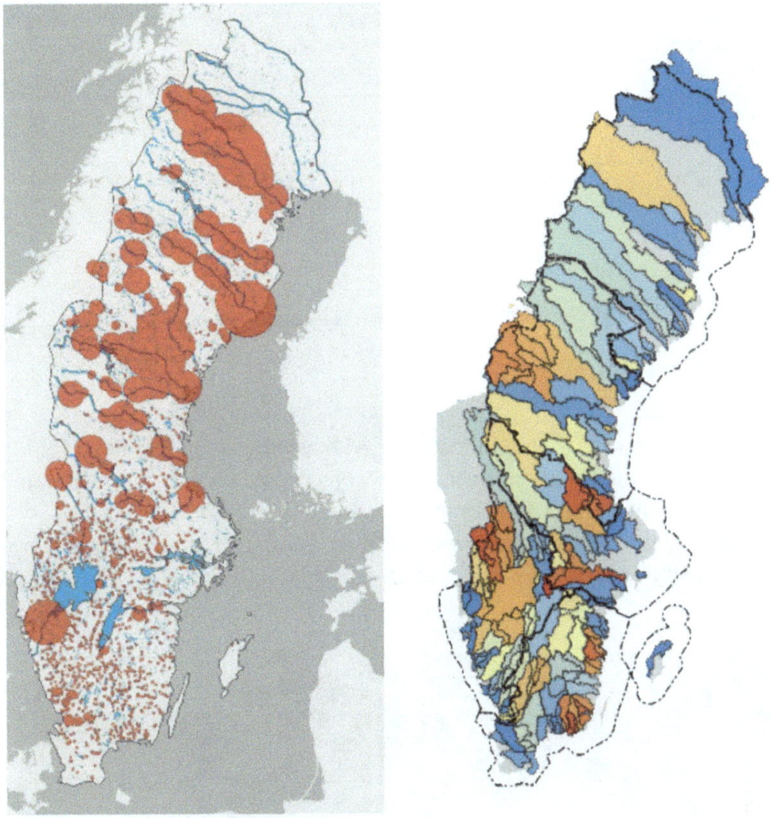

Fig. 61 Hydropower plants in operation in Sweden [18]

Vattenfall's oldest hydropower plant, Olidan on Göta River, was inaugurated in 1910. Älvkarleby hydropower plant celebrated its centenary in June 2015, while Porjus hydropower plant on Lule River marked its centenary in early 2015.

Vattenfall's research and development laboratory is located just a stone's throw from Älvkarleby hydropower station. Advanced three-dimensional computer simulations of various hydroelectric facilities are conducted there, while precise physical models are built to replicate real-life situations.

Despite sophisticated computer simulations, nothing beats physical models. A model can verify that flows, waves, vortices and water levels behave as expected before large-scale construction begins.

The Älvkarleby Laboratory is unique in Sweden. Only it is able to build and conduct tests on such models. Assignments come not only from Vattenfall but also from other power companies, universities and international clients. The 280 m^2 model of Slussen in Stockholm made the national news in 2012.

Table 4 Distribution of hydropower assets by category and their contribution to hydropower production in Sweden [18]

Installed power	Percentage of total production	Number of hydropower plants
> 10 MW	93.5%	208
1.5 MW–10 MW	3.9%	187
125 kW–1.5 MW	2.1%	680
< 125 kW	0.5%	1030
Total	100%	2105

5.2 *TS3 Case Study*

Älvkarleby hydropower plant is located in Älvkarleby, approximately 8 kilometres from the mouth of the Dalälven river. It was constructed by the Swedish State Power Board (now Vattenfall AB) and commissioned in 1915 (Figs. 62 and 63).

The establishment of hydropower stations marked a major leap forward for Swedish industry. Älvkarleby HPP became highly significant for industrialization

Fig. 62 Map with old and new Älvkarleby hydropower plants

throughout central Sweden as the power grid was expanded. Many mills and iron-works sourced their electricity here and the supply was sufficient to electrify Uppsala, Västerås, Gävle and Enköping.

The small community of Älvkarleby, located a few kilometres from the mouth of the Dalälven river, embraces both yesterday's technology and activities focused on future solutions. The hydropower station, which dates back to 1915, came at a time when knowledge of electricity had advanced and the discovery of alternating current made long-distance electricity transmission via power lines possible. This was considered "high tech" for its era, and great attention was paid to the architectural design of the facility. Both the machine hall and the switchgear building were constructed in the National Romantic style, giving the impression of a medieval castle, with steep roofs and facades clad in hand-made, hard-fired brick (Fig. 64).

The average water discharge in the Dalälven river at its mouth in the Baltic Sea is approximately 353–379 m^3/s. This number is based on long-term measurements and makes the river one of Sweden's largest waterways in terms of water flow. However, the discharge varies significantly throughout the year, with a recorded maximum of 2450 m^3/s and a minimum of 40 m^3/s. Therefore, the Dalälven river is heavily regulated and features many power stations, thus affecting the natural flow patterns (Fig. 65).

Älvkarleby was Vattenfall's third major project after the 135 MW Olidan hydropower station in Trollhättan and the 417 MW Porjus hydropower station. The aim was to supply electricity to central Sweden. The hydropower plant, designed by Erik Josephson, initially had a total capacity of 70 MW. The facility consists of five double horizontal camelback Francis turbines in the original turbine hall, each with a maximum flow of 90 m^3/s, resulting in a total flow of 450 m^3/s through the

Fig. 63 Old and new Älvkarleby hydropower plants: aerial view (in flow direction)

entire station. The hydropower plant has a head of 23 m and a water discharge of 700 m^3/s. Älvkarleby is a run-of-river (RoR) hydropower plant with a very limited storage capacity in the upstream reservoir. Therefore, a new hydropower station was built between 1988 and 1991, next to the old station. A sixth unit was added, giving the facility a total capacity of 125 MW. This new Älvkarleby hydropower plant is equipped with a vertical Kaplan turbine. The maximum flow is 250 m^3/s and the head 20 m (Fig. 66).

Around 1990, all dams were replaced with new ones in roughly the same locations, as the concrete of the original dams hasd severely degraded. The hydropower station was also expanded with the addition of new units.

Fig. 64 Old Älvkarleby hydropower plant: upstream view and side view

Fig. 65 Old Älvkarleby machine hall equipped with five horizontal hydropower units commissioned in 1915

Fig. 66 Älvkarleby spillway and dam

Today, the Älvkarleby hydropower plant has a different significance compared to a hundred years ago. While the mission to produce renewable electricity still remains, the plant now acts as a regulator within an energy system that increasingly consists of wind power production.

In keeping with the importance of the river as a sport fishery, Vattenfall conducts studies on the downstream migration of fish at its research facility in Älvkarleby.

6 The Fifth Training School on Sustainable Hydropower (TS4)

The fifth Training School on Sustainable Hydropower (TS4) aimed to assess hydropower assets specific to the Central European area, especially in Austria. TS4 was organized from 19-22 May 2026 in Salzburg. TU Wien acted as the local organizer.

6.1 Hydropower in Austria

Hydropower is an essential part of Austria's energy system. It provides a large portion of base load along the big rivers (e.g. Danube, Salzach, Inn, Drava), plus flexibility services via the big storage and pumped storage hydropower plants (PSHPs) in the alpine regions. The average electricity generation in Austria is stable at approximately 70 TWh per year (Fig. 67). More than 10 years ago the system was dominated by hydropower and fossil fuel generation, but in recent years the production of volatile renewable energy sources, i.e. wind and solar increased significantly. While they are increasingly replacing the share of fossil production, hydropower stays relatively constant. The latter is heavily influenced by annual precipitation restriction [7, 17].

The same trend is also evident in the installed capacity shown in Fig. 68. Over the last 10 years, the overall capacity has increased by 50% to over 30 GW, driven primarily by solar power. In the same period, hydropower increased by only 0.5 GW. In view of these two statistics and bearing in mind the Austrian history, hydropower was and continues to be the backbone of the national electricity grid [7, 17] (Fig. 69).

Due to the geographic topology differences between the flat eastern part of the country an the one hand, and the mountainous western and southern parts on the other, the distribution of solar, wind and hydropower plants across Austria is far from uniform. Wind is dominant the east, whereas solar and hydropower are spread out all over the country. 450 large HPP's (> 10 MW) operate in the country. Run-of-river HPPs on major rivers such as the Danube, Salzach, Inn and Drava provide the country with essential base load, while large storage and pumped storage plants are located in western and southern regions such as Salzburg, Tyrol, Vorarlberg and Carinthia. They are particularly important to provide flexibility and peak-load services to the national grid and beyond [40].

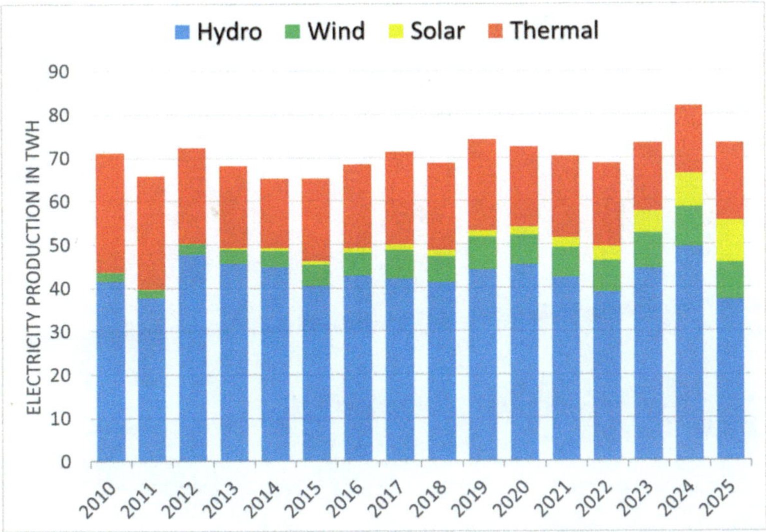

Fig. 67 Electricity production per generation type in TWh in Austria from 2010 to 2025 [17]

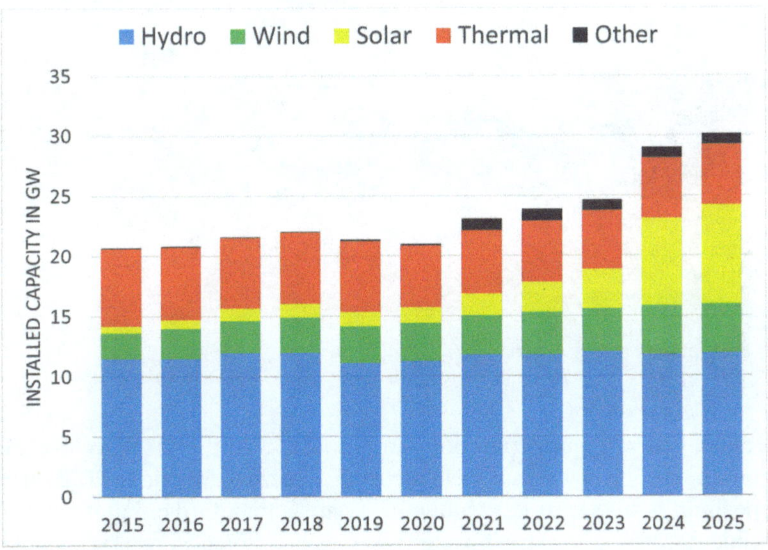

Fig. 68 Installed capacity per generation type in GW in Austria from 2015 to 2025 [7]

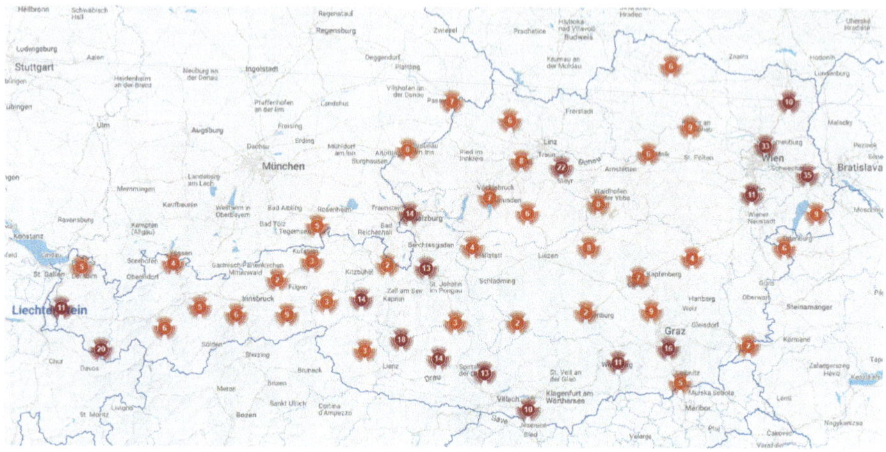

Fig. 69 Distribution of large hydropower plants (> 10 MW) across Austria [40]

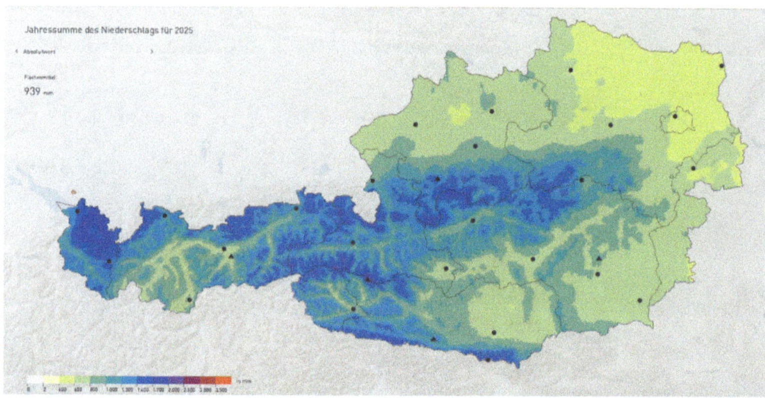

Fig. 70 Precipitation across Austria in 2025 [24]

The annual energy production of hydropower is directly linked to levels of precipitation. Figure 70 shows an typical distribution across Austria in 2025. It is clear that deposition is higher in the mountainous areas than in the flatter eastern parts of the country. 26% of Austria's water reserves are stored in reservoirs, lakes and glaciers [9].

6.2 TS4 Case Studies

Two hydropower plants were selected as case studies in TS4, namly Hintermuhr PSHP and Sohlstufe Lehen RoR HPP.

6.2.1 Sohlstufe Lehen HPP

The Sohlstufe Lehen hydropower plant (HPP) is a run-of-river (RoR) facility located on the Salzach River in the city of Salzburg, close to the historic old town. It is the final hydropower plant of the Salzach cascade before the river joins the Inn River. The plant was designed specifically to blend harmoniously into a densely populated urban environment, and its iconic 'swan-neck' hydraulic piers are widely recognized. The powerhouse's architecture and the four curved, arched weir fields were carefully designed to reflect the natural flow of water. This creates dynamic, curved profiles above the water surface that are reminiscent of swan necks (Fig. 71). A distinctive feature of the plant is the set of six-meter-high hydraulic levers for operating the gates. These structures serve functional and aesthetic purposes, reducing the visual impact of the mechanical equipment. Overall, the hydropower plant successfully combines heavy industrial engineering with refined urban architectural design.

Built between 2010 and 2013 on an existing riverbed step, the plant was designed to address riverbed erosion while generating renewable energy. With a design discharge of 250 m^3/s, a head of 6.6 m, and two Kaplan turbines, it has a capacity of approximately 13.7 MW and produces around 81 GWh of electricity per year. In addition to generating electricity, the project has significantly improved flood protection in the city. The weir system is designed to safely pass major flood events and manage sediment transport, which is essential in the sediment-rich Salzach River. Due to its sensitive urban location and complex geological conditions, careful planning, extensive communication with local residents and special architectural design were required for the project. Ecological measures were also implemented, including two fish passage systems (see Fig. 72) and the restoration of nearby river areas to improve ecological connectivity in line with EU water policy [46, 49].

Fig. 71 Sohlstufe Lehen RoR HPP (view from downstream) [46, 49]

Fig. 72 View of the two type of the fish passage systems installed at the Sohlstufe Lehen RoR HPP

6.2.2 Hintermuhr HPP

The Hintermuhr hydropower plant, commisioned in 1991, is a pumped storage facility located in the Austrian region of Salzburg. As it was the first cavern power station to be built in Salzburg, most of its mechanical and electrical components were installed underground in order to minimize surface impact and preserve the alpine environment of the surrounding national park. The facility uses water from the Rotgülden reservoir and its catchment area. This water is conveyed through a six-kilometre-long pressure tunnel and a penstock, utilizing a gross head of around 600 m to drive the turbines. Between 2006 and 2008, the plant was upgraded to a pumped storage facility by installing a new 68 MW reversible pump turbine alongside the existing 36 MW Pelton turbine. This conversion enables the plant to function as a large-scale energy storage system, providing peak-load generation and grid balancing services in both turbine and pump modes. This kind of flexibility in the energy system allows for better and more efficient integration of renewable energy sources. Additionally, the reservoir system supports hydrological management, including partial flood control and flow regulation. With an installed capacity of around 104 MW, the facility is the largest operated by Salzburg AG [47, 48] (Fig. 73).

Fig. 73 The Hintermuhr hydropower plant [47, 48]

7 Training School Outputs

Each of the five PEN@Hydropower training schools produced a variety of outputs. Firstly, different types of hydropower plants across Europe were compared with each other, providing insights into their technological, environmental, and regional characteristics. Secondly, the Hydropower Sustainability Standard (HSS) topics selected by the various stakeholder groups were analyzed and compared with each other, highlighting common priorities and differing sustainability perspectives. Thirdly, the training schools documented the personal experiences of both trainers and participants by way of interdisciplinary collaboration, field visits and knowledge exchange, demonstrating the educational and professional value of the initiative. Fourth, the authors provide an overview of the potential future implementation and expansion of the initiative into university curricula.

The following sections will address, each comparison in greater detail.

7.1 Comparison of Test Cases

One of the main objectives of the training schools was to compare different European regions and identify how hydropower plants are operated, which sustainability topics are prioritized and how regional conditions influence hydropower development. In line with this approach, the following analysis directly compares the selected hydropower plants with each other. More detailed descriptions of each power plant have been provided in the previous sections (2, 3, 4, 5, and 6), which discussed training schools individually.

The hydropower plants (HPPs) are grouped by asset type, namely run-of-river (RoR), storage and pumped storage hydropower plants (PSHPs), as listed in Table 5. Particular attention was paid ensuring a balanced representation of the various types of hydropower plants in the comparison.

The listed power plants are assigned to plant categories, then the plants of each category are compared with each other.

The first category to be addressed are run-of-river (RoR) hydropower plants (HPPs) (see Table 6). The three assessed power plants are located in different European regions, namely Romania, Sweden, and Austria. While the Timisoara SHPP and the original Älvkarleby HPP were commissioned over 100 years ago, Sohlstufe Lehen HPP and the newer Älvkarleby HPP expansion represent more modern hydropower developments.

The comparison of the three RoR hydropower plants reveals both common characteristics and regional differences in relation to the HSS criteria. From an environmental perspective, the more recently built Sohlstufe Lehen HPP is most closely aligned with modern biodiversity and river restoration practices, while the older plants exhibit varying degrees of ecological adaptation and flood-control capability. All three plants are highly relevant from the perspective of the local population, but Timisoara SHPP and Älvkarleby HPP stand out due to their historical and cultural significance. In terms of governance, the Austrian plant has more advanced and formalized management and safety standards, whereas the older facilities rely more on established operational structures and long-term experience. In economic terms, all three hydropower plants demonstrate medium to high long-term viability and continue to contribute significantly to the regional energy supply and security.

The second category comprises storage power plants (see Table 7). Three power plants in this category were also visited and assessed. These power plants are located in Romania and Greece, and were built in different decades. Ilarion and Polyphyto HPPs are both located on the same river cascade. Again, comparing the HPPs reveals common sustainability characteristics and regional differences. All storage hydropower projects demonstrate environmental flow compliance with environmental flow requirements and integrated water management measures. However, the Polyphyto HPP distinguishes itself through its large reservoir system and its strong focus on wetland protection and flow regulation. The Greek hydropower plant provides multiple benefits relating to the water–energy–food nexus, tourism and cultural

Table 5 Classification of TSs test cases by hydropower asset types

Type of HPP	Test cases
RoR	Timisoara SHPP (TS1), Älvkarleby HPP (TS3), Sohlstufe Lehen HPP (TS4)
Storage	Turnu-Ruieni HPP and Zervesti SHPP (TS1), Ilarion HPP + SHPP and Polyphyto HPP (TS2-1)
Pumped storage	Foz Tua PSHP (TS2-2), Hintermuhr PSHP (TS4)

Table 6 Comparison of RoR HPP test cases assessed during the TSs

HPP country year of commissioning	Timisoara SHPP Romania 1910	Älvkarleby HPP Sweden 1915	Sohlstufe Lehen HPP Austria 2013
Environmental	Moderate to minimal reservoir impact; likely lacks modern fish passage technology; high flood control capability	High reservoir impact; focus on fish bypasses and ecological measures; high flood control capability	Strong alignment with biodiversity best practices; high-quality fish bypass and river restoration measures; integrated flood control
Social and cultural	Very high cultural heritage value and strong integration into the urban identity and recreational landscape of the city	Recognized site with strong links to local history and sports activities	Good integration into the urban environment through public recreation areas and bridge infrastructure
Governance	Managed by local authorities; a clearer long-term strategy is required	Regularized management with established operational procedures	High governance standards with modern infrastructure safety measures
Economic	Very high long-term viability as one of the oldest operational plants in Europe	Very high long-term viability as one of the oldest operational plants in Europe	Economic viability and contribution to regional energy security

heritage integration. In contrast, the Romanian Turnu-Ruieni HPP primarily enhances regional energy security and decentralized electricity supply, while Zervesti SHPP delivers environmental flow. From a governance perspective, all projects have stable management. Restrictions on multipurpose use dominate Ilarion and Polyphyto HPP, while Turnu-Ruieni HPP is important for grid stability and restoration. In economic terms, storage HPPs are characterized by high operational efficiency, enhanced grid flexibility and significant contributions to regional energy stability, flood protection, tourism and farming activities.

The third and final category consists of pumped storage hydropower plants (PSHPs) (see Table 8). Two relatively modern hydropower plants were examined and assessed during the training schools. Both facilities are located in hilly and mountainous regions surrounded by national parks. The assessment of these PSHPs revealed both shared sustainability characteristics and regional differences relating to environmental, social, governance and economic aspects.

From an environmental perspective, both projects implement environmental flow management and mitigation measures. However, Foz Tua PSHP needs to focus more strongly on restoring vegetation and landscape impact compensation. The two projects differ in terms of their social and cultural dimensions. Foz Tua PSHP is closely connected to the UNESCO heritage debate and the architectural integration of the project into the surrounding landscape, whereas Hintermuhr PSHP

Table 7 Comparison of storage HPP test cases assessed during TSs

HPP country year of commissioning	Turnu-Ruieni HPP + Zervesti SHPP Romania 1983	Ilarion HPP + SHPP Greece 2014	Polyphyto HPP Greece 1974
Environmental	High head reservoir storage; environmental flow compliance	Cascade integration on the river; environmental flow compliance	Focus on flow regulation and artificial wetland protection; large reservoir; complex cascade
Social and Cultural	High contribution to regional energy security; decentralized power for local grid	Multipurpose use across the water-energy-food nexus (energy, water and irrigation); protection of cultural heritage; integration into landscape reflect a careful analysis of local heritage	Strong benefit sharing with local municipalities; multipurpose: power, water, irrigation, tourism
Governance	Managed by a national company; provides grid stability and restoration	Managed by a national company; multipurpose use of the river cascade; careful monitoring.	Managed by a national company; multipurpose use of the river cascade; careful monitoring
Economic	High operational efficiency; provides grid flexibility and black-start capability; high economic viability within a green portfolio	High efficiency and infrastructure safety; advanced flow regulation and operation; provides grid flexibility	Provides grid flexibility and flood protection; tourism and farming activities

contributes mainly to regional tourism, local communities and energy security. In terms of governance, both projects demonstrate high governance standards and strong stakeholder involvement. Foz Tua PSHP emphasizes community-based stewardship through regional development initiatives, while Hintermuhr PSHP focuses on transparency and decarbonization strategies. In economic terms, both pumped storage hydropower plants (PSHPs) provide substantial storage capacity and grid flexibility, thereby supporting the integration of renewable energy sources into the electricity system.

7.2 Comparison of HSS Topics

A key objective of the Training Schools, as noted in Sect. 7.1, is to examine power plants in different European regions from the perspectives of various stakeholders.

Table 8 Comparison of pumped storage hydropower plant (PSHP) test cases assessed during TSs

HPP Country Year of commissioning	Foz Tua PSHP Portugal 2017	Hintermuhr PSHP Austria 1991
Environmental	Mitigation of forest and land loss; ongoing monitoring of vegetation restoration; environmental flow management.	Utilizing existing water resources; environmental flow management and mitigation measures
Social and Cultural	High-profile UNESCO heritage debate; architectural integration to minimize visual impact	Strong community and tourism benefit and regional energy security
Governance	Strong community-based stewardship through Regional Development Association of the Tua Valley (ADRVT); high governance standards.	Strong commitment on transparency, decarbonization, safety standards and community integration; federal ownership
Economic	Provides grid flexibility; big storage which enables the integration of renewable energies	Provides grid flexibility; big storage which enables the integration of renewable energies

In this section, the selected Hydropower Sustainability Standard (HSS) topics are compared across the different stakeholder groups. As described in Sect. 1, participants worked solely with publicly available information and did not have access to operators' internal plant documentation. Although the HSS can be applied to all phases of a hydropower project, including planning, preparation, implementation and operation, the assessments carried out in the Training Schools covered only the operational phase with the exception of one group in TS2-1, which applied the HSS to the preparation stage. This can be chosen in cases where a project is close to transitioning to the next stage in its life cycle (e.g. from preparation to implementation or from implementation to operation), and the beneficiary wishes to be certified for the next stage of the life cycle.

For completeness' sake, all topics of the operational phase are listed below [32]:

- O-1 Communications and Consultation;
- O-2 Governance;
- O-3 Environmental and Social Issues Management;
- O-4 Hydrological Resource;
- O-5 Asset Reliability and Efficiency;
- O-6 Infrastructure Safety;
- O-7 Financial Viability;
- O-8 Project Benefits;
- O-9 Project Affected Communities and Livelihoods;
- O-10 Resettlement;
- O-11 Indigenous People;
- O-12 Labour and Working Conditions;
- O-13 Cultural Heritage;

- O-14 Public Heath;
- O-15 Biodiversity and Invasive Species;
- O-16 Erosion and Sedimentation;
- O-17 Water Quality;
- O-18 Reservoir Management;
- O-19 Downstream Flow Regime;
- O-20 Climate Change Mitigation and Resilience.

One group in TS2-1 used topics from the preparation phase. These are listed below [32]:

- P-13 Project Affected Communities and Livelihoods;
- P-17 Cultural Heritage.

These preparation phase topics are analogous to O-9 and O-13 respectively [32].

Table 9 lists all the topics chosen by the stakeholder groups, organized by training school. The most popular topic is O-6, followed by O-3 and O-5. An analysis of the table shows that some topics, e.g. O-2, O-7, O-9 and O-19, are mainly treated by a particular stakeholder group. Comparing the topics of individual TSs reveals more variety. Apart from TS2-1, O-6 was consistently selected.

The first evaluation focuses on the topics chosen by each group. Group 1 (Owners, operators) identified three common topics across the Training Schools. In Timisoara (TS1), the group highlighted issues such as limited communication with stakeholders, as well as inconsistencies in safety measures and labour management practices. In Greece (TS2-1), the multipurpose use of hydropower projects – covering water supply, irrigation, tourism and flood control – was recognized as a key benefit linked to reservoir management. In Foz Tua (TS2-2), Group 1 emphasized project benefits including a mobility plan for railways and tourism, job creation and reduction of the fire risk. Other topics addressed stakeholder engagement, such as the establishment of development agencies and the implementation of landscape protection initiatives. In Sweden (TS3), the owners and operators tended to focus more on technical aspects, including monitoring and maintenance, as well as biodiversity measures (related to flow and fish) and water quality concerns (seasonal temperature variations and suspended sediments). The Salzburg TS placed a strong focus on hydropeaking mitigation, flow regimes and labour conditions for the PSHP, while highlighting issues such as sedimentation and good asset reliability for the RoR HPP.

Meanwhile, Group 2 (Financiers, developers, insurers), selected four topics across the Training Schools. O-6 and O-7 were each chosen four times.

TS1, TS2-2, TS3 and TS4 all highlighted monitoring of infrastructure and dam safety, along with associated risk assessments. The financial viability topic, which was addressed in Greece, Portugal, Sweden and Austria, focused on reporting, market research and analysis, funding and offsets. In Thessaloniki, the second topic for Group 2 emphasized the reliability and efficiency of assets in relation to the flexible operation of HPPs, maintenance practices and market conditions. In Uppsala, the same topic emphasized reliable maintenance planning, established protocols and research support. For financiers and developers, the impacts of climate change are

Table 9 Selected hydropower sustainability standard topics in the training schools

	Group 1 (Owners, operators)	Group 2 (Financiers, developers, insurers)	Group 3 (Social stakeholders)	Group 4 (Environmental stakeholders)	Group 5 (Government, policies)	Most common topic in TS
TS1 (Timisoara) Turnu-Ruieni HPP, Zervesti SHPP, Timisoara SHPP	O-1, O-12	O-6	O-6 , O-9	O-3, O-4	O-2, O-5, O-6	O-6
TS2-1 (Thessaloniki) Ilarion HPP, Polyphyto HPP	O-8, O-18	O-5, O-7	O-3, O-8	O-17, O-19	P-13, P-17	O-8
TS2-2 (Porto) Foz Tua PSHP	O-3, O-8	O-6, O-7, O-20	O-9, O-15	O-3, O-15	O-2, O-6	O-3, O-6
TS3 (Uppsala) Älvkarleby HPP	O-5, O-15, O-17	O-5, O-6, O-7, O-18, O-20	O-3, O-6, O-12, O-13	O-15, O-16, O-17, O-19, O-20	O-8, O-20	O-5, O-6, O-15, O-17
TS4 (Salzburg) Hintermuhr PSHP, Sohlstufe Lehen HPP	O-4, O-5, O-12, O-16	O-6, O-7	O-1, O-3, O-6, O-9, O-12	O-15, O-16, O-17	O-2, O-3, O-6	O-3, O-6, O-12, O-16
Most common topic in groups	O-5, O-8, O-12	O-5, O-6, O-7, O-20	O-3, O-6, O-9	O-3, O-15, O-16, O-17, O-19	O-2, O-6	O-3, O-5, O-6

particularly relevant. Evaluations in Porto and Älvkarleby considered greenhouse gas (GHG) emissions and the increasing frequency of extreme events, highlighting potential risks. Adaptation measures are incorporated into planning, making reservoir management increasingly important. In Salzburg the wrong planning, thus lower cash flow was pointed out for one HPP, whereas the other has good financial performance.

Social stakeholders (Group 3) also chose several some topics several times, including Infrastructure Safety, Environmental and Social Issues Management and Project-Affected Communities and Livelihoods. Each of these topics was chosen three times. In TS1, the group highlighted monitoring, flood protection and safety procedures, as well as benefits such as local job creation, tourism and healthcare. In Greece, flood protection and the conservation of cultural heritage were emphasized for O-3, while the main benefit for local communities was considered to be the multipurpose use of the site, covering water supply, irrigation and tourism. In the Tua and Douro Valley, farmers and winegrowers benefited from compensation measures and biodiversity improvements through dedicated monitoring programmes, and the preservation of conservation areas was also prioritized. In Sweden, the focus was on maintaining high labour and safety standards, mitigating social and environmental impacts, and supporting local traditions and tourism. Both hydroelectric assets of Salzburg have high public acceptance, good integration, stakeholder engagement and labour conditions.

Similar to Group 2, environmental stakeholders (Group 4), selected the most common topics across the Training Schools. In TS1, the key issues were changing flow regimes, environmental flows, water quality monitoring and biodiversity. The HPPs visited either met or fell short of these criteria. In Greece, downstream flow regimes are regulated by irrigation, water supply, and environmental flow requirements, and the river that supplies Thessaloniki with fresh water is continuously monitored. At Foz Tua PSHP, which is located within a national park and adjacent to the Douro UNESCO World Heritage Site, biodiversity monitoring is essential. Stakeholders from the fields of ecology, agriculture and tourism are actively involved in discussions. In Sweden, although no typical fish pass has been installed, the operator and local fishers monitor fish and vegetation habitats, including the spawning of native species. The impacts of climate change, erosion, sedimentation and water quality are also treated as important considerations. The hydropower plants in Salzburg have reduced their surface impact. The PSHP is mitigating hydro peaking while ensuring high water quality. The RoR HPP supports fish migration with two fish ladders but struggles with high sediment concentrations and regulation.

Government and policy stakeholders (Group 5) selected the fewest common topics across the Training Schools. Governance and infrastructure safety were each chosen three times. In TS2-1, the group focused on topics from the preparation phase, examining the potential loss of cultural heritage during the construction of Polyphyto HPP, while also recognizing the benefits of multipurpose use. In TS1, the group highlighted grid services and asset efficiency at the Turnu-Ruieni HPP, but also noted governance irregularities and shortcomings in safety measures. The TS2-2 use case addressed social and environmental responsibilities, where multiple certifications demonstrate ethical business practices and corporate governance, although no public financial reporting was available. Although infrastructure safety is technically well documented and assessed, improved communication could enhance its effectiveness. In TS3, Group 5 identified project benefits including flood protection and local job creation. The facility also contributes to reducing GHG emissions despite limited measures to strengthen the resilience of freshwater ecosystems. In Salzburg, both power plants demonstrate good compliance and operational management, as well as high safety standards, and are socially and environmentally integrated.

The second evaluation focuses on the topics selected in each of the individual Training Schools. As Table 9 shows, O-6 was the only topic chosen more than once in TS1 (Timişoara), namely by Groups 2, 3 and 5, which can be seen as a reflection of the broad range of issues addressed. In TS2-1 (Thessaloniki), only one topic was selected by multiple groups. While most groups emphasized the multipurpose use of the sites, Groups 1 and 3 explored more specific and dedicated topics. In TS2-2 (Porto) the most frequently selected topics were Environmental and Social Issues Management and Infrastructure Safety. Owners and operators (G1) and environmental stakeholders (G4) chose O-3, while both Group 2 and 5 addressed O-6. The situation in TS3 (Uppsala) was notably different, with four topics being selected by multiple stakeholder groups. Groups 1 and 2 looked at asset reliability and efficiency; groups 2 and 3 focused on infrastructure safety; and Groups 1 and 4 shared two topics related to water quality and biodiversity. A similar pattern was observed in Salzburg, where groups also selected four common topics. As in Timisoara, Infrastructure Safety was chosen by groups 2, 3 and 5. Groups 1 and 3 selected Labour and Working Conditions, Groups 1 and 4 selected Erosion and Sedimentation, and the social and environmental stakeholder groups selected Environmental and Social Issues Management.

It is noteworthy that Owners and Operators (G1) and Social or Environmental Stakeholders (G3 and G4) shared topics in four out of five Training Schools. Groups 2 and 5 also selected the same topic (O-6) in three of the five editions.

Interestingly, Owners and Operators always selected topics that were different to those seleted by Governance and Policy (G5). Only once they did choose the same topics as Financiers, Developers and Insurers (G2). The same applies to Social and Environmental Stakeholders (G3).

When considering the limited availability of publicly accessible data for the visited sites, particularly in Sweden, the overlap of topics within a Training School is likely to be influenced by data availability.

Furthermore, the frequent selection of Infrastructure Safety (O-6) appears to be largely driven by the requirement to publicly communicate this subject to ensure protection and raise awareness among the general public. It should be noted that participants' professional backgrounds did not significantly influence their choice of topics. This is particularly evident in Sect. 7.4. The explanations provided by the guide during the visit also influenced topic selection.

7.3 *Personal Reflections and Statements*

As prevoiusly mentioned in Sect. 1, the training schools were attended by 111 participants from 31 countries. To gain a deeper understanding of participants' perspectives and collect valuable feedback on the course, an anonymous questionnaire was handed out to all attendees. The questionnaire covered the following topics:

- Participant background and motivation;
- Training school evaluation;
- Professional impact and networking;
- Experience and feedback;
- Willingness to provide a testimonial for this book.

To make it easier to assess all the answers, the questions were assigned to the categories of the SWOT (Strengths, Weaknesses, Opportunities and Threats) analysis (see the categorization in Table 10). Some questions allow for interpretation and can result in a response that is categorized as either a strength or a weakness, or an opportunity or a threat, and vice versa.

Table 11 presents the survey results using the same structure as the questionnaire and the assigned SWOT categories. This approach allows for a clearer interpretation of the responses and reduces uncertainties related to the categorization of individual questions, particularly where answers may reflect both positive and negative aspects depending on the participants' perspectives. Some questions are easy to assess (score between 1 (bad) and 5 (good)), while others allow for personal more statements. The latter are evaluated outside of Table 11.

The survey revealed that participants were primarily motivated to attend the Training School (TS) in order to broaden their perspective, deepen their knowledge of sustainable hydropower and expand their professional network. Overall, expectations were largely met, with the general framework and quality of the TS being rated as good. Participants highlighted the effectiveness of the trainers and speakers, the usefulness of the training materials and the well-organized logistics. They also appreciated the balance between theoretical sessions and practical activities, as well as the appropriate duration and pacing. Depending on their professional background, many participants noted that the knowledge gained could be applied in their future work, and that the TS had enhanced their understanding of the current challenges and innovations in sustainable hydropower. Participants especially valued the practical applications, such as the site visits and implementation of the Hydropower Sustainability Standard. Overall, attendees recommended the TS to their colleagues and peers, reflecting the programme's strong positive impact To further highlight personal opinion, including those of trainer, editors asked participants and speakers to provide short testimonials/statements.

Testimonial of Training School Teacher: "In times where so many -seemingly-new teaching concepts are proliferating, it is especially satisfying to see again the basics work best for teaching bright minds: the combination of in-the-room-teaching and the opportunity to test the fresh insights and newly learned tools on-site. That

Table 10 Survey questions assigned to SWOT categories

SWOT category	Survey questions
Strengths	How would you rate the overall quality of the training school?
	Did the training school meet your expectations in terms of learning about sustainable hydropower practice?
	Which topics or sessions did you find most valuable?
	Were the guest trainers and speakers effective in delivering their content?
	Did the training materials support learning effectively?
	How satisfied were you with the organization and logistics?
	Was the balance between theoretical sessions and practical activities appropriate?
	How would you rate the overall duration and pacing of the training school?
	What was the highlight of your experience during the training school?
	How likely are you to recommend the Training School to a colleague or within your community?
Weaknesses	How would you rate the overall quality of the training school?
	Did the training school meet your expectations in terms of learning about sustainable hydropower practice?
	Was the balance between theoretical sessions and practical activities appropriate?
	Which content should be expanded and/or given more time?
	How would you rate the overall duration and pacing of the training school?
	Were the guest trainers and speakers effective in delivering their content?
Opportunities	What motivated you to attend this training school?
	Did the training school help you expand your professional network?
	Did the training school enhance your awareness of current challenges and innovations in sustainable hydropower?
	Do you have any suggestions for improving future Training Schools?
	Which other content should be incorporated?
	Any other comments or suggestions?
Threats	How likely are you to apply the knowledge gained in your current or future work?
	How satisfied were you with the organization and logistics?
	If the training school were offered at a cost, what is the maximum attendance fee you would consider reasonable?
	Please share any challenges you experienced during the training school (if any)

Table 11 Survey results assigned to SWOT categories

SWOT category	Survey results
Strengths	The framework as a whole was rated as good in terms of both quality and overall performance
	Expectations in sustainable hydropower learning were largely met
	The most valuable topics were those related to sustainability and the HSS
	Guest trainers and speakers were effective in delivering their content
	The training materials supported effectively the TS
	Organization and logistics were good
	Theoretical sessions and practical activities were balanced
	The overall duration and pacing were appropriate
	The attendee's professional network was significantly expanded
	Overall, the knowledge gained will be applied in future work. However, this depends on the attendee's background and affiliation
	Current challenges and innovations in sustainable hydropower were explained
	Participants emphasized the importance of site visits and networking opportunities
	Participants recommend the TS to colleagues and peers
	Overall, there were no major challenges
Weaknesses	More time should be set aside for preparing the assessment
Opportunities	Networking opportunities and expanding knowledge in sustainable hydropower were the main motivations for attending the TS
	Coverage of additional specific topics and details was recommended
	A reasonable attendance fee was considered to be either <500 € or 500–1000 €
Threats	–

combination is at the core of the COST Action Training Schools, and the learning that results from them is deep. What I take away is equally the energy and the connections that form among participants, trainers, experts, organisers—a community that keeps talking long after the week ends. I am deeply grateful to have been part of it".

Testimonial of a participant from academia: "Participating in the Training School was an extremely valuable experience for me. The lectures on the Hydropower Sustainability Standard, as well as the practical application at an existing hydropower plant, significantly broadened my perspective and made me realize how many different aspects are involved in the hydropower sector. In addition, during this week I had the opportunity to establish many new connections with fellow participants from various academic and professional backgrounds. Our conversations allowed me to gain insights into their work and expertise. Overall, the Training School was highly enriching for me, both professionally and personally. For the practical application of the standard, we were divided into five groups, each focusing on a specific part of the Standard, such as Social Aspects or Environmental Aspects. While this approach allowed us to gain deeper insight into one particular area, it also meant that the other aspects were only addressed to a limited extent. One possible way to further broaden the overall perspective would be to rotate the groups, allowing participants to engage more closely with all areas of the Standard. However, this would require significantly more time, which may limit the practical feasibility of this suggestion".

Testimonial of a participant from industry: "Coming from an O&M role in hydropower plants, my day-to-day work is usually focused on operational issues, reliability and keeping the units running properly. The Training School was particularly interesting for me because it exposed me to a more academic and sustainability-oriented perspective that I do not normally deal with in daily operations. I especially appreciated the discussions with researchers and people from different technical backgrounds, since they helped me look beyond the immediate plant-level problems and understand hydropower from a more holistic perspective. It was also a very good opportunity to better understand how sustainability in hydropower is viewed from different stakeholder perspectives, which is something we do not usually analyze in depth during daily O&M activities. Overall, I found the experience very valuable and enriching".

The testimonials from trainers, academic participants, and industry representatives demonstrate that the Training Schools successfully combined theoretical learning with practical application, creating a highly enriching educational environment. The program fostered interdisciplinary exchange, strengthened professional networks, and encouraged collaboration between academia and industry. Participants particularly valued the opportunity to examine hydropower sustainability from multiple stakeholder perspectives, while the practical site assessments contributed to a deeper understanding of real-world challenges and solutions related to sustainable hydropower development and operation.

The feedback collected through the testimonials strongly overlaps with the results obtained from the participant surveys, confirming the overall positive impact of the Training Schools on knowledge transfer, practical skill development, interdisciplinary collaboration and the strengthening of the European sustainable hydropower community. At the same time, the feedback provided valuable and constructive suggestions for further improvement, particularly a possible expansion of topics and a broader exposure to all aspects of the Hydropower Sustainability Standard during the practical assessment activities.

7.4 Outlook on Implementation Possibilities

The PEN@Hydropower members' capacity-building activities in sustainable hydropower led to the initiation of theoretical knowledge and practical skills transfer nuclei in European universities. Starting with the 2025–2026 academic year, the Swiss Federal Institute of Technology of Lausanne (EPFL) in Switzerland and the Brno University of Technology (BUT) in the Czech Republic will introduce

Table 12 Selected topics for HSS assessment during the TU Wien excursion 2026

Group 1 (Owners, operators)	Group 2 (Financiers, developers, insurers)	Group 3 (Social stakeholders)	Group 4 (Environmental stakeholders)	Group 5 (Government, policies)
O-4, O-5, O-17	O-7, O-8	O-12, O-14, O-18	O-15, O-19	O-17, O-19

courses in sustainable hydropower. Over forty master's students have enhanced their theoretical knowledge of sustainable hydropower using materials provided by PEN@Hydropower.

In the future, the existing framework of the Training School could be expanded or integrated into university curricula. As discussed in the previous chapter on the current state of European hydropower education, there are currently no lectures that explicitly combine sustainability and hydropower. At a time when holistic approaches are becoming increasingly important, introducing or continuing a dedicated hydropower lecture or training programme would be highly valuable for the sector.

Feedback from colleagues indicates that many incorporate visits to hydropower plants into their courses, and these activities are consistently highly valued by students, as reflected in the Training School survey results (see Sect. 7.3). Combining theoretical education with practical, on-site experience enhances learning, as it allows students to observe the scale and complexity of hydropower facilities first-hand. Experiencing these facilities in person rather than through slides, images or diagrams provides a more comprehensive understanding of technical and sustainability aspects, helping to bridge the gap between theory and practice. This integrated approach could serve as a model for future hydropower education initiatives.

Similar to the site visits in the PEN@Hydropower Training School, such excursions can be used to illustrate sustainability assessments. As part of their biennial multi-day site visit, the Fluid-Flow Machinery research group at TU Wien incorporated a Hydropower Sustainability Standard (HSS) assessment into their programme. As in the training school, a specific site was selected for the assessment, preceded by a theoretical lecture covering sustainability in general and the HSS in particular. The same stakeholder groups were involved to mirror practical conditions. However, a notable difference is that all participants were mechanical engineering students, unlike the diverse professional backgrounds of Training School attendees. The visited site was unique in that its primary purpose is water supply rather than energy generation. This choice was deliberate, reflecting the objective of broadening the application of the HSS to water infrastructure beyond hydropower. The selected and assessed topics for this year's (2026) TU Wien excursion are listed in Table 12.

A comparison of the selected topics and the results from the field trip revealed that professional background played only a minor role in both topic selection and assessment. It should be noted that the amount of publicly available data significantly influences topic selection. The explanations provided by the guide during the visit also influenced topic selection A comparison of the HSS topics selected in the context of the Training Schools with those of the TU Wien field trip reveals topic overlaps for stakeholder groups G1, G2 and G4. In the other two groups, only one topic was selected once in previous Training Schools. While there is no definitive evidence as to the reason, the influence of more detailed explanations and the different infrastructure purpose is evident.

8 Conclusions

This chapter outlines the contribution of the PEN@Hydropower community to addressing the identified capacity-building gap in sustainable hydropower. It details the framework that was developed, implemented and tested by the community to support education and training in this field. This framework consists of a series of training schools on sustainable hydropower, which aim to transfer theoretical knowledge and practical skills to early career investigators (ECIs), as well as master's and PhD students. The PEN@Hydropower Training School structure combines: (*i*) clearly defined learning objectives related to sustainable hydropower; (*ii*) a curriculum integrating the theoretical foundations of the Hydropower Sustainability Standard (HSS) with practical assessment activities at existing hydropower facilities; (*iii*) modern teaching approaches, including teamwork and interactive exercises; and (*iv*) individual and group-based assessment methods.

The PEN@Hydropower Training School on Sustainable Hydropower was designed as a four-day program. The first day introduced the PEN@Hydropower initiative, presented the activities of the Working Groups (WGs) in hydropower and organized participants into stakeholder groups. The second day focused on the theoretical background of the Hydropower Sustainability Standard (HSS), complemented by hands-on exercises. During the third day, participants visited hydropower facilities and prepared assessment reports based on the HSS, strengthening their practical skills. The fourth day concluded with group presentations of the assessment results and the awarding of participation certificates.

Between 2023 and 2026, PEN@Hydropower organized five Training Schools across Europe, ranging geographically from east (TS1 in Timişoara, Romania) to west (TS4 in Salzburg, Austria), and from north (TS3 in Uppsala, Sweden) to south (TS2-1 in Thessaloniki, Greece and TS2-2 in Porto, Portugal). Although all events followed the same general format, each Training School had its own specific characteristics. These differences were shaped both by the diverse professional backgrounds and institutional affiliations of the participants from across Europe and by the unique features of each hydropower asset assessed. In total, 111 Early Career Investigators (ECIs) received training in sustainable hydropower through PEN@Hydropower activities across Europe. Particular attention was given to gender balance, resulting in the selection of 37 young women pursuing careers in hydropower. Around 30 industry representatives participated in the events, including 25 participants directly selected from hydropower companies across Europe. Experience gained from organizing the Training Schools demonstrated that the most successful events were those in which participants from academia (universities, research centres, and NGOs) represented diverse professional backgrounds, while the balance between academic and industry participants was maintained at approximately two-thirds to one-third.

A key objective of the Training Schools was to examine hydropower plants in different European regions from the perspectives of various stakeholder groups. The selected HSS topics were compared across the stakeholder groups: Group 1—owners and operators; Group 2—financiers, developers and insurers; Group 3—social stakeholders; Group 4—environmental stakeholders; and Group 5—government and policies. The assessments conducted during the Training School were limited to the operational phase although the HSS can be applied to all phases of a project. The most selected topics are O-6 Infrastructure Safety, followed by O-3 Environmental and Social Issues Management and O-5 Asset Reliability and Efficiency. The prominence of Infrastructure Safety (O-6) appears to be largely linked to the need for transparent public communication to ensure public protection and awareness. The analysis of topic selection also highlighted a strong alignment of interests among the different stakeholder groups. It should be noted that the availability of publicly accessible information strongly influenced topic selection, alongside explanations provided by site guides during the technical visits.

The test cases selected for assessment included all major types of hydropower infrastructure, namely run-of-river, storage and pumped storage hydropower plants (PSHPs). The assessed facilities ranged from small to large installed capacities and had commissioning dates spanning from 1910 to 2017. Comparative analysis based on the HSS highlighted the challenges faced by hydropower infrastructure commissioned during the last century, as well as the lessons learned from facilities commissioned in the current century.

The materials and experience gained in capacity building on sustainable hydropower by PEN@Hydropower members led to the establishment of theoretical and practical knowledge transfer initiatives in European universities. Swiss Federal Technology Institute of Lausanne (EPFL) from Switzerland and Brno University of Technology (BUT) from Czech Republic have introduced courses in sustainable hydropower starting with the 2025–2026 academic year. More than fourty master's students enhanced their theoretical knowledge in sustainable hydropower based on materials provided by PEN@Hydropower. TU Wien from Vienna, Austria has started implementing the transfer of theoretical knowledge and practical skills in sustainable hydropower within the framework of their biennial multi-day excursion. As in the PEN@Hydropower Training School, a specific site was selected for the assessment, preceded by a theoretical lecture covering sustainability in general and the HSS in particular. The same stakeholder groups were involved to mirror practical conditions. In total twenty students applied the topic to a real water-related infrastructure. All above data demonstrates the ongoing development of sustainability educational capacities in sustainable hydropower across Europe.

The PEN@Hydropower community provided ECIs, master's students and PhD students with opportunities to build professional networks, strengthen their knowledge in sustainable hydropower and collaborate in interdisciplinary groups. Participants who have undergone training in sustainable hydropower will contribute to the implementation of sustainable concepts in support of the clean energy transition. Participants who have completed the training programme will represent a new generation of professionals who will contribute to the implementation of sustainable practices and concepts within the hydropower sector and the wider transition to renewable energy.

Acknowledgements The authors would like to express their sincere gratitude to the local organizing committees for their excellent organisation of the PEN@Hydropower Training Schools and for preparing the dedicated Training School descriptions of this chapter. Special thanks are extended to the hydropower plant owners for granting access to their facilities and supporting the activities during the Training Schools. The authors also acknowledge the valuable contribution of the Hydropower Sustainability Alliance, especially João Costa, as well as all presenters of the individual Training Schools. Particular appreciation is given to Jian-hua Meng and Barbara Fischer-Aupperle for their teaching and guidance on the Hydropower Sustainability Standard. The authors further thank all contributors who provided testimonials for this chapter. In addition, the authors would like to acknowledge VGBE Energy and all individuals involved in promoting and disseminating the initiative. Sincere thanks are also extended to all participants in the Training Schools, whose engagement and enthusiasm greatly contributed to the success of the program. Finally, the authors gratefully acknowledge the support of the PEN@Hydropower community and the COST Association.

Competing Interests The authors have no conflicts of interest to declare that are relevant to the content of this chapter.

References

1. Administraţia Naţională Apele Române - ABA Banat (2023) Planul de Management al Riscului la Inundaţii - Spaţiul Hidrografic Banat, Timişoara, Romania. https://rowater.ro/. Cited 18 Feb 2026
2. Administraţia Naţională de Meteorologie, (ANM) (2025) Harta cu precipitaţiile in Romania 2024 [Map]. ANM INSPIRE Catalogue. Bucureşti, Romania. https://www.meteoromania.ro/. Cited 18 Feb 2026
3. Agência Portuguesa do Ambiente (APA) (2022) Barragens de Portugal. https://apambiente.pt/prevencao-e-gestao-de-riscos/barragens-de-portugal. Cited 6 Feb 2026
4. Aldescu V–C (2010) Studiul inundaţiilor pe râul Timiş amenajat în conceptul "mai mult spaţiu pentru râuri". Phd thesis, University Politehnica Timişoara, Politehnica Publishing House, Timişoara, Romania
5. Andrade C, Contente J, Santos J A (2021) Climate change projections of aridity conditions in the Iberian Peninsula. Water (Basel), p 13. https://doi.org/10.3390/w13152035
6. Argyrakis I (2016) Hydroelectric power plants of PCC SA in Greece. In: Proceedings of the seminar developing albania hydroelectric potential, IENE workshop, vol 3, Tirana, Albania, 3 June 2016. https://www.iene.eu/microsites/developing-albania-hydroelectric-potential/articlefiles/2nd_Session/HPP_OF_PPC_Argyrakis.pdf. Cited 12 Feb 2026
7. Austrian Power Grid - Installed capacity per production type. https://markt.apg.at/en/transparency/generation/installed-power-plant-capacity/. Cited 11 Feb 2026
8. Bakanos PI, Katsifarakis KL (2020) Optimizing current and future hydroelectric energy production and water uses of the complex multi-reservoir system in the Aliakmon River. Greece Energies 13(24):6499. https://doi.org/10.3390/en13246499

9. Bundesministerin für Landwirtschaft, Regionen und Tourismus (2018). Wasserland Österreich. ISBN: 978-3-85174-075-2
10. Calvin K, Dasgupta D, Krinner G, Mukherji A, Thorne PW, Trisos C, et al. (2023) IPCC 2023: climate change 2023: synthesis report. In: Lee H and Romero J (eds) Contribution of working groups I, II and III to the sixth assessment report of the intergovernmental panel on climate change, IPCC, Geneva, Switzerland (2023). https://doi.org/10.59327/IPCC/AR6-9789291691647
11. Catona V (1934) Monografia Uzinei Electrice Timişoara (1884–1934). Tipografia Românească, Timişoara, Romania
12. Cojocar M (2008) Hidroconstructia, vol I. Hydropower constructions, 2nd edn. Bucureşti, Romania
13. De Luis M, Gonzalez-Hidalgo J, Longares L, Stepanek P (2009) Seasonal precipitation trends in Mediterranean Iberian Peninsula in second half of XX century. Int J Climatol 29:1312–1323. https://doi.org/10.1002/joc.1778
14. Direção Geral de Energia e Geologia (DGEG) (2026). https://www.dgeg.gov.pt/. Cited 18 Feb 2026
15. Direcção Geral de Recursos Marinhos (DGRM) (2018) Report on the eel stock, fishery and other impacts in: Portugal. https://www.dgrm.pt/. Cited 6 Feb 2026
16. Drăghici IA, Rus I-D, Cococeanu A, Muntean S (2022) Improved operation strategy of the pumping system implemented in timisoara municipal water treatment station. Sustainability 14(15):9130. https://doi.org/10.3390/su14159130
17. E-Control - Elektrizitätsstatistik. https://www.e-control.at/statistik/e-statistik. Cited 11 Feb 2026
18. Energiforsk (2023) Potential for increased output and balancing from Sweden's hydropower. Report 2023:938. https://www.sverigesingenjorer.se/globalassets/aktuelltpress/potential-for-vattenkraft_slutrapport_forord.pdf. Cited 18 Feb 2026
19. European Commission (2021) The European pillar of social rights action plan. Luxembourg: Publications Office of the European Union. https://data.europa.eu/doi/10.2767/45. Cited 18 Feb 2026
20. European Environment Agency (2023) Natura 2000 data - the European network of protected sites. Munţii ŢarcuTarcu Mountains (ROSCI0126 - SCI). https://natura2000.eea.europa.eu/Natura2000/sdf/#/sdf?site=ROSCI0126&release=55. Cited 18 Feb 2026
21. Fernandez B, Carramolino R, Kavvadias K, Adamovic M, Gonzalez H (2018) The water-power nexus of the Iberian Peninsula power system – WATERFLEX project. https://doi.org/10.2760/739963
22. Georgiopoulos D and Dimou X K (2017) Hydroelectric plants 2015-2040 – development and prospects. In: 3rd National conference of dams and reservoirs, plant management and growth prospects Athens, Greece, 12–15 October 2017. https://www.researchgate.net/publication/320419653. Cited 12 Feb 2026
23. Georgopoulou E, Mirasgedis S, Sarafidis Y, Giannakopoulos C, Varotsos KV, Gakis N (2024) Climate change impacts on the energy system of a climate-vulnerable mediterranean country (Greece). Atmosphere 15(3):286. https://doi.org/10.3390/atmos15030286
24. GeoSphere Austria - Klimamonitoring. https://klimaportal.geosphere.at/klimamonitoring/. Cited 11 Feb 2026
25. Giannakopoulos C, Le Sager P, Bindi M, Moriondo M, Kostopoulou E, Goodess CM (2009) Climatic changes and associated impacts in the Mediterranean resulting from a 2°C global warming. Glob Planet Chang 68:209–224. https://doi.org/10.1016/j.gloplacha.2009.06.001
26. Guran MI (2022) Evoluţia în timp a amenajărilor de îmbunătăţiri funciare în România. Politehnica Publishing House, PhD thesis
27. Hellenic Association for Energy Economics and National Bank of Greece (2025) Greek Energy Market Report. https://www.haee.gr/. Cited 12 Feb 2026
28. Hydropower Sustainability Alliance - Eastmain-1 Hydroelectric Development. https://www.hs-alliance.org/published-assessments/eastmain-1/. Cited 24 May 2026

29. Hydropower Sustainability Alliance (HSA) (2021) Hydropower Sustainability Standard. https://www.hydropower.org/sustainability-standard. Cited 18 Feb 2026
30. Hydropower Sustainability Alliance. https://www.hs-alliance.org/. Cited 22 May 2026
31. Institutul Naţional de Statistică (INS) (2026) Anuarul Statistic al României [Date despre consumul de energie]. Bucureşti, Romania. https://insse.ro/cms/ro/content/statistica-energiei. Cited 18 Feb 2026
32. International hydropower Association (2020) Hydropower Sustainability Assessment Protocol. ISBN 978-0-9566228-1-5
33. International Hydropower Association Limited (IHA) (2022) Hydropower status report: sector trends and insights
34. International Hydropower Association Limited (IHA) (2026) Resources Hub. https://www.hydropower.org/. Cited 6 Feb 2026
35. International Renewable Energy Agency (IRENA) (2025) Renewable energy statistical profile: Greece. https://www.irena.org/-/media/Files/IRENA/Agency/Statistics/Statistical_Profiles/Europe/Greece_Europe_RE_SP.pdf. Cited 12 Feb 2026
36. Kaldellis JK (2008) Critical evaluation of the hydropower applications in Greece. Renew Sust Energ Rev 12:218–234. https://doi.org/10.1016/j.rser.2006.05.003
37. Liuba C (2003) The Hydropower Plant Ruieni as provider of tehnological services for the national power system. In: Power systems conference (PSC). Timişoara, Romania, pp 6–7
38. Ministry of Energy and Environment (2024) National Plan for Energy and Climate (NECP): Greece. https://ypen.gov.gr/energeia/esek/. Cited 12 Feb 2026
39. Muntean S, Câmpian VC, Cuzmoş A, Dumbravă C, Brebu N, Augustinov L (2009) Experimental investigations into a Francis turbine with low specific speed. UPB Sci Bull Ser D Mech Eng 71(2):3–12
40. Oesterreichs Energie: Österreichs E-Wirtschaft - Kraftwerkskarte. https://oesterreichsenergie.at/kraftwerkskarte. Cited 11 Feb 2026
41. Opriş M (1987) Timişoara. Tehnică Publishing House, Bucureşti, Romania, Mica monografie urbanistică
42. Panţu H (2009) Modernizarea sistemului hidrotehnic Timiş – Bega. PhD thesis, University Politehnica Timişoara, Politehnica Publishing House, Timişoara, Romania
43. Pop S, Ban I, Cojocar M (2012) Hidroconstructia. Subsidiary Bistra - Poiana Marului 30 years 1982 – 2012. Bucureşti, Romania
44. Ramião JP, Carvalho-Santos C, Pinto R, Pascoal C (2023) Hydropower contribution to the renewable energy transition under climate change. Water Resour Manag 37:175–191. https://doi.org/10.1007/s11269-022-03361-4
45. Rasilla DF, Garmendia C, García-Codron JC (2013) Climate change projections of streamflow in the Iberian peninsula. Int J Water Resour Dev 29:184–200. https://doi.org/10.1080/07900627.2012.721716
46. Salzburg AG - Kraftwerksgruppe Flachgau Tennengau. https://www.salzburg-ag.at/content/dam/web18/dokumente/unternehmen/erzeugung/Kraftwerksgruppe-Flachgau-Tennengau.pdf/. Cited 11 Feb 2026
47. Salzburg AG - Kraftwerksgruppe Lungau. https://www.salzburg-ag.at/content/dam/web18/dokumente/unternehmen/erzeugung/Kraftwerksgruppe-Lungau.pdf/. Cited 11 Feb 2026
48. Salzburg AG - Wasserkraftwerk Hintermuhr. https://www.salzburg-ag.at/ueber-die-salzburg-ag/unternehmen/erzeugung/erzeugungsanlagen/wasserkraftwerk-hintermuhr.html/. Cited 11 Feb 2026
49. Salzburg AG - Wasserkraftwerk Sohlstufe Lehen. https://www.salzburg-ag.at/ueber-die-salzburg-ag/unternehmen/erzeugung/erzeugungsanlagen/wasserkraftwerk-sohlstufe-lehen.html/. Cited 11 Feb 2026
50. Santos Silva C, Couceiro D, Castro MGR (2024) The role of pumped hydro storage in the Portuguese National Plan for energy and climate for 2030: A hybrid approach using Energy PLAN and machine learning. J Energy Storage 103(114194):4. https://doi.org/10.1016/j.est.2024.11419(APA)

51. Sistema Nacional de Informação de Recursos Hídricos (SNIRH) (2023) Boletim Armazenamento de Albufeiras. https://snirh.apambiente.pt/. Cited 6 Feb 2026
52. Stan S (2011) Turbinele Timişoarei. *Arhitectura* 4. Available at https://arhitectura-1906.ro/2011/08/turbinele-timisoarei/. Cited 18 Feb 2026
53. Teotónio C, Fortes P, Roebeling P, Rodriguez M, Robaina-Alves M (2017) Assessing the impacts of climate change on hydropower generation and the power sector in Portugal: a partial equilibrium approach. Renew Sust Energ Rev 74:788–799. https://doi.org/10.1016/j.rser.2017.03.002

Part III
Sustainable Hydropower Fields

The transition towards more sustainable and resilient energy systems is reshaping the role of hydropower worldwide. While hydropower has long been recognized for its contribution to renewable electricity generation and energy security, current challenges related to system flexibility and resilience, digitalization, climate change, environmental protection, new markets and policy frameworks, and societal acceptance are driving the sector towards new paradigms of development and operation. In this evolving context, sustainable hydropower is increasingly understood as a multidimensional concept that extends beyond energy production alone, requiring integrated approaches capable of balancing technical performance, ecological integrity, economic viability, and social expectations.

This part builds upon these emerging perspectives by bringing together a collection of chapters developed within the framework of the COST Action PEN@Hydropower. The part reflects the interdisciplinary nature of the network and its effort to foster collaboration among researchers, industry stakeholders, policymakers, and early-career scientists working on the future of hydropower systems across Europe.

A key value of this part lies in its cross-disciplinary perspective. The contributions presented here move beyond conventional hydropower analysis by combining insights from engineering, environmental sciences, climate studies, energy systems, governance, and digital technologies. Together, the chapters provide a broader understanding of how hydropower systems are evolving in response to the accelerating clean energy transition and the growing need for sustainable water and energy management strategies.

The part also highlights the importance of regarding hydropower as part of a larger and interconnected system. Future hydropower development will increasingly depend on its ability to interact effectively with variable renewable energy sources, changing hydrological conditions, environmental restoration objectives, and societal priorities. Addressing these interconnected challenges requires not only technological innovation, but also new forms of collaboration and integrated decision-making. The work developed within PEN@Hydropower demonstrates how interdisciplinary research networks can contribute to this transition by creating platforms for knowledge exchange, harmonization of methodologies, and development of shared research agendas.

The five chapters included in this part reflect the core themes explored within the PEN@Hydropower Working Groups. Together, they provide both a review of current knowledge and a discussion of emerging research directions for sustainable hydropower systems.

The first chapter examines the evolving role of hydropower and pumped storage hydropower in future low-carbon energy systems, with particular attention to flexibility services, energy storage, and system resilience. The second chapter focuses on technological innovation and digitalization, including advances in monitoring systems, predictive maintenance, and digital tools that can support more efficient and adaptive hydropower operation. The third chapter addresses environmental sustainability and climate adaptation, discussing topics such as ecological impacts, river restoration, environmental flows, sediment management, and the implications of changing hydrological regimes. The fourth chapter explores governance and policy dimensions including sustainability integration into hydropower planning and management. Finally, the fifth chapter synthesizes interdisciplinary perspectives including the integration of environmental and social considerations as well as stakeholders engagements by outlining an holistic framework for research and exploitation of sustainable hydropower.

By combining foundational perspectives with insights emerging directly from the PEN@Hydropower Working Groups, this section offers a timely overview of the challenges, opportunities, and research priorities shaping the future of sustainable hydropower in Europe and beyond.

Hydropower Role in Flexible Energy Synergies

Elena Vagnoni

Where water meets adaptability, power becomes resilient.

Abstract High shares of variable renewable energy sources change both the physics and the economics of power systems: variability and uncertainty increase; synchronous inertia decreases; and the demand for reserves, ramping, and storage expands across time scales from sub-seconds to seasons. In Europe, hydropower is simultaneously a mature renewable generator, a dispatchable resource, and a long-duration storage technology through pumped storage hydropower. This chapter synthesizes recent evidence and results from the COST Action PEN@Hydropower (CA21104) to explain hydropower's evolving role in flexible energy synergies. A system perspective including flexibility products, residual load, and storage requirements is combined here with an asset perspective with focus on the implication of flexible operation on the wear and tear and OPEX costs of hydropower equipment. In particular, fatigue accumulation and useful remaining lifetime of mechanical component are considered. Quantitative results from EnergyPLAN modeling for five European countries and from fatigue mapping of five hydropower plants are used to illustrate the magnitude of storage value and the mechanical cost of flexibility. The chapter concludes with an integrated roadmap linking modernization, digitalization, and environmental mitigation to enable sustainable flexibility provision in forward-looking power systems amid climate change challenges.

Keywords Flexibility · Storage · Ancillary services

E. Vagnoni (✉)
Hydraulic Machinery Platform, École Polytechnique Fédérale de Lausanne, Lausanne, Switzerland
e-mail: Elena.vagnoni@epfl.ch

E. Doujak et al. (eds.), *Rethinking Hydropower: A Holistic Vision for a Sustainable Future*, Springer Tracts in Civil Engineering,
https://doi.org/10.1007/978-3-032-33328-5_7

1 Introduction

Europe's electricity sector is experiencing a structural transition driven by decarbonization objectives and the accelerating deployment of wind and solar photovoltaic (PV) generation. The defining technical consequence is not only the reduction of fossil energy, but also a shift in how variability and uncertainty are managed [1, 2]. Wind and PV output follow meteorological conditions and can change quickly over multiple horizons [2]. In parallel, converter-interfaced generation displaces synchronous machines and reduces system inertia. Power systems that historically relied on inertial response and governor behavior of thermal fleets now need to maintain frequency stability, adequacy, and reliability with a different portfolio of assets [3, 4]. Evidence from high-renewable regions has highlighted the operational implications of this change. In systems with increasing shares of inverter-based generation, system operators observe reduced frequency response capability and greater difficulty in maintaining reserves and frequency containment margins [5, 6]. These observations are consistent with the underlying physics: when fewer synchronous machines are online, the stored kinetic energy in rotating masses is smaller, and the rate of change of frequency following disturbances can increase. At the same time, net-load ramps become more frequent and larger as the penetration of variable renewable energy sources (VRES) increases. Consequently, flexibility becomes a primary design variable in future power systems [7]. Hydropower is uniquely positioned in this landscape. It is renewable, dispatchable within operational constraints, and capable of shifting energy over longer durations when configured as storage or pumped-storage. Available literature, as well as the work performed and presented in PEN@Hydropower, depict the ability of hydropower to contribute across the full time-horizon of grid services and emphasizes its role in stabilizing high-VRES systems [8, 9]. Hydropower can provide inertia, frequency containment and restoration, ramping, intraday balancing, and long-duration energy shifting. Because few other technologies can cover this spectrum, hydropower is increasingly framed as a cornerstone flexibility technology for deeply decarbonized scenarios. However, flexibility comes at a cost. Providing fast ramping, frequent start-stop cycles, and operation away from best-efficiency point (BEP) increases mechanical stress and hydraulic instabilities. Consequently, the energy transition contains an underappreciated tension: the same flexibility demanded by the system can shorten the lifetime of critical assets unless modernization and digitalization are deployed. In parallel, flexibility patterns may intensify ecological pressures, increasing the need for environmental mitigation and multi-stakeholder governance [9, 10]. This chapter provides evidence on the crucial role of hydropower flexibility in the energy system, explicitly linking (i) power-system flexibility needs, (ii) hydropower technology capabilities and asset integrity, and (iii) roadmap towards the future role of hydropower to enable the success of the energy transition. It builds primarily on available literature and outputs from the PEN@Hydropower COST Action: a European survey of hydropower flexibility participation; EnergyPLAN-based analyses of storage needs for five countries; and fatigue lifetime assessments of flexible operation for five hydropower plants.

2 Flexibility in Decarbonized Power Systems

2.1 Flexibility and Power-System Services

Flexibility requirements are driven by both predictable variability of daily generation and consumption and uncertainty due to forecast errors across timescales. At high penetration, curtailment becomes frequent unless storage and transmission expand. These results are consistent across modeling paradigms, including unit-commitment simulations, capacity expansion models, and integrated energy system models. An important nuance is that the marginal value of flexibility is not smooth: it tends to increase at points where the system transitions from "rare surplus" to "frequent surplus," or where deficits become multi-day during low-wind events. This is precisely the range where long-duration storage technologies become critical [1, 2].

A convenient quantitative lens for flexibility is residual load, defined as $L_{res}(t) = L(t) - G_{VRES}(t)$, where $L(t)$ is demand and $G_{VRES}(t)$ is wind/solar generation. The statistical properties of $L_{res}(t)$, its variance, ramp-rate distribution, duration curve, and autocorrelation shape the need for operational reserves and storage [1, 2]. When VRES penetration increases, negative residual load episodes become more frequent, representing potential curtailment if storage or exports are insufficient. Positive residual load peaks represent periods when dispatchable generation or storage discharge must provide energy [5].

System operators translate residual-load dynamics into layered services. At the fastest horizon, inertial response and fast frequency response aim to reduce the rate of change of frequency following contingencies. Primary control (FCR) stabilizes frequency by automatic governor response within seconds. Secondary control (aFRR) restores frequency and reserves within minutes; tertiary reserves (mFRR) handle sustained imbalances and forecast errors. At slower horizons, ramping and load following cover continuous net-load changes while intraday balancing manages within-day uncertainty and long-duration storage addresses multi-day deficits and curtailment. Hydropower can contribute across these horizons thanks to its storage capacity and by adjusting discharge rapidly. In pumped storage hydropower (PSH), the same unit can provide upward reserve through generation and downward reserve by pumping depending on its current mode, subject to mode change constraints [6].

2.2 Flexibility Metrics and Constraints

Operational indicators include start-up time, minimum stable generation, ramp rate, and operating range. Several studies and real-world scenarios report hydropower start-up times typically in the 2–20 min range and often below 3 min for certain configurations. Minimum stable operation is often around 35–45% of rated power. Ramp rate can be expressed as $R = dP/dt$. Yet ramping is constrained by waterway dynamics, cavitation risk, vibration, and environmental ramping, hydropeaking and thermopeaking, limits downstream [11, 12].

Storage is characterized by power capacity and energy capacity $E_{storage}$. For a storage unit, $E_{storage} = P_{storage} \cdot T_{storage}$ where $T_{storage}$ is the duration at rated power. Energy storage spans a portfolio: batteries (short to medium duration), PSH (multi-hour to multi-day), storage hydropower and synthetic fuels (seasonal). PSH and storage hydropower seasonal capacity and operations are governed by water availability constraints with bounds on minimum and maximum water level in the reservoirs [13–15]. PSH is consistently identified as a mature long-duration storage technology with large global deployment and high cycle life of several decades. Comparative assessments position PSH as cost-effective for multi-hour to multi-day storage, particularly where suitable topography and water availability exist. PSH has unique characteristics: large power and energy capacity at comparatively low marginal storage cost in suitable sites, high cycle life, and the ability to provide synchronous services when generating. Many studies highlight that while batteries can dominate short-term balancing, PSH and other long-duration resources become valuable at high VRES shares and under electrification-driven demand growth. PSH also interacts with geography and environmental regulation. New greenfield projects may face permitting hurdles; therefore, modernization and repurposing of existing infrastructure (adding pumping capability, upgrading turbines, variable-speed retrofits) can be a practical pathway. Variable-speed technology is often highlighted as improving pumping-mode controllability and enabling ancillary service provision in pumping mode. The economic viability depends on market designs that reward flexibility and on regulatory frameworks that recognize the system value of storage [14].

2.3 The Stability Value of Synchronous Hydropower

Hydropower units are typically synchronous machines and therefore contribute inertia and active and reactive power. In high inverter-based systems, these attributes become valuable: inertia limits RoCoF and improves frequency nadir [3, 4]. Furthermore, reactive power supports voltage stability and protection performance. While synthetic inertia can be provided by converters, synchronous machines provide inherent inertial response through their operation in spinning reserve and as synchronous condenser. Thus, hydropower value extends beyond energy and active power reserves to stability services [14].

Flexibility products create system value by reducing curtailment, avoiding peaking investments, and maintaining reliability. Yet providing these services may increase wear and reduce lifetime. Therefore, a complete flexibility assessment must couple system models, estimating value, with asset models, which estimates wear.

3 European Hydropower Flexibility: Evidence from PEN@Hydropower Working Group 1

Working Group WG1 conducted a survey across fifteen European countries to assess hydropower participation in flexibility services and to identify barriers and priorities. The survey results comprehensively presented in [9] show that hydropower participates in multiple ancillary services across all timescale, from sub-seconds regulation to seasonal adequacy, in more than 70% of the considered European countries, and thus plays a pivotal role in both short-term and long-term flexibility. Respondents highlight that short-term services such as frequency regulation and rapid balancing are technically demanding because they require frequent transients and operation away from BEP, leading to increased wear. The survey also reveals significant differences across countries. In regions with high VRES shares, hydropower is increasingly used for short-term balancing and reserve provision. In countries with large reservoirs, hydropower can also provide seasonal balancing and adequacy support. Hydropower storage and PSH emerges as a key technology for long-duration storage; however, investment barriers such as regulatory uncertainty and insufficient remuneration are persistent.

4 EnergyPLAN Modeling and Sensitivity Interpretation

4.1 Introduction to the EnergyPLAN Study

Energy system modeling software plays an important role in the analysis and planning of modern energy systems. These tools help researchers and decision makers to understand how different energy sources interact within the electricity, heating, transport, and industrial sectors. With the increasing integration of renewable energy, energy systems become more complex and require detailed analysis to ensure reliability, efficiency, and sustainability [16].

Software of this kind is especially useful for assessing the role and potential of renewable energy resources. By using energy simulation tools, it is possible to evaluate how energy sources such as hydropower interact with other renewable sources, support grid stability, and reduce greenhouse gas emissions. Among the available tools, EnergyPLAN is widely recognized as an effective software for integrated energy system analysis [17]. Developed by Aalborg University, EnergyPLAN allows users to simulate the operation of entire energy systems with hourly resolution over one year. The software can model electricity production, heating demand, transport energy use, storage technologies, and sector coupling approaches. This makes it particularly suitable for studying renewable integration and evaluating different future energy scenarios. EnergyPLAN is also valuable because it combines technical, economic, and environmental analyses in a single framework. Researchers can compare different development pathways, assess green-house emissions, analyze energy balances, and investigate the impacts of policies or infrastructure changes. Due to its flexibility and detailed modelling capability, EnergyPLAN has been widely used in scientific studies related to renewable energy planning, including hydropower assessment and sustainable energy transition strategies. In storage studies, the model essentially decides when storage charges and discharges, based on the availability of surplus electricity and the need to meet demand, while minimizing system cost or respecting predefined dispatch priorities. For PSH, the model represents pumping power limits, storage energy capacity, and round-trip efficiency. Key outcomes such as CEEP and dispatchable fossil generation are emergent properties of these constraints. A useful way to interpret EnergyPLAN outputs for PSH is through a decomposition of curtailment drivers. Curtailment occurs when (i) VRES output exceeds demand; (ii) storage charging power is insufficient to absorb surplus; (iii) storage energy capacity is saturated; and (iv) exports are constrained. Increasing pumping power addresses (ii); increasing storage duration addresses (iii). However, if surpluses are frequent and long, both constraints bind, which can explain nonlinear benefits of expanding storage.

In many systems, PSH also provides negative load flexibility: during pumping, it increases demand and can stabilize the net-load profile. In operational terms, this can reduce the required down-ramping of other generators and can mitigate negative prices by absorbing surplus. When storage is large enough, it can also reduce the frequency of curtailment episodes, enabling higher VRES penetration without compromising system balance. While EnergyPLAN is deterministic, scenario analysis can emulate uncertainty by exploring different weather years or by varying VRES profiles. In a planning context, storage sizing should be robust to such variability, therefore, multi-year simulations or probabilistic approaches are desirable.

While the survey described in Sect. 3 establishes the importance of hydropower flexibility, system-level modeling is required to quantify storage needs and to compare the role of hydropower storage across countries. A dedicated analysis performed within WG1 of PEN@Hydropower evaluated storage requirements in Greece, Hungary, Italy, Poland, and Portugal [18]. The objective was to assess how storage, particularly PSH, supports high-VRES scenarios by reducing CEEP and fossil import reliance. EnergyPLAN was used to perform an hourly simulation of energy balances under specified technology portfolios. Selected multi-day periods are analyzed using stacked area and cumulative line charts that illustrate charging and discharging behavior. Publicly available dataset including forward-looking scenarios to 2040 were used as input. For instance for Poland, available data indicates that demand composition includes 196.9 TWh/year of load and BEV charging components; dispatchable capacity includes biomass/biogas (6 GW), natural gas (14.9 GW), and nuclear (5.6 GW); and VRES capacity includes PV (45 GW), onshore wind (20 GW), offshore wind (17.9 GW), and run-of-river hydropower (0.6 GW). Baseline storage includes utility batteries (5.0 GW, 20 GWh, 4 h), small-scale batteries (1.7 GW, 3.8 GWh, 2 h), and PSH (1.7 GW, 7.6 GWh, 4.47 h). These scenarios provide a transparent foundation for interpreting how increasing storage duration and power affects system metrics. Across countries, the work reports that PSH contributes between approximately 2.7

and 11.2% of annual demand coverage depending on scenario assumptions [4]. The benefit of increasing storage duration is most visible in systems with large surplus episodes. Italy's PV-dominant system shows lower marginal value of storage beyond 6 h in some assumptions, whereas wind-heavy systems exhibit higher marginal value for longer storage durations. These results emphasize that storage planning must be context-specific and tied to the VRES portfolio and interconnection assumptions. While EnergyPLAN is valuable for strategic scenario comparison, it does not explicitly model transmission constraints, reservoir optimization with water value, cascaded hydropower systems, or ecological flow constraints. Future work should integrate these constraints into optimization frameworks and incorporate probabilistic uncertainty. Nevertheless, EnergyPLAN provides a robust baseline quantification for policy discussion and highlights the system-level importance of hydropower.

4.2 Country Case Deep-Dive and Interpretation

This section provides a deeper narrative interpretation of the five-country EnergyPLAN modeling, emphasizing why national context matters and how hydropower-VRES synergies manifest differently across systems. Full results can be appreciateted in [18]

Poland. The Poland 2040 dataset used in the work performed combines large wind (onshore and offshore) with a substantial PV build-out, while retaining dispatchable capacity from natural gas, biomass/biogas, and nuclear. In such systems, surplus episodes are driven strongly by wind output, which can persist over multi-day periods. The combination of high surplus potential and electrification-driven demand growth implies that long-duration storage is valuable not only for intraday shifting but also for absorbing extended wind-driven surpluses and returning energy during subsequent deficits. PSH complements batteries: batteries can manage diurnal PV-driven fluctuations and fast balancing, while PSH can provide bulk shifting and reduce the system's reliance on fossil dispatch. The sensitivity results can be interpreted as follows: increasing PSH power increases the rate at which surplus can be absorbed, thus reducing instantaneous curtailment, while increasing duration increases the total energy that can be shifted. In practice, the optimal combination depends on the expected surplus duration distribution, which is shaped by wind regime statistics and interconnection constraints.

Portugal. Portugal's system is characterized by significant wind and an increasing PV share, combined with existing hydropower and interconnections within the Iberian Peninsula. The key synergy mechanism is the coordination of hydropower dispatch with PV diurnal patterns and wind variability. During many hours, PV creates a midday net-load valley; pumping in these hours can reduce curtailment and store energy for evening peaks. Wind-driven surpluses can occur overnight and during stormy periods; PSH can absorb part of this surplus if sufficient pumping power is available. Importantly, Portugal's hydropower fleet may include both reservoir and run-of-river units; reservoir units can provide energy shifting directly, whereas run-of-river units are constrained by inflows. Therefore, PSH and reservoir operations must be coordinated to maximize flexibility without violating environmental flows.

Italy. Italy is often presented as a PV-heavy system with strong diurnal structure. Under such conditions, storage value is driven by the day-night cycle: PV surpluses appear around midday, and deficits occur in evening and early morning hours. Batteries with 2–6 h duration are effective for this pattern; the result notes that the marginal value of increasing storage beyond around six hours can be lower in some Italian assumptions. However, this does not imply that long-duration storage is irrelevant: multi-day deficits can still occur during low solar periods in winter or when electrification increases demand peaks. In addition, Italy's geography includes Alpine hydropower systems with reservoir storage that can provide multi-day balancing. The key message is that a single "duration threshold" should not be generalized; rather, the value of long-duration storage depends on the residual-load distribution, seasonal patterns, and interconnection availability.

Greece. Greece's system features strong solar potential and increasing wind deployment, with interconnections that can mitigate some variability through imports and exports. PSH can provide both balancing and congestion relief in isolated subsystems. Storage planning in such contexts must consider network constraints and the location of VRES relative to demand centers. While EnergyPLAN does not explicitly model transmission, the qualitative implication remains: storage located strategically can reduce curtailment and improve reliability, especially where network reinforcement is slow.

Hungary. Hungary's 2030 scenarios show that storage needs depend strongly on assumptions regarding VRES penetration and dispatchable capacity. In systems where VRES penetration is moderate or where interconnections and dispatchable capacity provide sufficient balancing, the marginal value of storage duration beyond a certain point can be limited. Yet as decarbonization deepens, the same system can transition to a regime where surplus becomes frequent and long-duration storage becomes more valuable. Therefore, the appropriate storage strategy is dynamic: it should evolve over time as the system moves from moderate VRES integration to high VRES dominance. Across all cases, the same conceptual conclusion emerges: PSH provides value through both power and energy capacity, but the relative importance of each depends on the time structure of surpluses and deficits. The results of this work provide a foundation for more in depth national studies that incorporate reservoir constraints, hydrology, and network constraints.

5 Mechanical Integrity Under Flexible Operation: Fatigue and Lifetime

Mechanical integrity under flexible operation is increasingly critical for modern hydraulic machinery, as the transition from steady baseload operation to highly dynamic regimes fundamentally alters the stress environment experienced by components. In flexible operation, machines such as Francis pump-turbines, Pelton turbines, and multistage pumps are subjected to frequent transients, off-design conditions, and rapid load variations. These operating modes significantly modify the structural response of mechanical components, introducing a complex combination of high-cycle fatigue and low-cycle fatigue mechanisms. Start-up and shut-down events are particularly severe, as they induce rapid changes in flow velocity and pressure, often leading to hydraulic transients such as water hammer, pressure waves, and flow oscillations. These transient phenomena generate large stress amplitudes over short durations, contributing predominantly to low-cycle fatigue. At the same time, part-load and off-design operation can lead to cavitation, vortex formation, and pressure pulsations, which produce high-frequency stress fluctuations associated with high-cycle fatigue. Additionally, increased frequency of regulation requires more frequent actuation of servomechanisms, thereby accelerating wear and fatigue in mechanical and control components [19, 20].

To quantify the impact of such flexible operation on lifetime consumption, a practical and data-driven methodology based on the concept of damage accumulation was introduced within in WG1 of PEN@Hydropower [21]. Central to this approach is the use of Palmgren–Miner's linear damage rule, which expresses cumulative fatigue damage as follows:

$$D = \sum_i \frac{n_i}{N_i},$$

where n_i is the number of cycles experienced at a given stress level and N_i is the number of cycles to failure at that specific level. Although simplified, this formulation provides a useful engineering approximation for aggregating fatigue effects across varying operating conditions. The key outcome of this work lies in the introduction of damage maps, which relate operating points, defined by variables such as head, discharge, and rotational speed, to a damage rate expressed as damage per unit time. These maps are derived from experimental strain gauge measurements and fatigue characterization of critical components. By combining these maps with high-resolution operational data, it becomes possible to continuously integrate damage over time, effectively translating real operating histories into quantitative lifetime consumption.

For improved interpretability and operational relevance, fatigue damage is often expressed in terms of equivalent operating hours at the best efficiency point (BEP). This normalization allows operators and engineers to compare the relative severity of different operating regimes using a familiar reference condition. One of the most striking findings of this analysis is the disproportionate contribution of start-up and shut-down cycles to overall damage. For example, in a Francis pump-turbine, a single start-up event can be the equivalent of approximately 119 equivalent hours of operation at BEP. In the case of a Pelton turbine, a start-up event may produce damage equivalent to about 7.2 times the damage that occurs at BEP operation over the same time interval. These results clearly indicate that transient events dominate fatigue accumulation when they occur frequently, as is typical in flexibility-intensive systems such as those supporting renewable energy integration. Consequently, even if the machine spends a relatively small fraction of time in transient modes, the cumulative damage associated with these events can far exceed that of steady-state operation.

The impact of flexible operation varies across machine types, reflecting differences in hydraulic design, operating principles, and susceptibility to transient phenomena. The study shows that, when start-up and shut-down cycles are included, total fatigue damage can increase by approximately two orders of magnitude ($\sim 10^2$) for Francis pump-turbines and by more than three orders of magnitude ($> 10^3$) for multistage pumps. Pelton turbines exhibit smaller relative increases, but the effects remain significant and cannot be neglected in lifetime assessments. These differences are further elucidated through cumulative damage plots, which compare scenarios with and without transient events. Such plots typically show a dramatic divergence over time, with damage trajectories under flexible operation rising steeply due to the inclusion of start-stop cycles. Complementary analyses of operating mode distributions, i.e., the fraction of time spent in start-up, shut-down, part-load, and full-load conditions, provide insight into the operational drivers behind these damage patterns.

Given these findings, mitigating fatigue damage under flexible operation requires a combination of engineering design improvements and optimized operational strategies. From a design perspective, turbine runners and hydraulic passages can be refined to reduce flow instabilities, cavitation, and pressure pulsations under off-design conditions. Structural components can be reinforced or redesigned to better withstand transient loads, and materials with improved fatigue resistance can be employed. On the operational side, optimizing start-up and shut-down trajectories is crucial to minimize the severity of transient stresses; this may involve controlling ramp rates, synchronizing valve and guide vane movements, and avoiding critical operating zones. Advances in control systems and governors also play a key role, enabling smoother regulation and reducing unnecessary actuation cycles. Furthermore, hybridization with energy storage systems such as batteries can alleviate the burden on hydraulic machines by absorbing high-frequency load fluctuations, thereby reducing the need for frequent start-stop operations.

Finally, predictive maintenance strategies are essential for managing lifetime consumption in this context. By leveraging continuous monitoring, sensor data, and digital twin models, it is possible to track damage accumulation in real time and anticipate potential failures before they occur. This enables condition-based maintenance planning, reduces unplanned outages, and supports more efficient asset management. In an energy system increasingly characterized by variability and flexibility requirements, such integrated approaches are indispensable to ensure the long-term mechanical integrity and reliability of hydraulic machinery.

6 Impact of Flexibility on Hydropower Sustainability

Flexible operation of hydropower plants, while essential for supporting variable renewable energy sources, can exacerbate environmental pressures, particularly in riverine ecosystems. One of the primary concerns is hydropeaking, which refers to rapid fluctuations in river discharge caused by frequent changes in turbine operation. These fluctuations can significantly alter downstream flow regimes, affecting aquatic habitats, sediment transport, and river morphology. Hydropeaking, thermopeaking, and downstream water supersaturation are identified in literature as key environmental risks associated with flexible hydropower operation. Supersaturation, often caused by air entrainment during spill or turbine operation, can lead to gas bubble disease in fish and other aquatic organisms [10, 12].

To mitigate these impacts, several technical and operational measures can be implemented. Ramping rate limitations constrain the rate of change of discharge, thereby smoothing transitions between operating states and reducing ecological stress. Environmental flow requirements ensure that a minimum discharge is maintained at all times to preserve ecosystem functions. Re-regulation basins downstream of hydropower plants can further attenuate flow fluctuations by temporarily storing excess water during peak releases and subsequent gradual release. These mitigation measures must be explicitly incorporated into dispatch optimization and system planning models as operational constraints, ensuring that flexibility provision does not compromise environmental sustainability.

Climate change introduces an additional layer of complexity by increasing hydrological variability and uncertainty. Changes in precipitation patterns, snowmelt dynamics, and extreme weather events can alter inflow regimes, affecting both energy production and water availability [22]. In this context, adaptive reservoir management becomes essential. This involves dynamic rule curves, probabilistic inflow forecasting, and real-time optimization to balance competing objectives such as energy generation, flood control, irrigation, and ecological preservation. Integrated water-energy governance frameworks are required to coordinate these objectives across sectors and stakeholders.

In regard of sustainability enhancement, institutional and regulatory frameworks also play a crucial role in the ways flexibility can be enhanced while facilitating sustainability aspects. From a governance perspective, aligning hydropower flexibility with environmental objectives requires a shift from purely energy-centric optimization to integrated resource management. This includes the adoption of ecosystem-based management principles, the use of environmental indicators in operational decision-making, and the integration of hydropower into broader river basin management plans. Coupled hydro-ecological models and decision-support systems, can enhance the ability to assess trade-offs and design adaptive strategies under uncertainty. In Europe, the opportunity set to improve flexibility provision includes both new projects and modernization of existing assets, including conversion of reservoirs to pumping operation, addition of pumping capability, and variable-speed upgrades [23, 24]. Modernization includes improved turbine designs that extend stable operating ranges, upgrades to governors and control systems to improve response while limiting transients, and retrofits to reduce cavitation and vibration. Modernization can also include digitalization: sensor deployment, advanced condition monitoring, and digital twins that integrate operational data into physics-based models to support predictive maintenance and wear-conscious operation. The interplay between modernization, digitalization, and market incentives is increasingly recognized as central: without remunerated flexibility products, investment in modernization may not occur at the necessary scale. Furthermore, integration of modernization options into system planning models would be an asset to valorize the long-term value of these modernizations by considering the forecasts on energy needs and power systems scenarios. Many power-system models treat hydropower as a simplified dispatchable resource with limited constraints. In reality, the flexibility envelope is plant-specific, and modernization changes constraints and costs. Bridging the gap requires multi-scale modeling that connects system-level dispatch needs with plant-level dynamics.

Concession renewals represent key opportunities to modernize infrastructure and incorporate advanced environmental mitigation measures, e.g. in Switzerland [25]. However, they also involve complex negotiations among operators, regulators, local communities, and environmental organizations. These processes must balance economic efficiency, environmental protection, and social acceptance. Strengthening stakeholder engagement, improving transparency, and adopting multi-criteria decision-making approaches can facilitate more sustainable outcomes.

7 Market Design and Investment for a Forward-Looking and Flexible Hydropower Technology

Hydropower provides a wide range of flexibility services to the power system, including frequency regulation, reserve provision, and peak load balancing. However, current electricity market designs often fail to fully capture and remunerate the value of these services. As highlighted in Sect. 5, short-term flexibility provision imposes significant mechanical wear and accelerates fatigue damage. These costs are typically borne by plant operators but are not adequately reflected in market revenues, leading to a misalignment between system needs and economic incentives [8].

To address this issue, market designs can be modernized and evolve to incorporate wear-conscious mechanisms. One approach is to introduce compensation schemes that explicitly account for cycling costs, linking remuneration to the intensity and frequency of operation. Alternatively, new market products can be defined to incentivize smoother operation patterns that reduce mechanical stress, such as ramping services or constrained flexibility products. In addition, system services that are inherently provided by hydropower–such as inertia, voltage support, and short-circuit strength–should be explicitly valued and procured through dedicated markets or ancillary service mechanisms [26].

Investment in hydropower infrastructure and modernization of existing assets, is highly sensitive to revenue stability and regulatory risk. Hydropower storage, including PSH, plays a critical role in long-duration energy storage and system balancing, but its economic viability depends on sufficient price spreads and predictable market conditions. Uncertainty in future market revenues, coupled with lengthy permitting processes and environmental constraints, can hinder investment decisions [15]. Ultimately, aligning market incentives with system needs is essential to unlock the full potential of hydropower flexibility.

8 Integrated Roadmap and Conclusions

The evidence gathered within the PEN@Hydropower framework supports the development of a comprehensive and integrated roadmap for the sustainable deployment of hydropower flexibility. This roadmap must address multiple dimensions simultaneously, including system modeling, operational strategies, asset management, environmental protection, and market design.

Comprehensive system dynamics: System-level models must be enhanced to better represent hydropower dynamics, including hydraulic constraints, reservoir interactions, and climate-induced uncertainties. Incorporating stochastic inflow scenarios and multi-scale temporal resolution allows for more accurate assessment of flexibility potential and system resilience.

Operational optimization: Operational strategies should be optimized with explicit consideration of wear and fatigue, leveraging methodologies such as damage mapping and equivalent operating hours to minimize lifetime consumption while meeting system requirements.

Environmental inclusion: Environmental monitoring and mitigation measures must be systematically implemented and co-designed with stakeholders, to ensure that flexibility provision does not compromise ecological integrity.

Enhanced monitoring: The deployment of advanced monitoring systems and digital twins enables real-time assessments of asset condition and predictive maintenance. These technologies integrate sensor data, physical models, and machine learning techniques to provide actionable insights into component health and remaining useful life.

Modern governance: Market and regulatory frameworks should properly value flexibility services and support investment in modernization and new capacity. This can involve the introduction of wear-conscious compensation mechanisms, the explicit valuation of system services, and the provision of long-term investment signals through policy instruments.

Hydropower occupies a unique position in the energy system, as it can provide flexibility across all temporal scales–from fast frequency response to seasonal storage. It is therefore expected to play a central role in meeting future flexibility demand in Europe. Energy system simulations using tools such as EnergyPLAN demonstrate that hydropower with PSH significantly reduces renewable curtailment and reliance on fossil-based generation across diverse national contexts.

Achieving sustainable and flexible hydropower therefore requires a coordinated approach that integrates technological innovation, regulatory reform, and environmental stewardship.

Competing Interests The author has no conflicts of interest to declare that are relevant to the content of this chapter.

References

1. International Energy Agency (2018) Status of power system transformation 2018. https://www.iea.org/reports/status-of-power-system-transformation-2018
2. International Renewable Energy Agency (2018) Power system flexibility for the energy transition, IRENA. https://www.irena.org/publications/2018/Nov/Power-system-flexibility-for-the-energy-transition
3. ENTSO-E—European network of transmission system operators for electricity (2020) Inertia and rate of change of frequency. https://www.entsoe.eu/Documents/SOC%20documents/Inertia%20and%20RoCoF_v17_clean.pdf
4. ENTSO-E—European network of transmission system operators for electricity (2020) The inertia challenge in Europe: present and long-term perspective. https://eepublicdownloads.blob.core.windows.net/public-cdn-container/tyndp-documents
5. ENTSO-E—European network of transmission system operators for electricity (2009) Policy 1: load-frequency control and performance
6. ENTSO-E—European network of transmission system operators for electricity (2023) Frequency Containment Reserves Cooperation. https://www.entsoe.eu/
7. International Energy Agency (2025) The Value of Demand Flexibility
8. International Energy Agency (2021) Hydropower special market report. https://www.iea.org/reports/hydropower-special-market-report
9. Vagnoni E, others (2024) The new role of sustainable hydropower in flexible energy systems and its technical evolution through innovation and digitalization. Renew Energy. https://www.sciencedirect.com/science/article/pii/S0960148124009005
10. Terese E, Venus, others (2025) The power of hydropeaking: trade-offs between flexible hydropower and river ecosystem services in Europe. Ecol Econ 233:108583
11. XFLEX HYDRO consortium (2024) Technical white paper. https://www.xflexhydro.com/knowledge/technical-white-paper. Accessed 20 May 2026
12. Hayes DS, others (2022) Hydropeaking: processes, effects, and mitigation. In: Encyclopedia of Inland Waters (Second Edition), Mehner T, Tockner K (eds), 2nd ed. Oxford, Elsevier, pp 134–149. https://doi.org/10.1016/B978-0-12-819166-8.00171-7, https://www.sciencedirect.com/science/article/pii/B9780128191668001717
13. Quaranta E, others (2025) Pumped storage hydropower operation for supporting clean energy systems. Nat Rev Clean Technol. https://www.nature.com/articles/s44359-025-00057-x
14. Quaranta E, others (2024) Clean energy technology observatory: hydropower and pumped storage hydropower in the European Union—2024 status report on technology development, trends, value chains and markets. European Commission, Joint Research Centre. https://doi.org/10.2760/8354439, https://publications.jrc.ec.europa.eu/repository/handle/JRC139225

15. Yang W, others (2025) Pumped storage hydropower operation for supporting clean energy systems. Nat Rev Clean Technol 1:454–473. https://doi.org/10.1038/s44359-025-00057-x, https://www.nature.com/articles/s44359-025-00057-x
16. Lund H, others (2021) EnergyPLAN—advanced analysis of smart energy systems. Smart Energy 1:100007. https://doi.org/10.1016/j.segy.2021.100007
17. Lund H, others (2014) 4th Generation District Heating: Integrating smart thermal grids into future sustainable energy systems. Energy 68:1–11. https://doi.org/10.1016/j.energy.2014.02.089
18. Vagnoni E (2024) Report on generation/accumulation/consumption of electric energy and the role of hydropower in future flexible power systems, COST action CA21104–Pan-European network for sustainable hydropower (PEN@Hydropower), 2024, deliverable 1.1. https://static1.squarespace.com/static/635d45f550e75155f79c7d18/t/670cda6af839d91019493d66/1728895596262/PEN%40Hydropower+deliverable+1.1.pdf. Accessed 20 May 2026
19. Nobilo M, others (2026) Lifetime analysis of hydro turbines with focus on fatigue damage in a renewable energy system—a review. Renew Sust Energy Rev 228:116578
20. Seydoux M, others (2024) On the prediction of the induced damage by the start-up sequence of Francis turbines: on operational resilience framework. Renew Energy 120587. https://doi.org/10.1016/j.renene.2024.120587
21. Den Dekker M (2024) Short term scientific mission report, COST action CA21104–Pan-European network for sustainable hydropower (PEN@Hydropower), 2024, STSM report within PEN@Hydropower COST action. https://static1.squarespace.com/static/635d45f550e75155f79c7d18/t/690640dd924cb353be156706/1762017501230/MatthijsDenDekker+STSM+PEN%40hydropower+Report.pdf. Accessed 20 May 2026
22. European Environment Agency (2025) Europe's Environment 2025, European Environment Agency. https://doi.org/10.2800/1121467, https://www.eea.europa.eu/en/europe-environment-2025
23. Quaranta E, others (2021) Assessing the energy potential of modernizing the European hydropower fleet. Energy conversion and management, vol 246, pp 114655. https://doi.org/10.1016/j.enconman.2021.114655, https://doi.org/10.1016/j.enconman.2021.114655
24. Kougias I, others (2019) Analysis of emerging technologies in the hydropower sector. Renew Sustain Energy Rev 113:109257. https://doi.org/10.1016/j.rser.2019.109257, https://doi.org/10.1016/j.rser.2019.109257
25. Office fédéral de l'énergie OFEN (2022) Procédure de renouvellement des concessions de force hydraulique, Département fédéral de l'environnement, des transports, de l'énergie et de la communication (DETEC), Rapport, Enquête auprès des services cantonaux, publié le 20 juillet 2022. https://pubdb.bfe.admin.ch/fr/publication/download/10987
26. Yang W, others (2018) Burden on hydropower units for short-term balancing of renewable power systems. Nat Commun 9(1):2633. https://doi.org/10.1038/s41467-018-05060-4, https://doi.org/10.1038/s41467-018-05060-4

Technological Evolution Through Innovation and Digitalization

Dogan Gezer and Pavel Rudolf

Reimagining hydropower through innovation and digitalization

Abstract Hydropower is facing new challenges due to the increasing need for flexibility, environmental compliance and integration into modern energy systems. In this context, innovation and digitalization have become key enablers for improving the performance and efficient role of hydropower plants. This chapter focuses on technological evolution through innovation and digitalization as supported by the literature and activities outcome of Working Group 2 (WG2) within the PEN@Hydropower COST Action. The chapter follows a holistic approach, considering different perspectives such as plant operation, system integration, environmental impacts and stakeholder needs. Key topics include digital twins, data-driven monitoring, predictive maintenance, new designs of components and system concepts. In addition, the interaction between hydropower and the water-energy-environment nexus is discussed, with examples related to sediment management, ecological flows and hybrid systems. Finally, current limitations and future directions are outlined, highlighting the need for practical, scalable and integrated solutions in hydropower.

Keywords Digitalization · Innovation · Water–energy nexus · System integration · Digital twin

D. Gezer (✉)
TUBITAK Marmara Research Center, ODTU Kampusu, Ankara, Turkey
e-mail: dogan.gezer@tubitak.gov.tr

P. Rudolf
Faculty of Mechanical Engineering, V. Kaplan Department of Fluid Engineering, Brno University of Technology, Brno, Czech Republic
e-mail: rudolf@fme.vutbr.cz

E. Doujak et al. (eds.), *Rethinking Hydropower: A Holistic Vision for a Sustainable Future*, Springer Tracts in Civil Engineering,
https://doi.org/10.1007/978-3-032-33328-5_8

1 Introduction

Hydropower has long been a key component of electricity systems, providing reliable and controllable energy production. In recent years, however, its role has been evolving due to increasing expectations related to flexibility, environmental performance and integration with other intermittent energy and water systems [1, 2]. These changes require not only improvements in individual components, but also a broader reconsideration of how hydropower plants are designed, operated and interconnected with the larger power systems.

In this context, innovation and digitalization have emerged as important drivers of the transformation. Developments such as advanced monitoring systems, data-driven operation and digital twins are enabling new ways to understand and operate hydropower plants. At the same time, technological innovations in turbines, new control techniques such as variable speed control, refurbishment approaches and small-scale solutions are expanding the range of feasible applications.

This chapter builds on the work carried out within Working Group 2 (WG2) of the COST Action PEN@Hydropower [3], which focuses on technological evolution through innovation and digitalization [4]. The content is based on a series of regular meetings, multidisciplinary workshops and round-table discussions, bringing together perspectives from academia, industry and system operators. Rather than presenting a purely literature-based review, the chapter reflects the collective experience and discussions within WG2.

A key aspect of the approach adopted in this chapter is the consideration of interconnections. Instead of treating different topics in isolation, the chapter highlights how digitalization, technological innovation and environmental aspects are linked to each other and to system-level requirements. For this reason, topics such as system integration and flexibility are not addressed as separate sections, but are embedded within the relevant technological and operational discussions.

This chapter is structured as follows: first, hydropower is discussed as a complex and multidisciplinary system. This is followed by sections on digitalization and technological innovation. Subsequently, the interaction between hydropower and the water–energy–environment nexus is examined. Building on these elements, cross-domain interconnections are discussed, which is followed by an overview of current barriers and implementation challenges. Finally, insights from WG2 activities and future perspectives are presented.

2 Hydropower as a Complex System

Hydropower plants are often described in terms of their main components, such as turbines, generators and hydraulic structures. However, in practice, they operate as complex systems where multiple physical, technical and operational elements

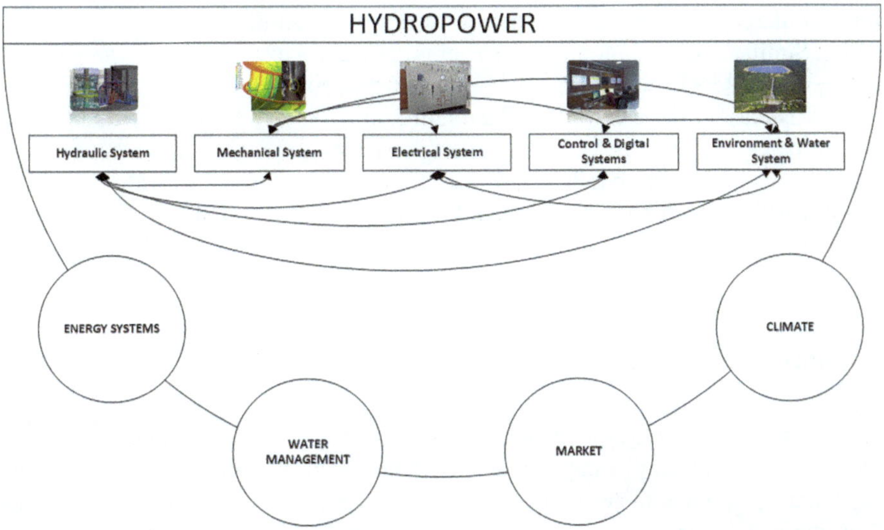

Fig. 1 Hydropower as a multi-domain system integrating hydraulic, mechanical, electrical, digital and environmental components

interact with each other. These interactions go beyond individual components and extend to the surrounding environment and the broader energy system.

From a technical perspective, hydropower combines hydraulic, mechanical and electrical subsystems that must operate in a coordinated manner. The hydraulic system governs water conveyance and flow conditions, the mechanical system is responsible for energy conversion and the electrical system ensures grid connection and power delivery. In addition, control and protection systems play a key role in maintaining stable and safe operation under varying conditions.

At the same time, hydropower plants are closely linked to environmental and social factors. Issues such as sediment transport, ecological flow requirements and water quality directly influence plant operation and design. These aspects are not only external constraints, but constitute integral parts of the system that must be considered together with technical performance.

Another important dimension is the interaction with the wider energy system. Hydropower is increasingly expected to provide flexible operation, support variable renewable energy sources and contribute to system stability in presence of high penetration of intermittent renewable energy resources. This requires plants to operate under more dynamic conditions, including frequent start-stop cycles and operation outside traditional design points. Recent large-scale blackouts in Europe, including events in Spain and the Balkan region, have highlighted the importance of inertia-providing generators such as hydropower units for maintaining system stability.

Hydropower as a multi-domain system is depicted in Fig. 1. These multiple layers make hydropower highly multidisciplinary. As a result, improvements in one part of the system may have consequences for others. For example, changes in operation aimed at increasing flexibility can affect mechanical loading or environmental conditions. Similarly, digital monitoring systems can improve operational awareness but require reliable data and integration across different subsystems.

For this reason, the technological evolution of hydropower cannot be addressed through isolated solutions. A system-level perspective is required, where different domains are considered together. This perspective forms the basis for the following sections, which examine digitalization, technological innovation and their interactions with environmental and system-level aspects.

3 Digitalization in Hydropower Systems

Digitalization has become one of the main drivers of change in hydropower, enabling new ways of monitoring, analyzing and operating plants. While traditional systems relied mainly on local measurements and operator experience, recent developments allow the use of large volumes of data, advanced analytics and remote access to support decision-making.

Rather than being limited to individual technologies, digitalization introduces a systemic transformation in how hydropower plants are operated and managed. It enables the integration of information across hydraulic, mechanical and electrical subsystems, supporting a more comprehensive understanding of plant behavior. In this context, digitalization is not only a tool for improving efficiency, but also a key enabler for flexibility, reliability and system-level coordination.

At the same time, the implementation of digital solutions depends on the availability of reliable data, the integration of heterogeneous systems and the ability to translate data into actionable insights. These aspects highlight the importance of combining technological development with practical operational considerations.

Modern digital hydropower architecture, as shown in Fig. 2, is composed of multiple layers such as a physical layer including site signals, data layer with input/output modules with controllers, a digital layer which benefits from the data collected by analytics, and finally the cloud/remote layer involving remote control and forecasting.

3.1 Data Acquisition and Monitoring

The increasing availability of data from different parts of the plant is a key enabler of digitalization. Sensors installed on mechanical, electrical and hydraulic components provide continuous information on operating conditions. This data can be used not only for basic supervision, but also for identifying patterns, detecting anomalies and improving overall performance.

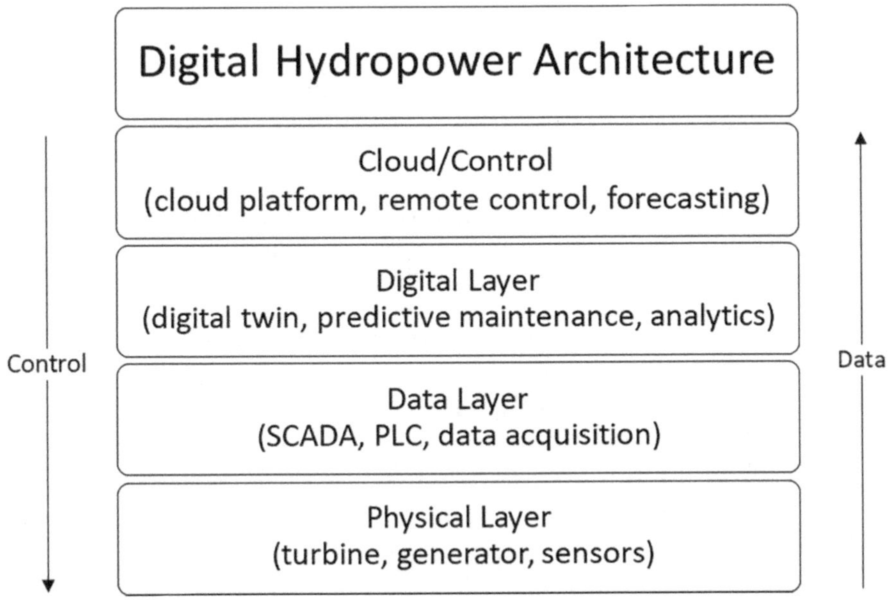

Fig. 2 Layered digital architecture of hydropower systems, including data acquisition, analytics and cloud-based operation

The integration of data from different subsystems is particularly important, as it enables a more comprehensive understanding of plant behavior. Instead of analyzing individual signals in isolation, combined datasets allow operators to identify correlations between hydraulic conditions, mechanical response and electrical performance.

Field measurement remains a critical aspect of digitalization, as accurate flow, pressure and water quality data are essential for model calibration and validation. Various sensor technologies, including ultrasonic and electromagnetic flow meters as well as pressure transducers, are increasingly used in combination with communication systems such as LoRa, GSM and WiFi. However, challenges related to sensor accessibility, calibration and measurement uncertainties under dynamic and harsh operating conditions remain significant.

3.2 Digital Twins

Digital twins represent one of the most advanced applications of digitalization in hydropower. By combining physical models with real-time data, digital twins provide a dynamic representation of plant behavior under different operating conditions.

This enables operators to compare measured and simulated values, detect deviations and better understand system dynamics. In addition, digital twins support commissioning processes, transient analysis and the assessment of operating limits. These capabilities are particularly valuable in complex systems where direct experimentation is limited or not feasible. However, the application of digital twin approaches in hydropower faces several interconnected challenges. Data quality and availability remain a primary concern, as many facilities lack modern sensor infrastructure, and retrofitting aging equipment with comprehensive monitoring systems is both costly and technically demanding. Even where instrumentation exists, data is often noisy, sparse, or inconsistently formatted across the civil, mechanical, electrical, and hydrological subsystems that must be integrated. This integration itself poses a significant challenge due to multi-physics complexity, i.e. coupling fluid dynamics, structural mechanics, electromechanical behavior, and hydrology into a single coherent model, demanding substantial computational resources and rare cross-disciplinary expertise. Moreover, operating a digital twin of a critical infrastructure also poses a cybersecurity challenge.

3.3 Predictive Maintenance

Data-driven monitoring and predictive maintenance approaches are increasingly used to assess equipment condition and prevent failures. Techniques based on machine learning and statistical analysis can detect early signs of faults, such as changes in vibration patterns, temperature trends, or hydraulic conditions.

Compared to traditional threshold-based methods, predictive approaches provide a more flexible and adaptive way to evaluate system health. This enables optimized maintenance planning, reduced unplanned outages and extended component lifetime.

However, the application of predictive maintenance in hydropower involves several unique features:

1. operating conditions vary substantially across plants due to differences in head, flow regime, turbine type and age, making it difficult to generalize models from one facility to another
2. sensor instrumentation is often incomplete, providing insufficient coverage of the parameters most relevant to emerging faults
3. degradation processes such as fatigue crack growth, winding insulation breakdown, or cavitation damage evolve over timescales of several months to years, requiring models that can capture slow trends amid noisy operational signals.

Compared to traditional threshold-based methods, predictive approaches provide a more flexible and adaptive way to evaluate system health. This enables optimized maintenance planning, reduced unplanned outages and extended component service life.

3.4 Cloud-Based Architectures and Remote Operation

Cloud-based architectures are increasingly used to support data processing, storage and analysis in hydropower systems. By centralizing data handling, these approaches enable scalable analysis across multiple units and facilitate the integration of different data sources collected from various control systems.

Such architectures also support remote monitoring and, in some cases, remote control of hydropower plants. This reduces the need for on-site presence and enables centralized operation of multiple facilities. However, these advantages must be balanced with considerations related to communication reliability, cybersecurity and system resilience.

In addition, cloud-based platforms facilitate the integration of short-term forecasting tools, enabling operators to estimate inflow variations, electricity demand and market prices. This supports more proactive and optimized operation, particularly under dynamic system conditions.

4 Innovation in Hydropower Technologies

In addition to digitalization, technological innovation continues to play an important role in the evolution of hydropower. While many existing plants are based on well-established designs, there is an increasing need to adapt these systems to new operational and environmental requirements. This has led to a renewed focus on both the development of new technologies and the refurbishment of existing infrastructure [5].

Rather than being limited to individual components, innovation in hydropower increasingly reflects a system-oriented perspective. It encompasses turbine and component design, small-scale and distributed solutions, refurbishment strategies and environmental technologies. These developments are driven by the need to improve efficiency under variable operating conditions, increase flexibility and ensure compatibility with environmental constraints.

At the same time, innovation is closely linked to the changing role of hydropower in modern energy systems. More dynamic operation, integration with variable renewable energy sources and the utilization of previously untapped hydraulic potential are determining the direction of technological development. This requires solutions that are not only efficient, but also adaptable to different contexts and operational requirements.

Environmental considerations are increasingly a feature of technological innovation. Developments such as fish-friendly turbine designs, improved sediment management solutions and technologies supporting ecological flow requirements are becoming part of the design process These approaches aim to reduce environmental impacts while maintaining energy production. Rather than treating environmental constraints as external limitations, they are increasingly considered as design parameters that influence both system configuration and operation. Overall, innovation in hydropower technologies is moving towards more adaptable, efficient and context-aware solutions. This includes both the development of new concepts and the improvement of existing systems, with a focus on meeting the evolving demands of energy systems and environmental requirements.

4.1 Turbine and Component Innovations

One of the main areas of innovation is related to turbine design and operation. Advances in hydraulic design, materials and control strategies are enabling turbines to operate more efficiently over a wider range of conditions. In particular, there is growing interest in improving performance under part-load operation, which is becoming more common due to changing operational requirements.

This also impacts the design stage by increased use of simulation tools, not only computational fluid dynamics (CFD) and structural mechanics, but more often also coupled approaches of fluid-structure interaction (FSI), fatigue analysis and fracture propagation analysis. Regarding manufacturing and materials, 3d rapid prototyping and added manufacturing becomes increasingly relevant at least for some constructional parts. New coatings that are resistant to cavitation and sediment erosion are being developed and applied to turbine components (applied by HVOF, plasma, electric arc, cold spray). Last but not least, significant effort is being invested to design fish-friendly turbines that do not harm fish passing through the turbine blade channels. Innovations are not limited to the turbine itself, but also include auxiliary components such as guide vanes, bearings and sealing systems. Improvements in these elements contribute to increased efficiency, reduced wear and enhanced reliability under variable operating conditions.

Recent developments also include innovations in generator technologies and control strategies. Advances in synchronous generator design, including improved excitation systems and enhanced thermal management, contribute to higher efficiency and operational reliability under dynamic conditions. Simultaneously, variable speed operation concepts are attracting increasing attention, as they allow hydropower units to operate more flexibly across a wider range of hydraulic conditions. These approaches enable better adaptation to fluctuating system requirements, particularly in systems with high shares of variable renewable energy.

In addition, pressure recovery system (PRS) turbines, such as crossflow designs equipped with diffusers, offer simple inline configurations with pressurized outflow. These solutions are particularly suitable for integration into water distribution networks, as demonstrated in pilot applications such as the San Giovannello site in Palermo.

4.2 Small-Scale and Distributed Hydropower

Small-scale and low-head hydropower technologies have gained increasing attention in recent years. Concepts such as pumps operating as turbines, in-pipe energy recovery systems and modular micro-hydropower units offer opportunities to generate energy from the existing water infrastructure.

These solutions are particularly relevant in water distribution networks, irrigation systems and industrial applications, where unused hydraulic potential can be exploited. Compared to conventional large-scale hydropower, these systems can often be installed more easily and integrated with minimal environmental impact. Low-head hydropower technologies, typically operating in the head range of 1.5–5 m, are particularly relevant for flat landscapes such as those found in the Netherlands, where conventional hydropower applications are limited, but also in most river weirs all around Europe with untapped hydro-energy potential.

4.3 Refurbishment and Modernization

Another important approach is the refurbishment and modernization of existing hydropower plants. A significant portion of the current hydropower fleet was designed and installed decades ago under different technical and environmental conditions. As a result, upgrading these plants has become a key strategy for improving performance and extending operational lifetime.

Refurbishment may include replacing or upgrading turbines, improving control systems and optimizing operational strategies. In addition to increasing efficiency, these measures can support more flexible operation and reduce mechanical stress under dynamic conditions such as part load operation, frequent starts/stops or fast ramping up and down.

5 Integration Within the Water–Energy–Environment Nexus

Hydropower is inherently linked to water systems and its operation cannot be fully understood without considering its interaction with other water-related uses and environmental processes. In recent years, this interaction has been described increasingly within the framework of the water–energy–environment nexus, which emphasizes the need to address multiple objectives simultaneously (Fig. 3).

Rather than focusing on single objectives such as energy production, the nexus perspective highlights the need to balance technical performance, environmental sustainability and water management requirements. This introduces new challenges, as

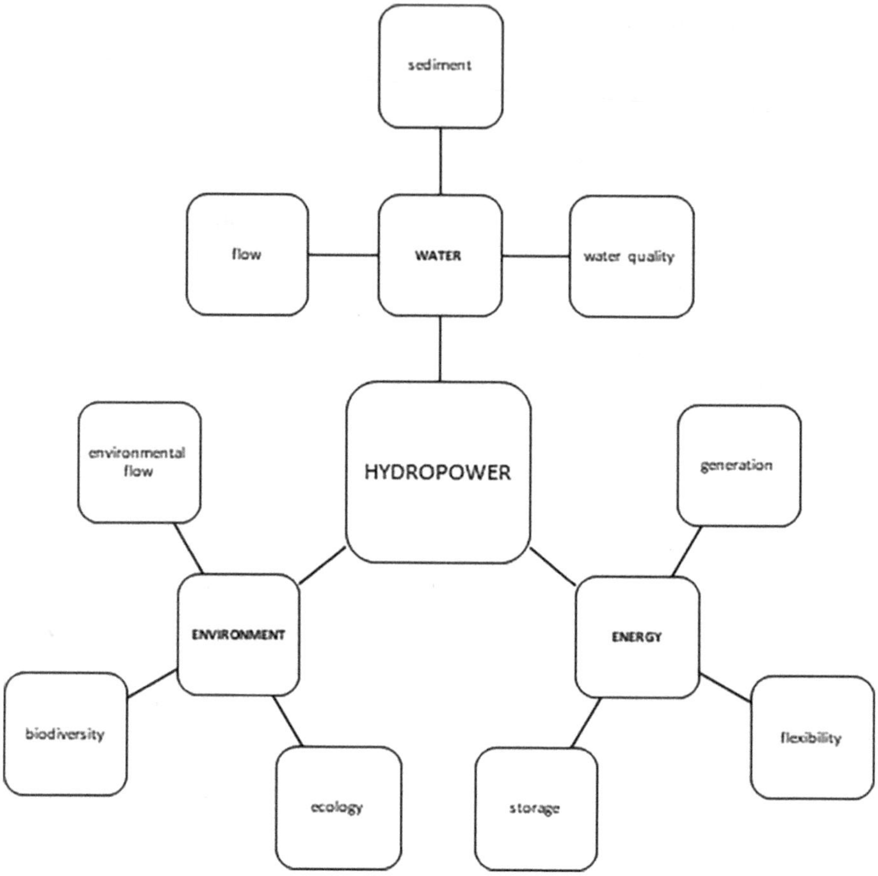

Fig. 3 Hydropower within the water–energy–environment nexus, highlighting interactions between technical, environmental and resource-related processes

decisions related to hydropower operation can have direct implications for hydraulic conditions, ecosystem dynamics and water resource utilization.

At the same time, the increasing integration of hydropower into existing water infrastructure and multi-purpose water systems expands the scope of potential applications. This includes not only large-scale power generation, but also distributed energy recovery solutions and hybrid water-energy systems. As a result, hydropower must be considered as part of a broader system where energy, water and environmental aspects are tightly interconnected. Climate change is expected to further influence hydropower operation, with increasing variability in inflow patterns. While average precipitation may increase in some regions such as Northern Europe, reductions are projected in the Mediterranean, along with more frequent extreme and flash flow events.

One important aspect of this nexus is the integration of energy production into existing water infrastructure. In addition to conventional hydropower plants, there is growing interest in utilizing hydraulic potential in water distribution networks, irrigation systems and industrial processes, taking advantage of the residual pressure in the water system. Energy recovery solutions in pipelines and pressure reduction points offer opportunities to generate electricity without the need for new large-scale infrastructure. These approaches can contribute to improving overall system efficiency while minimizing environmental impact. At the same time, their implementation requires careful consideration of hydraulic constraints and service requirements. This leads to green cities concept with distributed generation.

6 Interconnections Across Domains

The previous sections have addressed digitalization, technological innovation and the water–energy–environment nexus from different perspectives. However, in practice, these aspects are not independent of each other. Hydropower systems operate at the intersection of multiple domains, where changes in one area can directly influence others. Understanding these interconnections is therefore essential for evaluating the overall performance and role of hydropower. Rather than being driven by a single objective, modern hydropower operation involves balancing multiple requirements. These requirements are generally managed as a trade-off. These include efficiency, flexibility, environmental constraints and system-level demands. As a result, decisions in one domain can create cascading effects across others, highlighting the need for an integrated approach. The increasing complexity of hydropower systems is further increased by their evolving role within the energy system and their interaction with water management processes. This requires not only technological solutions, but also a broader understanding of cross-domain dependencies and trade-offs.

One of the most evident interactions exists between digitalization and plant operation. Advanced monitoring systems, data analytics and digital twins enable a more detailed understanding of system behavior. This supports more flexible and adaptive operation, including operation under varying load conditions and outside traditional design ranges. At the same time, the effectiveness of these tools depends on data quality, system integration and the ability to interpret results within the operational context.

6.1 Technology–Environment Interaction

There is a strong link between technological innovation and environmental performance. Improvements in turbine design, flow control and sediment management can enhance efficiency while reducing environmental impacts.

Conversely, environmental constraints such as ecological flow requirements or sediment dynamics may limit certain operational strategies, requiring adjustments in both design and operation.

6.2 Hydropower Within the Energy System

Hydropower is increasingly required to support variable renewable energy sources and provide flexibility services. This leads to more dynamic operating regimes, including frequent start-stop cycles and operation away from optimal design conditions.

Such changes affect mechanical loading, maintenance requirements and overall system reliability, demonstrating how system-level requirements directly influence plant-level behavior.

Looking towards 2050 scenarios, it is increasingly recognized that batteries alone will not be sufficient to address long-term flexibility needs and that hydropower-based storage solutions will remain essential.

Small-scale pumped storage hydropower systems have also been identified as viable solutions for islands and remote communities, where they can provide grid stability, enhance energy autonomy and reduce reliance on fossil fuel-based generation.

6.3 Integration with Water Systems

The integration of hydropower within water systems creates additional interdependencies. Energy recovery solutions must comply with hydraulic constraints and service requirements, while water management decisions directly affect energy production.

These interactions highlight the need for coordinated approaches across energy and water domains.

6.4 Need for Integrated Approaches

Hydropower cannot be optimized based on a single objective. Trade-offs between technical performance, environmental constraints and system requirements are often inevitable.

Addressing these challenges requires integrated and interdisciplinary approaches, where digitalization, innovation and environmental considerations are treated as elements of a unified system.

7 Barriers and Implementation Challenges

Despite the significant opportunities associated with digitalization and technological innovation, their implementation in hydropower systems remains challenging in practice. In reality, a combination of technical, economic and organizational barriers can limit the adoption and effectiveness of new solutions [7].

These challenges are particularly relevant in the context of existing hydropower plants, where modernization efforts must be balanced with operational constraints, legacy systems and economic feasibility. As a result, the transition towards more digitalized and flexible hydropower systems depends not only on technological development, but also on system integration and practical implementation.

7.1 Data Availability and Quality

One of the main challenges is related to data availability and quality. Many—particularly older—hydropower plants were not originally designed with advanced monitoring systems. As a result, available data may be limited, incomplete, or inconsistent. This directly affects the reliability of data-driven approaches such as predictive maintenance and digital twins, where the quality of input data plays a critical role in model accuracy and decision-making.

7.2 Legacy Infrastructure

A large share of the hydropower fleet consists of long-standing installations. Integrating modern digital solutions into these plants can be complex, requiring additional interfaces and system adaptations. In some cases, technical limitations of existing equipment may restrict the level of achievable integration, making it necessary to prioritize selective upgrades rather than full system replacement.

7.3 Interoperability and Standardization

Hydropower plants often include components from different vendors and generations. Ensuring seamless communication between these systems can be difficult, especially in the absence of common standards. This creates challenges for data integration, system interoperability and the implementation of centralized monitoring or control solutions.

7.4 Economic Constraints

Economic considerations play an important role, particularly for small and medium-sized hydropower plants. While large installations may justify investments in advanced technologies, smaller plants often face tighter budget constraints.

As a result, the cost-benefit balance of digitalization and modernization measures must be carefully evaluated, taking into account both short-term investments and long-term operational benefits.

7.5 Human and Organizational Factors

The adoption of new technologies requires changes in operational practices and depends on the acceptance of plant operators and maintenance personnel. Resistance to change, lack of familiarity with digital tools and limited training can slow down implementation processes.

In addition, there may be a lack of specialized expertise needed to implement and maintain advanced digital solutions, particularly in smaller organizations.

Public perception also represents an important barrier, as hydropower is sometimes associated with negative environmental impacts due to past unsustainable practices. Improving communication and demonstrating modern, sustainable approaches are therefore essential.

7.6 Cybersecurity and Communication

The increasing use of remote monitoring and cloud-based systems introduces additional challenges related to cybersecurity and communication reliability. Ensuring secure and stable operation is essential for maintaining trust in digital solutions and for preventing potential disruptions in critical infrastructure. Overall, these barriers highlight that the implementation of innovation and digitalization is not only a technical issue, but just as much an organizational and systemic task. Addressing these challenges requires a coordinated effort, practical solutions and consideration of the specific conditions of each plant.

8 Insights from WG2 Activities

The content presented in this chapter is based on a structured series of activities carried out within Working Group 2 (WG2) of the COST Action PEN@Hydropower.

These activities bring together perspectives from academia, industry and system operators, providing a comprehensive basis for the analysis presented in this chapter.

WG2 activities have been organized in different formats, each addressing specific aspects of hydropower innovation and contributing complementary insights:

- **Regular Meetings:** A series of technical meetings where topics such as digitalization, monitoring systems, turbine technologies and environmental aspects were presented and discussed. These meetings enabled continuous knowledge exchange and the identification of key technological trends.
- **Round-Table Discussions:** Interactive sessions bringing together experts from different disciplines to exchange views on cross-cutting topics, including system integration, environmental challenges and future directions. These discussions contributed to a broader understanding of stakeholder perspectives and system-level interactions.
- **Workshops and Events:** Multidisciplinary workshops and scientific events aimed at fostering collaboration between academia, industry and stakeholders. These activities supported the integration of technical knowledge with practical experience and facilitated cross-disciplinary dialogue.
- **Short-Term Scientific Missions (STSMs):** Targeted research activities promoting collaboration between institutions. These missions contributed to specific topics such as digitalization frameworks, innovation pathways and cross-domain analysis and supported the development of joint scientific outputs.
- **Training Schools:** Educational activities designed to disseminate knowledge and support capacity building in areas related to hydropower innovation, digitalization and sustainability.

Together, these activities provide a broad and complementary set of perspectives on the ongoing transformation of hydropower systems.

One of the main outcomes of these activities is the identification of digitalization as a key enabler for improving hydropower performance and operation. Across multiple WG2 meetings, it has been consistently observed that data-driven approaches, advanced monitoring systems and digital twins offer practical benefits in terms of reliability, efficiency and operational flexibility. At the same time, discussions have highlighted the importance of integrating these tools into existing systems and ensuring their usability in real operating environments.

Another important insight is related to the role of small-scale and distributed hydropower solutions. Presentations and discussions within WG2 have shown that significant energy recovery potential exists in water distribution networks and other non-traditional applications. These solutions are particularly relevant in the context of increasing energy efficiency and decentralized energy systems.

The interaction between hydropower and environmental processes has also been a recurring topic. Workshops and round-table discussions have emphasized the need to better integrate sediment management, ecological flow requirements and water quality considerations into both design and operation. These aspects are increasingly seen as integral parts of hydropower systems rather than external constraints.

In addition, WG2 activities have highlighted the importance of system-level thinking. The role of hydropower within modern energy systems is evolving, with increasing expectations related to flexibility, hybrid operation and integration into other energy sources. This has led to a broader understanding of hydropower as part of interconnected energy and water systems.

Finally, the discussions within WG2 have identified several practical challenges that need to be addressed. These include data-related limitations, integration of legacy systems, economic constraints for smaller installations and the need for skilled personnel. Addressing these challenges requires not only technological solutions, but also collaboration between different stakeholders and disciplines.

Overall, WG2 activities provide a comprehensive view of the ongoing transformation of hydropower, combining technical developments with practical experience and stakeholder perspectives. These insights form an important basis for the analysis and conclusions presented in this chapter.

9 Future Perspectives

Hydropower is expected to continue playing an important role in future energy systems, particularly in supporting flexibility, reliability and system stability. However, its contribution will increasingly depend on its ability to adapt to changing technical, environmental and operational requirements.

One of the main directions of development is the further integration of digital technologies into hydropower systems. Advances in data analytics, machine learning and digital twins are expected to improve the understanding of plant behavior and support more efficient and flexible operation. In the longer term, these developments may enable higher levels of automation and more autonomous plant operation, especially in combination with centralized monitoring and control systems. Pumped storage hydropower is expected to play a key role as a cornerstone technology for balancing variable renewable energy generation.

At the same time, technological innovation will remain essential. Improvements in turbine design, materials and refurbishment approaches are expected to enhance performance under a wider range of operating conditions. In addition, small-scale and distributed hydropower solutions may become more relevant, particularly in the context of energy recovery within existing water infrastructure.

Environmental considerations will also play an increasingly important role. Future hydropower systems will need to operate in closer alignment with ecological and water management requirements. This includes more effective sediment management, improved compatibility with ecological flows and greater attention to water quality. These aspects will likely influence both design and operational strategies.

Another key aspect is the continued integration of hydropower into interconnected energy and water systems. The interaction with variable renewable energy sources, hybrid system configurations and multi-purpose water use will require more coordinated and system-oriented approaches. This reinforces the importance of considering hydropower not as an isolated technology, but as part of a broader system. Cross-sector synergies, including integration with water supply networks, hydrogen production and smart digital systems, are expected to open new pathways for innovation. Cross-sector synergies, including integration with water supply networks, hydrogen production and smart digital systems, are expected to open new pathways for innovation.

Overall, the future of hydropower will depend on the ability to combine digitalization, technological innovation and environmental considerations in a coherent way. This requires not only advances in individual technologies, but also a holistic perspective that takes into account system-level interactions and practical implementation aspects.

Competing Interests The authors have no conflicts of interest to declare that are relevant to the content of this chapter.

References

1. Vagnoni E, Gezer D, Anagnostopoulos I, Cavazzini G, Doujak E, Hočevar M, Rudolf P (2024) The new role of sustainable hydropower in flexible energy systems and its technical evolution through innovation and digitalization. Renew Energy 230. https://doi.org/10.1016/j.renene.2024.120832
2. Chernobrova A, Moloshnyi O, Koleva R, Nycz A, Szakal R, Kandukuri ST, Schlanbusch R, Shanbhag VV, Siviero M, Quaranta E (2026) A comprehensive review of novel technologies for hydropower and strategic opportunities for Europe. Energy Rep 15. https://doi.org/10.1016/j.egyr.2025.109019
3. PEN@Hydropower COST Action Web Site, https://www.pen-hydropower.eu/. Lact access: Apr 09, 2026
4. Gezer D, Bilkova E, Lučin I, Chernobrova A, Rudolf P (2024) Advancing sustainable hydropower through digitalization and innovation: insights from the pan-European network EERA workshop "Revamping hydropower research", 2024 September
5. van Dijk M, Gezer D, Rudolf P (2025) Unlocking hydropower's potential: retrofitting infrastructure and harnessing unconventional sources for clean energy transitions. In: IOP conference series: earth and environmental science, vol 1442, p 012006. https://doi.org/10.1088/1755-1315/1442/1/012006
6. Karapici V, Trojer A, Lazarevikj M, Pluskal T, Chernobrova A, Nezirić E, Zuecco G, Alerci AL, Seydoux M, Doujak E, Rudolf P (2024) Opportunities of hidden hydropower technologies towards the energy transition. Energy Rep 12:5633–5647. https://doi.org/10.1016/j.egyr.2024.11.039
7. Butuner MA, Gezer D, Kandukuri S, Schlanbush R, Voglin IK (2025) Digitalization challenges for hydropower plants – insights from a drinking water recovery example, CIGRE SEERC 2025, 2025 June

Environmental Sustainability of Hydropower and Its Adaptation to Climate Change

Epari Ritesh Patro, Behnam Balouchi, Marie-Pierre Gosselin, Maria Ubierna Aparicio, and David Christian Finger

Sustainable hydropower in the 21st century demands a holistic, multi-stakeholder approach that integrates climate adaptation, sediment management, ecological integrity, and social equity.

Abstract Despite the ongoing technological innovation and the rapidly increasing penetration of new renewable energy sources, the hydropower sector managed to remain a central actor for power production worldwide while establishing itself as a strong support to compensate the intermittency induced by most of these new energy sources. However, the sustainability of hydropower is increasingly challenged by climate-induced hydrological shifts, sediment regime changes, and complex socio-environmental trade-offs. This chapter argues that sustainable hydropower in the 21st century demands a holistic, multi-stakeholder approach that integrates climate adaptation, sediment management, ecological integrity, and social equity. Drawing on case studies and modelling examples, the physical foundations, e.g. climate, hydrology, sediment dynamics, and their interconnections with ecosystems and society are highlighted. As a renewable and affordable, usually low-carbon source of energy, hydropower can significantly contribute to the fulfilment of the UN Sustain-

E. R. Patro (✉)
Water, Energy and Environmental Engineering Research Unit, University of Oulu, Oulu, Finland
e-mail: ritesh.patro@oulu.fi

B. Balouchi
Hydropower group, AFRY Norway AS, Oslo, Norway

M.-P. Gosselin
Norwegian Institute for Nature Research (NINA), Trondheim, Norway
e-mail: marie-pierre.gosselin@nina.no

M. U. Aparicio
Open Hydro Ltd., London, UK
e-mail: mubierna@openhydro.net

D. C. Finger
Department of Engineering, Reykjavik University, Reykjavik, Iceland
e-mail: davidf@ru.is

E. Doujak et al. (eds.), *Rethinking Hydropower: A Holistic Vision for a Sustainable Future*, Springer Tracts in Civil Engineering,
https://doi.org/10.1007/978-3-032-33328-5_9

able Development Goals. However, any hydropower plant also potentially induces important ecological, social, and economic impacts at the local scale, calling for a sustainable strategy in building and management, but also in existing hydropower infrastructures.

Keywords Hydropower · Climate change · Sediment dynamics · Environmental flows · River ecology · Adaptation · Sustainability

1 Introduction

As the global energy transition accelerates toward net-zero emissions, the role of hydropower is evolving from a traditional baseload source to a flexible, integrating force for variable renewables such as wind and solar. Understanding this evolving role is not merely technical, it is foundational to designing energy systems that are both sustainable and resilient [1]. Hydropower systems must not be seen in isolation. They must be understood as a dynamic component of an interconnected, decarbonizing energy landscape [2].

In this context, the sustainability of hydropower can no longer be assessed through a static lens. Climate change is altering the very resources on which hydropower depends through shifting precipitation patterns, growing hydrological variability, and intensifying extreme weather events [3]. These changes threaten the reliability, safety, and ecological balance of hydropower systems. Thus, adaptation is no longer optional; it is central to the long-term viability and sustainability of hydropower. Future engineers must be trained to anticipate, model and respond to climate-driven uncertainties, integrating adaptive thinking into every stage of planning, design, and operation.

Addressing these complex, interwoven challenges demands more than just technical expertise. It requires a holistic, multi-stakeholder framework that bridges disciplines and perspectives [4]. Climate science, hydrology, ecology, social sciences, and engineering must intersect in meaningful ways. Tomorrow's engineers will need to engage with diverse stakeholders, including local communities, indigenous groups, policymakers, ecologists and investors, to co-create solutions that are environmentally sound, socially equitable and economical. This chapter highlights the imperative of educating engineers to navigate this complexity, fostering system thinking and collaborative governance as core professional competencies.

Historically, hydropower plants were built in different river basins with diverse hydrological regimes that worked as a "weather portfolio", with long transmission lines transporting electricity in space from "wet" areas to "dry" areas [12]. Reservoirs store water (or electricity) from "wet" periods to "dry" periods, sometimes across multiple years, with multi-year regulation capacity. In the last two decades, however, there has been a shift away from large storage projects. The power-to-flooded area-ratio of new plants has gone up significantly due to environmental constraints [13];

generally these new plants were "run-of-river" plants that do not significantly modify the natural seasonal pattern of river flows. The world now faces the dual challenge of having to increase the availability of electricity while moving toward a zero-carbon future–and it must do so while minimizing any negative impacts on people and the environment.

Dams represent an extension of human influence over the environment. Currently, there are plans to construct over 3,700 hydropower dams globally in the next 30 years [14]. The most intensive development is planned in river basins like Congo, Amazon, and Mekong which host approximately one third of the world's riverine biodiversity [14–16].

Human-altered flow regimes count among the adverse effects of dams, having serious consequences and posing persistent threats to aquatic ecosystems and natural habitats in streams and rivers [17, 18]. Dams influence hydrological flow regimes at varying time scales, altering low-flow regimes [19] and decreasing peak flows on a seasonal basis [20], but also on a sub-daily time scale. This operational regime, called *hydropeaking*, adversely influences riverine ecosystems and vegetation in several ways [21, 22]. Another dam-induced flow regime change is the reduced sediment load carrying capacity of a river, since the reservoirs of the dam create sediment starvation in the downstream section. Various modellers have indicated both global and local reduction in hydropower generation capacity two to five decades into the future if no preventive measures are introduced [23–26].

In spite of these various confounding factors, hydropower is still the dominant supply source in many countries, although its relative participation in electricity production has decreased [27]. Significant questions arise as to whether this general approach can be sustained in the future. The challenge going forward is to find new approaches that allow the continued use of the existing hydropower base, together with new innovations in order to pave the way for a massive further expansion of VRES.

To address the changing nature of the water-energy nexus for hydropower from the "water for energy" perspective, it is necessary to identify upcoming research questions focusing on hydropower planning, management, and rehabilitation. A need for trade-off studies is an important first step toward identifying more balanced outcomes–specifically to explore ecology–energy trade-offs at the decision-relevant scale of an individual hydropower plant. Legal regimes, governance structures, and the challenges surrounding hydropower projects are diverse. Until this challenge of identifying and navigating trade-offs is confronted at multiple scales, broader basin-scale studies may struggle to credibly identify the trade-offs and ecological outcomes of complex, multi-billion-dollar hydropower portfolios.

2 Climate Change, Hydrology, and Hydraulics: The Physical Foundation

For engineers coming into the hydropower sector, a robust understanding of the physical environment is no longer sufficient. They must also grasp how that environment is changing. The foundations of sustainable hydropower–climate, hydrology, and hydraulics–are now dynamic, driven by anthropogenic climate change. Educating future engineers in these shifts is critical to designing infrastructure that is resilient, efficient, and ecologically integrated over its entire lifecycle.

2.1 Climate and Water Cycle Changes: Observed and Projected Climate Trends Relevant to Hydropower

Future hydropower engineers must be fluent in climate science as it pertains to water resources. Observed trends–such as increasing temperatures, shifting precipitation patterns, and the accelerated melting of glaciers and seasonal snowpacks–directly influence river regimes. In many regions, projections indicate greater hydrological variability: more intense rainfall events, longer droughts, and altered seasonal flows. These changes challenge the assumptions underpinning historical hydrological data, the very data traditionally used for design and planning. Engineers must learn to work with climate models, uncertainty ranges, and scenario-based projections to future-proof hydropower systems.

The type of hydropower is crucial for understanding the influence of climate change. Storage-type hydropower has inherent operational flexibility by shifting power production from the time of water availability to the moment of energy demand [3, 26, 28–31]. Run-of-river type, on the other hand, is quite responsive to the hydrological regime due to very negligible to zero water storage capacity [25, 29]. Pump-storage type produces electricity during peak demand and re-pumps during off-peak periods [31, 32]. This type of hydropower plays a key role in the integration of other intermittent renewable energy sources into the electricity market [11, 33].

2.1.1 Climate Changes

The translation of climate trends into river hydrology is a core competency for hydropower engineers. Changes in precipitation type (rain vs. snow), timing of snowmelt, glacial contribution, and evaporation rates alter the magnitude, frequency, and seasonality of river flows. This can lead to increased flood risks, reduced dry-season flows, or shifts in the annual hydrograph. Understanding these impacts requires moving beyond static flow duration curves to dynamic, climate-informed hydrological modelling.

There is general scientific consensus that the climate has warmed over the past century [34], consisting of an increase in temperature and a variation in the spatial and temporal patterns of solid and liquid precipitation [35, 36]. Variations in climate may affect almost all processes in the biosphere to some extent, including landscape, ecosystem, hydrological regimes, and water resource infrastructure [37]. As a result, quantifying the impacts of climate change on hydropower moved into focus, by proposing and formulating operational adaptation strategies within different hydroclimatic scenarios and electricity markets [23, 25, 26, 38–43]. The footprints of recent climatic trends can thus be observed in various global and local hydrological variables, such as an increase of the frequency and extent of extreme weather events, as well as a modification of seasonal hydrological cycles, particularly in snow- and glacier-fed basins [25, 44]. The latter can significantly alter regional hydroclimatology, which in turn can have a substantial and negative economic and environmental impact on hydropower. Generalisations are difficult given the dominance of local conditions; assessments of the future of hydropower need to be interpreted with care [45].

2.1.2 Meltwater

Mountain catchments are subjected to rapid and sustained environmental changes [46], and it is necessary to focus on the mountain cryosphere as it hosts most of the hydropower [30]. Meltwater produced from perennial and seasonal snow, ice and glaciers provides the most significant runoff contribution to the major tributaries of such catchments and is therefore highly vulnerable to climate change [47, 48]. During the last century, most glaciers across the globe have retreated in response to climate variability [49, 50].

In mountain areas, precipitation and temperature may combine and result in a shift from solid to liquid precipitation [25]; onset of snowmelt [51]; shorter snow season [37]; decline in peak snow accumulation [52, 53]; changes in flow regimes from glacial to nival or nival to pluvial [23]. From this follows a substantial impact on the seasonality of water availability with direct consequences for reservoir management and hydropower operation [43].

Many studies across Europe have observed negative trends in snow depth and snow duration [54–58], with pronounced correlation to elevation [59, 60]. These observed changes are mainly linked to a shift of precipitation from solid to liquid [61] and lead to more frequent and more intense melting [62] as a result of increasing air temperatures in spring and winter.

Numerous hydropower plants are built downstream of glaciers. Glacier shrinkage may provide a boost to hydropower installed capacity and new installations [26, 63]. The combined synoptic weather patterns with local effects driven by topography may increase meltwater–the so-called "peak water"–for some decades, but will eventually disappear in the future [49]. Thus climate change may hamper subglacial hydrology and present a key reason for the susceptibility of future hydropower [64]. In the first half of the century, ice ablation would provide a significant increase in water flows; however, as glaciers largely disappear, a decrease in stream flows would be seen, with the role of precipitation becoming paramount.

2.1.3 Extreme Events

Altered hydrology imposes new demands on hydraulic infrastructure. More extreme flows challenge spillway capacity, dam safety, and flood risk management. Sediment loads may increase due to glacial melt or intense erosion, which would increase the risk of turbine abrasion and reservoir sedimentation. Conversely, reduced flows can affect power generation efficiency, cooling capacity, and minimum ecological releases. Future engineers must be trained in adaptive design: incorporating larger safety margins, designing for sediment bypass, and creating flexible operating rules. The Reykadalur case study exemplifies these challenges. It highlights how changes in catchment hydrology and extreme weather events can impact plant operation and infrastructure resilience, providing a practical framework for students to evaluate design trade-offs under climate uncertainty.

Climate change under global warming increases the incidence of extreme natural events, namely hazards. Climate-dependent hazards–including flash floods, droughts, shallow landslides, rock avalanches, snow/ice avalanches, and glacial lake outburst floods–have an increased chance of concurrent occurrence [65]. The issue of a potentially increased impact of multiple hazards under present and prospective climate change conditions has so far been largely neglected, and risk assessment is typically pursued independently for different processes.

Water-based calamities such as heavy rainfall and torrential floods are repeatedly observed around the world. Dams have deep upstream and downstream effects on a river's natural hydrological regime, principally through their effects on five components: magnitude, timing, frequency, duration and rate of change [66]. In case of streamflow drought, hydropower production immediately decreases, adversely affecting suppliers and consumers. Drought-induced reduction in dam-based electricity supply can have negative effects on daily life and the economy, particularly in regions that rely heavily on hydropower, for example in Africa and South America [67].

2.1.4 Interconnection with Ecology and Society

Hydropower engineering does not exist in a physical vacuum; it interacts directly with ecological systems and human communities. Changes in flow regimes affect river temperature, dissolved oxygen, habitat connectivity, and aquatic biodiversity. These ecological shifts, in turn, impact fisheries, water quality, and ecosystem services upon which communities depend. Furthermore, socio-hydrological feedbacks are critical: water releases for power generation may conflict with agricultural needs, drinking water supply, or cultural practices under changing climate conditions.

During droughts, the water use domains of energy, water, food, health, and the environment are often competing intensely. Water markets and reallocation systems play a significant role in balancing trade-offs between sectors and improving resilience to the impacts of drought. Moreover, fish and riverine resources are critical factors for the livelihoods of many indigenous peoples, providing both financial security and food. It is these communities that are often most heavily impacted by hydropower and dam developments. Dams threaten zones of aquatic biodiversity in all the major rivers of undeveloped, developing, and developed countries [68].

2.1.5 Energy Systems and Electricity Mix

Anomalies in total electricity production can provide crucial information about changes in the share of different energy sources and the manner in which each electricity mix affects the whole electricity system. It is essential to accelerate the development of renewable energy and its share, as countries strive to reduce carbon emissions and air pollution [27]. The relative flexibility and storage capacity most hydropower plants (storage and pumped hydro) offer provides the necessary back-up to balance demand and supply [69].

Nevertheless, hydroelectricity itself is susceptible to climate because of its strong reliance on natural water resources and climate conditions. Climate change is projected to lead to an increase in hydropower generation in some parts of the world and to a decrease in others [70]. In the context of current changing climatic conditions, the demand for electricity often increases due to the soaring requirements of air conditioning and heating [26]. The complementarity between hydro and coal, wind, solar or other energy sources is highly recommended [71]. Although this integration has already been implemented, the mix of multiple sources is expected to grow further in the context of technology development combined with the projected stronger and more frequent hydrological extremes in the near future [72].

In current literature, hydropower models mostly simulate the physical impacts of changes in streamflow or outflow from reservoirs, whereas socio-economic feedbacks, changes of electricity prices, and demand load portfolios are not modelled. The modelling of anthropogenic river flow alteration is urgently required in future research, and a thorough investigation of human component impacts is suggested.

3 Integrating Sediment Dynamics and Climate Change for Sustainable Hydropower

Large reservoirs may be capable of controlling a wide range of floods, and consequently can reduce the magnitude and frequency of floods experienced by the downstream channel. The reduced flow may not transport sediment delivered to the river below the dam by tributaries, promoting channel aggradation and potentially increasing flooding risk. Thus, depending on the balance between the available transport energy and sediment supply, some river stretches below a dam will exhibit sediment deficit and others sediment surplus [73, 74]. Hydropower plants also act as sediment barriers, blocking natural riverine flows of water, sediment and critical nutrients that in turn impact fish and other aquatic organisms.

Hydropower infrastructure regulates not only water but also sediment. Because sediment transport governs reservoir longevity, channel stability, habitat structure, and downstream geomorphology, it is a central determinant of hydropower sustainability. Under climate change, shifts in sediment supply and transport capacity challenge design assumptions that were historically based on stationary conditions. For future engineers, understanding sediment dynamics is therefore essential to designing hydropower systems that remain technically reliable, ecologically responsible, and socially legitimate throughout their operational lifetime.

3.1 Flow Regulation and Morphodynamic Adjustment

Alterations to discharge regimes fundamentally modify sediment transport processes. Changes in flood magnitude, seasonal timing, and hydrograph shape influence shear stress distribution, bed mobility, and sediment continuity along the river corridor [75]. Reduced peak flows may decrease transport capacity and promote channel stabilization, vegetation encroachment, or localized deposition. In contrast, hydropeaking operations can induce rapid stage fluctuations that destabilize fine sediments and alter bed structure.

Such changes often translate into long-term morphological adjustment [76]. Braided or multi-thread systems may evolve toward simplified, more stable configurations, while sediment deficits downstream of reservoirs can lead to channel incision and coarsening. Conversely, altered hydrographs may increase suspended sediment concentrations if flow releases mobilize previously stored material. The direction and magnitude of response depend on the interaction between sediment supply, hydraulic forcing, and operational rules. Nevertheless, these responses are highly non-linear and inherently unpredictable, as they depend on complex interactions among system parameters, regulated by hydropower dams, human interventions, and episodic disturbances such as landslides [77].

Recognising these morphodynamic consequences is critical when evaluating hydropower sustainability. Channel adjustment affects not only infrastructure performance but also ecological habitat diversity and floodplain connectivity.

3.2 Climate Change and Sediment Regime Shifts

The sediments carried by a river are deposited in the reservoir, which in turn reduces the river's transport capacity. Although reservoirs are typically designed with dead storage volume to accommodate accumulated sediment, sedimentation also occurs in the useful volume of the reservoir [78], leading to loss in storage capacity that hampers normal operation. The main causes of sediment deposition at reservoirs are either of two types: physical and chemical [79]. The physical type is due to the decrease in water speed of tributaries flowing into the reservoir; essentially there is not enough energy for the transport of suspended sediments. The second type depends substantially on the interaction between the various chemical characteristics in the water.

The phenomenon is complicated by the simultaneous action of various factors (variable flow rates released by the dam, sediment loads, reservoir level variations, etc.) [80]. The dynamics of the sediment are affected by other actions associated with the management of reservoirs, such as manoeuvres to reduce the reservoir level carried out at high speed. This also adds a drag effect to the phenomena described above [81].

Climate change modifies sediment dynamics at the catchment scale. Intensified precipitation, glacier retreat, permafrost degradation, and increased hillslope instability can alter both the magnitude and timing of sediment delivery. In mountain and glacier-fed basins, enhanced meltwater can temporarily increase fine sediment transport, whereas long-term glacier retreat may eventually reduce coarse sediment input once easily erodible sources are exhausted. Similarly, increased rainfall intensity may accelerate erosion processes and slope failures. These changes interact with existing hydraulic infrastructure. Reservoirs may trap greater volumes of sediment than anticipated during the design stage, accelerating storage loss and reducing operational flexibility. Altered timing of sediment inputs may also complicate flushing or bypass strategies [82]. The combined effect of climatic and operational drivers reshapes sediment continuity at the basin scale, requiring adaptive and forward-looking management.

3.3 Implications for Infrastructure Performance

Sediment processes directly influence the technical and economic sustainability of hydropower systems. Reservoir sedimentation reduces active storage, limits flood buffering capacity, and constrains energy production over time. Elevated suspended sediment concentrations increase turbine abrasion, and as a consequence maintenance demands and operational costs rise. Deposition near intakes and in tailrace channels can impair hydraulic efficiency and lead to bigger safety risks.

The deposition of sediments in reservoirs can generate problems of stability and infrastructure security. Accumulation of sediments increases the possibility of localized erosions at the foot of the dam. The operating rules of the dam greatly influence the shape and location of the sediment deposits [83]. Indeed, following the flushing operations it is necessary to carry out inspections or maintenance of the parts of dam infrastructure since changes occur in the structure of the sedimentation profile, which can cause a greater accumulation of sediment close to the dam outlet [84].

Under non-stationary climatic conditions, sediment-related risks may intensify. Historical sediment records may no longer provide reliable design references, and extreme sediment loads can exceed expected operational thresholds. Consequently, sediment management must be integrated into climate-resilient design and operational standards. Approaches such as sediment bypass systems, controlled flushing operations, sediment augmentation downstream, and adaptive reservoir management offer potential pathways toward maintaining sediment continuity while protecting infrastructure. Evaluating these strategies requires balancing energy production objectives with long-term geomorphic stability and ecological function.

3.4 Modelling Across Scales: Capabilities and Uncertainty

Predicting sediment dynamics under changing climatic and operational conditions remains a complex hydraulic modelling challenge. Sediment transport is highly sensitive to hydraulic boundary conditions, grain-size distribution, bed roughness parameterization, and transport formulations [77].

At the catchment scale, hydrologic models estimate sediment yield based on projected precipitation patterns, land use, and erosion processes. Uncertainty arises from climate scenarios and the representation of erosion mechanisms. At the reach scale, hydraulic models simulate flow fields and transport capacity. Results depend on roughness coefficients, regulated and non-regulated discharge, mesh size, turbulence representation, and calibration procedures [85]. At the morphodynamic scale, two- and three-dimensional models simulate bed evolution and channel adjustment [86]. Outcomes are influenced by grid resolution, sediment transport functions, hiding–exposure corrections, and bed roughness predictors. Sensitivity analyses consistently demonstrate that parameter selection can substantially affect predicted morphological change.

When combined with climate projections, these uncertainties propagate across scales. Robust assessment, therefore requires scenario-based analysis rather than reliance on single deterministic predictions. It should be noted that, while 2D flow and sediment modelling is well suited for long-term and climate change assessments, it remains computationally demanding, often requiring the application of acceleration factors. In addition, uncertainties related to calibration procedures and data scarcity further affect model reliability and must be carefully addressed. Educating engineers to critically interpret model outputs and recognize structural limitations is essential for responsible hydropower planning under climate uncertainty.

3.5 *Interconnections with Ecology and Society*

Sediment dynamics form a direct link between physical processes and ecological integrity. Channel morphology governs habitat heterogeneity, spawning substrate availability, and riparian stability. Excessive fine sediment deposition can degrade water quality, while sediment deficits may cause incision and bank instability. For communities, altered sediment regimes influence flood risk, agricultural productivity, cultural landscapes, and downstream coastal stability. Reduced sediment delivery to deltas can exacerbate shoreline erosion and increase vulnerability to rising sea levels.

The suspended sediments have a direct effect on the respiratory and alimentary systems of fish and invertebrates and on the aquatic plant component [87]. There are several factors that intervene in reaching the lethality threshold (size of particles, dissolved oxygen, duration of exposure, background concentration). Turbidity induced by the sediments affects the ability of sunlight to penetrate the water column, with repercussions on autotrophic organisms, the entire food web and on the productivity of the ecosystem [88]. The deposition of sediment on spawning areas prevents the regular hatching of the eggs and their development as the filling of the interstitial spaces restricts water and gas circulation; furthermore, the deposition of sediment can prevent larvae from emerging from nests, trapping them in the sediment after hatching [89].

Sediment supply downstream can alter mesohabitat structure through the filling of pools and the formation of islands, ultimately reducing hydraulic-morphological diversity [90]. On the other hand, the release of sediment can in some cases improve the heterogeneity of downstream physical habitats [91]. Integrating sediment considerations into hydropower development therefore requires collaboration among engineers, ecologists, policymakers, and stakeholders. Basin-scale sediment budgeting, transparent communication of trade-offs, and adaptive governance frameworks are key elements of long-term resilience.

4 Ecological and Natural Resource Dimensions of Sustainable Hydropower

In addition to a climate crisis, the world is also experiencing a biodiversity crisis, the latter often being overshadowed by the former. Freshwater ecosystems are the most threatened ecosystems on the planet [92] with freshwater fish species (including migratory ones) having declined by 85% since the 1970s. In addition, according to the largest global assessment of freshwater animals on the IUCN Red List of Threatened Species, 24% of the world's freshwater fish, dragonfly, damselfly, crab, crayfish and shrimp species are at high risk of extinction [93]. Hydropower remains the world's largest source of renewable electricity and plays a leading part in achieving the green energy transition by providing a stable "baseload" that complements the variability of wind and solar [94]. However, hydropower itself has a significant ecological footprint due to the damming of river systems and modification of natural flow regimes. Achieving sustainable hydropower will only be possible by balancing the demand for low-carbon energy with the preservation of complex, well-functioning and healthy aquatic ecosystems–and the natural resources and ecosystem services that human communities depend on.

Riverine species are adapted to the natural flow regime [95] of the river ecosystem they inhabit and to the natural mosaic of habitats and their variability in this river ecosystem. Modifications to the hydrology and morphology of the river will impair river health, functioning, and ultimately resilience.

Riverine ecosystems have historically faced many different pressures, not only from hydropower but also from other human activities (e.g. agriculture, industrialisation). They are now experiencing a twofold crisis: the simultaneous impact of global climate change and the expansion of physical infrastructures. Already in 2015, [14] discussed the global boom in hydropower dam construction and recorded that almost 4,000 major dams with a capacity above 1 MW were either planned or under construction globally [14]. The global dataset GOODD for georeferenced dams registered over 38,000 dams [96]. As of 2020, 1.2 million instream barriers were identified in 36 European countries [97]. River systems are currently caught between two forces–climate stress and energy infrastructure stress–that create a degradation feedback loop. As an example, a recent study in Oregon, USA showed that an increase in stream temperature by 1.8 to 3.2 °C–and this is predicted for the end of the century–will reduce the amount of thermally suitable habitats for several salmonid species by 50% or more [98]. Anthropogenic structures lead to a fragmentation of the river's longitudinal continuity, preventing species from reaching potentially suitable habitats and refuges, and undermining the resilience of these ecosystems and freshwater species to adapt to climate change.

4.1 How Hydropower Affects River Ecology

The impacts of hydropower production on riverine ecosystems and on biodiversity are well documented [99–101]. The most visible and direct impacts of hydropower production are linked to the disruption of river hydromorphology, where longitudinal and lateral connectivity of the river are affected, leading to habitat fragmentation and barriers to fish migration. River flow is regulated by smoothing out seasonal high and low flows. While this is beneficial for flood control and irrigation, many species rely on natural flow cues for spawning and migration. Hydropeaking leads to fish stranding, habitat dewatering, and increased erosion [100]. Hydropeaking will be more frequent due to an increased demand for flexibility in hydropower, and this demands that such impacts are further researched and addressed [102].

Rivers are "conveyor belts" for sediment, and hydropower production modifies the natural sediment transport, leading to significant geomorphological changes (see Sect. 3). The upstream accumulation of sediment due to impoundment leads both to a reduction in reservoir storage capacity over time and to sediment deprivation in down-stream reaches, depriving habitats of the material needed to maintain them. Finally, the transition from lotic to lentic ecosystem due to impoundment changes the chemistry, temperature, and biochemical cycles in the river ecosystem.

Independent of the capacity of a hydropower plant, energy production has an impact on riverine ecosystems and their biodiversity, and such impacts can be significant regardless of the size of the hydropower plant. Impacts from small hydropower plants can be considerable, especially when these hydropower plants are built along the same river stretch.

4.2 *The Compounded Effect of Climate Change and Hydropower*

The impacts of hydropower operation alone have led to profound alteration of river ecosystems, in addition to other human pressures river ecosystems are facing. When considering the combined effect of hydropower operation and climate change, the synergy between these stressors may often result in a level of damage that is "greater than the sum of its parts", meaning existing damages will be exacerbated.

Rivers have natural season-dependent temperature cycles. Alteration in the temperature regime associated with climate change (e.g. warming of rivers) and hydropower operation (e.g. thermopeaking) can lead to thermal stress as well as disturbance to biological cues for riverine biota (e.g. fish migration and spawning) [103–105]. Rising water temperature in combination with decreased flows exacerbates pollution and eutrophication, leading to water quality problems and a general degradation in the ecological status [106].

Levees and other structures impair lateral connectivity, i.e. they prevent rivers from connecting with their floodplains. During extreme climate change-driven floods, which are forecasted to increase in frequency and magnitude [107], floodwater can no longer dissipate into adjacent wetlands, leading to catastrophic downstream surges. Climate change modifies rainfall and snowmelt patterns. This affects riverine ecosystems directly by modifying the timing of flow and nutrient pulses, which are then no longer aligned with the requirements of the life cycles of riverine biota; and indirectly by influencing hydropower production and thus patterns of flow releases downstream. This will push vulnerable populations of native species to a tipping point, beyond which they are unlikely to recover.

4.3 Successful Mitigation of Ecological Impacts Is Needed to Achieve Sustainable Hydropower

A variety of targeted measures exist to attempt to mitigate the effect of hydropower [101]. In the last decades, these very much focused on improving fish passage, both upstream and downstream, and on retrofitting fish passage facilities on older dams, with varying success [108–111]. Developments in turbine technologies have led to the design of turbines that decrease the mortality risk for fish, featuring wider spacing between blades and lower rotational speeds to reduce physical trauma and barotrauma.

However, efficient mitigation measures should be proactive and included already in the design stage of hydropower plants. The most important mitigation measure to maintain and/or restore a healthy aquatic ecosystem will always be: sufficient and adapted environmental flows to maintain suitable instream habitat; lateral and longitudinal hydrological connectivity; and maintaining the processes and functions that are necessary to meet habitat and life-cycle requirements [112, 113]. The 2007 Brisbane Declaration on Environmental Flows stated that "environmental flows are essential to freshwater ecosystem health and human well-being" [114]. The most inclusive definition associated with the Brisbane Declaration defines environmental flows as the "quantity, timing and quality of freshwater flows and levels necessary to sustain aquatic ecosystems, which, in turn, support human cultures, economies, sustainable livelihoods, and well-being" [114]. More than 300 methods for the determination of environmental flows exist, from simple hydraulic rating methods to habitat modelling and complex, transdisciplinary holistic methods [115].

Despite numerous calls to assert the implementation of environmental flows globally [113, 114], and despite the release of ecological flows being a key measure of the EC Water Framework Directive in the European Union [112], numerous challenges remain when it comes to implementation. One area that needs particular focus is the adaptation of environmental flows (both design and implementation) to reflect climate change-induced shifts.

The release of environmental flows means a loss of water for hydropower production and often leads to conflict between stakeholders on the trade-off needed between economic and energy gain on the one hand, and ecosystem conservation on the other. Adaptive management strategies are relevant, even more so under climate change scenarios. Some hydropower plants are not necessarily viable, and their removal can help restore rivers to free-flowing conditions, offering refuges and habitats to species. Replanting of riparian zones will contribute to the cooling of water temperature, providing thermal refuge to an array of species. Case-by-case assessment, based on sound science and on precautionary principles, will ensure a tailored mitigation regime with a higher probability of target achievement than a one-size-fits-all solution [116]. In addition, hydropower must be planned and operated at the watershed level, taking into account other pressures and land uses along riverine corridors, instead of being managed as isolated projects.

5 Operational Adaptation Strategies

Electricity and power markets are intimately linked to weather variables [117]. Temperature and humidity influence consumption (demand), production (supply), and transmission [118], resulting in an impact on prices. Energy demand may be influenced by climate change, and in particular the seasonality of this demand is governed by changes in temperature and humidity patterns [28]. Typical seasonal cycles may be altered due to a warmer future, resulting in, for example, decreasing demand in winter and increasing demand in summer [119]. For this reason, it is very important to understand how hydropower revenues will be impacted by shifts in electricity price seasonality.

But modifications in temperature trends and precipitation patterns are also expected to modify the entire hydrological cycle. Due to its high dependency on the water regime and water availability, hydropower is extremely exposed to climate change [119]. Climate change is expected to impact hydropower through two main factors: runoff and its seasonality, and sedimentation [25, 30, 120, 121].In mountainous areas, changes may overlap and result in an earlier beginning of snow and ice melting [51], impacting especially summer streamflow.

Many power plants are built in catchments dominated by glaciers. As a consequence of climate change, these are progressively shrinking [122, 123]. Initially, this would have a positive impact on runoffs; however, after complete depletion, this results in a net loss of yearly volume [49]. In these catchments, there would be a shift from a glacio-nival to a nival-pluvial regime, impacting seasonality and volumes available [23]. Finally, the retreat of the glaciers would support sediment formation [124], increasing the rate of sedimentation in reservoirs and ultimately reducing their capacities faster than the 0.8–1% we currently observe [63].

According to the Fifth Assessment Report of the IPCC, at world scale, projected changes include an increase in water resources at high latitudes, in tropical East Africa, and in Southeast Asia, and a decrease of water resources in many semi-arid and arid areas (e.g., the Mediterranean Basin, Western US, Southern Africa, and North-eastern Brazil) [36]. According to [125], Europe's gross hydropower potential is expected to decrease by 6% by the 2070s [125]. Areas around the Mediterranean will experience a reduction spanning from 20% to 50%, while Northern and Eastern European countries will see an increase between 15% and 30%, and Western and Central Europe will likely remain stable. Golombek et al. [126] found an overall reduction of 15% for Western European countries [126].

Moreover, hydropower can play an effective role in flood mitigation [121]. Further safety concerns include the melting of permafrost and the possibility of more frequent heavy rainfalls, resulting in more frequent slope instabilities and potential flood waves that may endanger hydropower [124, 127]. Amplified sediment yields from deglacierized catchments may put further stress on power generation by accelerating the reduction in storage capacity, reducing operating efficiency, and increasing turbine abrasion and infrastructure wear [128–131].

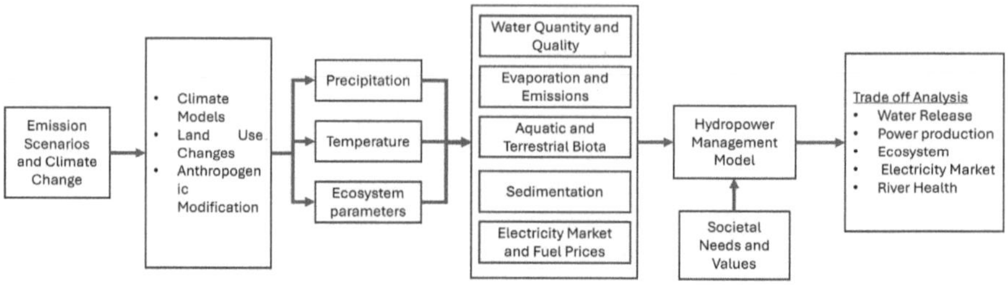

Fig. 1 This flowchart shows the assessment of hydropower generation under future climate scenarios, as adopted in literature. Blue arrows indicate increases, changes in seasonality and/or frequency and/or intensity of variables and parameters under consideration. Adapted from [151]

In available literature, the vast majority of future climate impact assessments on hydropower relies on the classical framework where climate model outputs are used as input for hydrological/water resources system models [132], applied either at local scale [64, 133] or at global scale. This modelling chain framework, which is used for many case studies around the world, is summarised in Fig. 1. Implications of changing hydrology on hydropower have been discussed both quantitatively and qualitatively [63, 134]. In these studies, the present hydropower operating rules are typically assumed to simulate system operations under future hydrological regimes, thus overlooking the adaptation potential of hydropower systems. Only in rare cases are adaptation measures including reoperation [25], structural measures [135], or change in hydropower functioning [25, 26, 28, 29] considered. To the authors' knowledge, there are no models so far that explore how hydropower operations might be affected by ecosystem, economic, sediment flux and political scenarios.

6 Way Forward: Redefining Sustainable Hydropower Under Climate Change

Sustainable hydropower under climate change depends on managing both water and sediment as an integrated system. Flow regulation alters sediment transport, while climate change modifies sediment supply; together, these processes drive morphological adjustments that influence infrastructure performance, ecosystem stability, and societal well-being.

Embedding sediment dynamics into climate adaptation strategies strengthens the physical, ecological, and social foundations of hydropower sustainability. At the same time, although 2D flow and sediment modelling is well-suited for long-term and climate change assessments, it remains computationally demanding and often requires the use of acceleration factors. Furthermore, uncertainties related to calibration procedures and data scarcity must be carefully addressed to ensure model reliability.

Reconciliation of multiple uses of water is one of the most important problems faced globally. The aim of sustainable water resource management and allocation is to deliver the quality and quantity of water needed for all usages by controlling it sustainably [136]. To enable hydropower sustainability, full integration of multiple water uses and proper water management must be established, with focus on strategy, planning and decision-making [137, 138]. The decision-making for water resource management in dams is primarily governed by multiple objectives in diverse strategic sectors, such as power generation, irrigation, navigation, flood control, recreational use, and consumptive uses [139, 140]. A decision-making process for hydropower sustainability is composed of a wide range of variables associated with hydrological factors, political and social interests, as well as cost/benefit relationships and water availability [141, 142]. In this sense, water must be managed to accommodate multiple uses such as supply, power generation, navigation, irrigation, industry and sustainability [143, 144]. These conflicting demands require the use of an integrated hydrological, geomorphological, hydraulic and water quality modelling system [145–147], enabling the methodology used to perform resource allocation and satisfy multiple uses through processes that optimise objectives often subjected to restrictions [148–150].

By adopting integrated water–sediment–ecological management approaches and acknowledging these modelling limitations, future engineers can contribute to the development of hydropower systems that remain resilient in an era of accelerating environmental change.

7 Conclusion

Holistic frameworks are necessary to point the way to sustainable and reliable hydropower in the future. Climate change and sediment dynamics challenge historical design assumptions, while ecological integrity and social equity must become core priorities, not afterthoughts. Meeting these demands requires a holistic, multi-stakeholder approach that unites engineering, natural sciences, and communities. Educating future engineers to navigate trade-offs and manage water and sediment as an integrated system will be essential in order to assure that hydropower remains a sustainable pillar of the energy transition–a system that respects the rivers and the people it serves.

Acknowledgements The authors wish to thank the working group members and all contributors who provided input and feedback throughout the preparation of this chapter.

Competing Interests The authors have no conflicts of interest to declare that are relevant to the content of this chapter.

References

1. Prieler M, Fazeni-Fraisl K, Lindorfer J, Markova D, Finger DC (2022) Decarbonizing the European Energy Sector: Frameworks, Examples and How Education Plays a Key Role. In Vasconcelos C, Calheiros CSC (eds) Enhancing environmental education through nature-based solutions. Integrated science, vol 4. Springer, Cham. https://doi.org/10.1007/978-3-030-91843-9_22
2. Stefanò D, Finger DC (2025) Advancing education on energy and landscapes. EGU general assembly 2025, Vienna, Austria, 27 Apr–2 May 2025, EGU25-17707. https://doi.org/10.5194/egusphere-egu25-17707
3. Finger D, Heinrich G, Gobiet A, Bauder A (2012) Projections of future water resources and their uncertainty in a glacierized catchment in the Swiss Alps and the subsequent effects on hydropower production during the 21st century. Water Resour Res 48:W02521. https://doi.org/10.1029/2011WR010733
4. Voegeli G, Finger DC (2021) Disputed dams: mapping the divergent stakeholder perspectives, expectations, and concerns over hydropower development in Iceland and Switzerland. Energy Res & Soc Sci 72:101872. https://doi.org/10.1016/j.erss.2020.101872
5. Moran EF, Lopez MC, Moore N, Müller N, Hyndman DW (2018) Sustainable hydropower in the 21st century. Proc Natl Acad Sci. https://doi.org/10.1073/pnas.1809426115
6. IEA (2015) WEO-2015 special report: energy and climate change. International Energy Agency. https://doi.org/10.1038/479267b
7. Schillinger M, Weigt H, Barry M, Schumann R (2017) Hydropower operation in a changing environment. In: International conference on the European energy market, EEM. https://doi.org/10.1109/EEM.2017.7981887
8. Puspitarini HD, François B, Zaramella M, Brown C, Borga M (2020) The impact of glacier shrinkage on energy production from hydropower-solar complementarity in alpine river basins. Sci Total Environ. https://doi.org/10.1016/j.scitotenv.2020.137488
9. Nautiyal H, Goel V (2020) Sustainability assessment of hydropower projects. J Clean Prod. https://doi.org/10.1016/j.jclepro.2020.121661
10. Berga L (2016) The role of hydropower in climate change mitigation and adaptation: a review. Engineering 2(3):313–318. https://doi.org/10.1016/J.ENG.2016.03.004
11. Gaudard L, Madani K (2019) Energy storage race: Has the monopoly of pumped-storage in Europe come to an end? Energy Policy. https://doi.org/10.1016/j.enpol.2018.11.003
12. McCully P (1996) Silenced rivers the ecology and politics of large dams. https://doi.org/10.2307/2624501
13. Towler BF (2014) The future of energy. Elsevier. https://doi.org/10.1016/C2013-0-19049-6
14. Zarfl C, Lumsdon AE, Berlekamp J, Tydecks L, Tockner K (2015) A global boom in hydropower dam construction. Aquat Sci 77(1):161–170. https://doi.org/10.1007/s00027-014-0377-0
15. Dugan PJ, Barlow C, Agostinho AA, Baran E, Cada GF, Chen D, Cowx IG, Ferguson JW, Jutagate T, Mallen-Cooper M, Marmulla G, Nestler J, Petrere M, Welcomme RL, Winemiller KO (2010) Fish migration, dams, and loss of ecosystem services in the Mekong basin. Ambio. https://doi.org/10.1007/s13280-010-0036-1
16. Winemiller KO, McIntyre PB, Castello L, Fluet-Chouinard E, Giarrizzo T, Nam S, Baird IG, Darwall W, Lujan NK, Harrison I, Stiassny MLJ, Silvano RAM, Fitzgerald DB, Pelicice FM, Agostinho AA, Gomes LC, Albert JS, Baran E, Petrere M, Sáenz L (2016) Balancing hydropower and biodiversity in the Amazon, Congo, and Mekong. Science 351(6269):128–129. https://doi.org/10.1126/SCIENCE.AAC7082
17. Chen Q, Zhang X, Chen Y, Li Q, Qiu L, Liu M (2015) Downstream effects of a hydropeaking dam on ecohydrological conditions at subdaily to monthly time scales. Ecol Eng. https://doi.org/10.1016/j.ecoleng.2014.12.017

18. Fruhmann C, Tuerk A, Kulmer V, Gubina AF (2019) Balancing environmental benefits and damages of small hydropower generation in policy-making: assessing the implementation of a contradicting EU policy framework in Austria and Slovenia. Int J Sustain Energ. https://doi.org/10.1080/14786451.2018.1452741
19. Heggenes J, Alfredsen K, Bustos AA, Huusko A, Stickler M (2018) Be cool: a review of hydro-physical changes and fish responses in winter in hydropower-regulated northern streams. Environ Biol Fishes. https://doi.org/10.1007/s10641-017-0677-z
20. Mailhot A, Talbot G, Ricard S, Turcotte R, Guinard K (2018) Assessing the potential impacts of dam operation on daily flow at ungauged river reaches. J Hydrol Reg Stud. https://doi.org/10.1016/j.ejrh.2018.06.006
21. Bejarano MD, Sordo-Ward Á, Alonso C, Nilsson C (2017) Characterizing effects of hydropower plants on sub-daily flow regimes. J Hydrol. https://doi.org/10.1016/j.jhydrol.2017.04.023
22. Moreira M, Hayes DS, Boavida I, Schletterer M, Schmutz S, Pinheiro A (2019) Ecologically-based criteria for hydropeaking mitigation: a review. Sci Total Environ. https://doi.org/10.1016/j.scitotenv.2018.12.107
23. Bongio M, Avanzi F, De Michele C (2016) Hydroelectric power generation in an Alpine basin: future water-energy scenarios in a run-of-the-river plant. Adv Water Resour 94:318–331. https://doi.org/10.1016/J.ADVWATRES.2016.05.017
24. Gaudard L, Gabbi J, Bauder A, Romerio F (2016) Long-term uncertainty of hydropower revenue due to climate change and electricity prices. Water Resour Manage. https://doi.org/10.1007/s11269-015-1216-3
25. Patro ER, De Michele C, Avanzi F (2018) Future perspectives of run-of-the-river hydropower and the impact of glaciers' shrinkage: the case of Italian Alps. Appl Energy 231:699–713. https://doi.org/10.1016/j.apenergy.2018.09.063
26. Ranzani A, Bonato M, Patro E, Gaudard L, De Michele C (2018) Hydropower future: between climate change, renewable deployment, carbon and fuel prices. Water 10(9):1197. https://doi.org/10.3390/w10091197
27. IRENA (2018) Global energy transformation: a roadmap to 2050. International renewable energy agency
28. Bonato M, Ranzani A, Patro ER, Gaudard L, De Michele C (2019) Water-energy nexus for an Italian storage hydropower plant under multiple drivers. Water 11(9):1838. https://doi.org/10.3390/w11091838
29. Gaudard L, Avanzi F, De Michele C (2018) Seasonal aspects of the energy-water nexus: the case of a run-of-the-river hydropower plant. Appl Energy 210:604–612. https://doi.org/10.1016/j.apenergy.2017.02.003
30. Kumar A, Schei T, Ahenkorah A, Caceres Rodriguez R, Devernay J-M, Freitas M, Hall D, Killingtveit Å, Liu Z (2011) Hydropower. IPCC special report on renewable energy sources and climate change mitigation, pp 437–496. https://doi.org/10.1016/j.rser.2005.06.003
31. Lueken R, Apt J (2014) The effects of bulk electricity storage on the PJM market. Energy Syst. https://doi.org/10.1007/s12667-014-0123-7
32. Chang MK, Eichman JD, Mueller F, Samuelsen S (2013) Buffering intermittent renewable power with hydroelectric generation: a case study in California. Appl Energy 112:1–11. https://doi.org/10.1016/j.apenergy.2013.04.092
33. Guittet M, Capezzali M, Gaudard L, Romerio F, Vuille F, Avellan F (2016) Study of the drivers and asset management of pumped-storage power plants historical and geographical perspective. Energy. https://doi.org/10.1016/j.energy.2016.04.052
34. IPCC (2007) Climate change 2007–the physical science basis: working group I contribution to the fourth assessment report of the IPCC. Cambridge University Press
35. Blackshear B, Crocker T, Drucker E, Filoon J (2011) Hydropower vulnerability and climate change—a framework for modeling the future of global hydroelectric resources. Middlebury college environmental studies senior seminar

36. Stocker TF, Qin D, Plattner GK, Tignor M, Allen SK, Boschung J, Nauels A, Xia Y, Bex V, Midgley PM (2013) IPCC. Climate change 2013: the physical science basis. Cambridge University Press
37. Beniston M, Farinotti D, Stoffel M, Andreassen LM, Coppola E, Eckert N, Fantini A, Giacona F, Hauck C, Huss M, Huwald H, Lehning M, López-Moreno JI, Magnusson J, Marty C, Morán-Tejéda E, Morin S, Naaim M, Provenzale A, Vincent C (2018) The European mountain cryosphere: A review of its current state, trends, and future challenges. Cryosphere. https://doi.org/10.5194/tc-12-759-2018
38. Beniston M, Stoffel M (2014) Assessing the impacts of climatic change on mountain water resources. Sci Total Environ 493:1129–1137. https://doi.org/10.1016/j.scitotenv.2013.11.122
39. François B, Hingray B, Creutin JD, Hendrickx F (2015) Estimating water system performance under climate change: influence of the management strategy modeling. Water Resour Manage. https://doi.org/10.1007/s11269-015-1097-5
40. Li FF, Qiu J (2016) Multi-objective optimization for integrated hydro-photovoltaic power system. Appl Energy. https://doi.org/10.1016/j.apenergy.2015.09.018
41. Mideksa TK, Kallbekken S (2010) The impact of climate change on the electricity market: a review. Energy Policy 38(7):3579–3585. https://doi.org/10.1016/j.enpol.2010.02.035
42. Schaefli B (2015) Projecting hydropower production under future climates: a guide for decision-makers and modelers to interpret and design climate change impact assessments. Wiley Interdiscip Rev Water 2(4):271–289. https://doi.org/10.1002/wat2.1083
43. Schaefli B, Manso P, Fischer M, Huss M, Farinotti D (2019) The role of glacier retreat for Swiss hydropower production. Renew Energy. https://doi.org/10.1016/j.renene.2018.07.104
44. Avanzi F, Yamaguchi S, Hirashima H, De Michele C (2015) Bulk volumetric liquid water content in a seasonal snowpack: modeling its dynamics in different climatic conditions. Adv Water Resour 86:1–13. https://doi.org/10.1016/j.advwatres.2015.09.021
45. Van Vliet MTH, Vögele S, Rübbelke D (2013) Water constraints on European power supply under climate change: impacts on electricity prices. Environ Res Lett 8(3). https://doi.org/10.1088/1748-9326/8/3/035010
46. Gobiet A, Kotlarski S, Beniston M, Heinrich G, Rajczak J, Stoffel M (2014) 21st century climate change in the European Alps—a review. Sci Total Environ. https://doi.org/10.1016/j.scitotenv.2013.07.050
47. Barnett TP, Adam JC, Lettenmaier DP (2005) Potential impacts of a warming climate on water availability in snow-dominated regions. Nature 438(7066):303–309. https://doi.org/10.1038/nature04141
48. Rees HG, Collins DN (2006) Regional differences in response of flow in glacier-fed Himalayan rivers to climatic warming. Hydrol Process. https://doi.org/10.1002/hyp.6209
49. Huss M, Jouvet G, Farinotti D, Bauder A (2010) Future high-mountain hydrology: a new parameterization of glacier retreat. Hydrol Earth Syst Sci 14(5):815–829. https://doi.org/10.5194/hess-14-815-2010
50. Zemp M, Hoelzle M, Haeberli W (2009) Six decades of glacier mass-balance observations: a review of the worldwide monitoring network. Ann Glaciol. https://doi.org/10.3189/172756409787769591
51. Huss M, Farinotti D, Bauder A, Funk M (2008) Modelling runoff from highly glacierized alpine drainage basins in a changing climate. Hydrol Process 22(19):3888–3902. https://doi.org/10.1002/hyp.7055
52. Laternser M, Schneebeli M (2003) Long-term snow climate trends of the Swiss Alps (1931–99). Int J Climatol 23(7):733–750. https://doi.org/10.1002/joc.912
53. Mote PW (2003) Trends in snow water equivalent in the Pacific Northwest and their climatic causes. Geophys Res Lett 30(12). https://doi.org/10.1029/2003GL017258
54. Durand Y, Giraud G, Laternser M, Etchevers P, Mérindol L, Lesaffre B (2009) Reanalysis of 47 years of climate in the French Alps (1958–2005): climatology and trends for snow cover. J Appl Meteorol Climatol. https://doi.org/10.1175/2009JAMC1810.1
55. Falarz M (2008) Changes in extreme nival conditions in Poland during the second half of the 20th century. Meteorol Z. https://doi.org/10.1127/0941-2948/2008/0293

56. Micu D (2009) Snow pack in the Romanian Carpathians under changing climatic conditions. Meteorol Atmos Phys. https://doi.org/10.1007/s00703-009-0035-6
57. Pons MR, San-Martín D, Herrera S, Gutiérrez JM (2010) Snow trends in Northern Spain: analysis and simulation with statistical downscaling methods. Int J Climatol. https://doi.org/10.1002/joc.2016
58. Scherrer SC, Appenzeller C, Laternser M (2004) Trends in Swiss Alpine snow days: the role of local- and large-scale climate variability. Geophys Res Lett. https://doi.org/10.1029/2004GL020255
59. Rabatel A, Letréguilly A, Dedieu JP, Eckert N (2013) Changes in glacier equilibrium-line altitude in the western Alps from 1984 to 2010: evaluation by remote sensing and modeling of the morpho-topographic and climate controls. Cryosphere. https://doi.org/10.5194/tc-7-1455-2013
60. Terzago S, Fratianni S, Cremonini R (2013) Winter precipitation in Western Italian Alps (1926–2010). Meteorol Atmos Phys. https://doi.org/10.1007/s00703-012-0231-7
61. Serquet G, Marty C, Dulex JP, Rebetez M (2011) Seasonal trends and temperature dependence of the snowfall/precipitation-day ratio in Switzerland. Geophys Res Lett. https://doi.org/10.1029/2011GL046976
62. Klein G, Vitasse Y, Rixen C, Marty C, Rebetez M (2016) Shorter snow cover duration since 1970 in the Swiss Alps due to earlier snowmelt more than to later snow onset. Clim Change. https://doi.org/10.1007/s10584-016-1806-y
63. Terrier S, Jordan F, Schleiss AJ, Haeberli W, Huggel C, Künzler M (2011) Optimized and adapted hydropower management considering glacier shrinkage scenarios in the Swiss Alps. In: Dams and reservoirs under changing challenges–proceedings of the international symposium on dams and reservoirs under changing challenges–79th annual meeting of ICOLD. https://doi.org/10.1201/b11669-63
64. Schaefli B, Hingray B, Musy A (2007) Climate change and hydropower production in the Swiss Alps: quantification of potential impacts and related modelling uncertainties. Hydrol Earth Syst Sci 11(3):1191–1205. https://doi.org/10.5194/hess-11-1191-2007
65. Turner SWD, Voisin N, Fazio J, Hua D, Jourabchi M (2019) Compound climate events transform electrical power shortfall risk in the Pacific Northwest. Nat Commun. https://doi.org/10.1038/s41467-018-07894-4
66. Song X, Zhuang Y, Wang X, Li E, Zhang Y, Lu X, Yang J, Liu X (2020) Analysis of hydrologic regime changes caused by dams in China. J Hydrol Eng 25(4):05020003. https://doi.org/10.1061/(ASCE)HE.1943-5584.0001891
67. EIA (2019) Annual energy outlook 2019 with projections to 2050. U.S. Energy information administration. DOE/EIA-0383(2012)
68. Wisser D, Fekete BM, Vörösmarty CJ, Schumann AH (2010) Reconstructing 20th century global hydrography: a contribution to the Global Terrestrial Network-Hydrology (GTN-H). Hydrol Earth Syst Sci. https://doi.org/10.5194/hess-14-1-2010
69. Gaudard L (2015) Pumped-storage project: a short to long term investment analysis including climate change. Renew Sustain Energy Rev. https://doi.org/10.1016/j.rser.2015.04.052
70. Teotonio C, Fortes P, Roebeling P, Rodriguez M, Robaina-Alves M (2017) Assessing the impacts of climate change on hydropower generation and the power sector in Portugal: a partial equilibrium approach. Renew Sustain Energy Rev 74:788–799. https://doi.org/10.1016/j.rser.2017.03.002
71. Engeland K, Borga M, Creutin JD, François B, Ramos MH, Vidal JP (2017) Space-time variability of climate variables and intermittent renewable electricity production—a review. Renew Sustain Energy Rev. https://doi.org/10.1016/j.rser.2017.05.046
72. De Michele C, Avanzi F (2018) Superstatistical distribution of daily precipitation extremes: a worldwide assessment. Sci Rep 8(1):1–11. https://doi.org/10.1038/s41598-018-31838-z
73. Ho M, Lall U, Allaire M, Devineni N, Kwon HH, Pal I, Raff D, Wegner D (2017) The future role of dams in the United States of America. Water Resour Res 53(2):982–998. https://doi.org/10.1002/2016WR019905

74. Schmitt RJP, Bizzi S, Castelletti A, Kondolf GM (2018) Improved trade-offs of hydropower and sand connectivity by strategic dam planning in the Mekong. Nat Sustain. https://doi.org/10.1038/s41893-018-0022-3
75. Balouchi B, Mirzaahmadi A (2025) Assessing the effect of a morphological acceleration factor on 2D morphodynamic modeling in a braided river. In: Proceedings of the 41st IAHR world congress–innovative water engineering for sustainable development, Singapore. https://doi.org/10.64697/978-90-835589-7-4_41WC-P2046-cd
76. Schwedhelm H, Siegfried T, Casas-Mulet R, Dallmeier A, Rüther N (2026) A novel framework for analysing river planform characteristics in data-scarce regions using remote sensing data. Earth Surf Process Landf
77. Balouchi B, Rüther N, Schwarzwälder K (2024). Temporal variation of braided intensity and morphodynamic changes in a regulated braided river using 2D modeling and satellite images. River Res Appl 40(11). https://doi.org/10.1002/rra.4268
78. ASDSO Task Group (2016) The cost of rehabilitating our nation's dams: a methodology. Estimate & proposed funding mechanisms. Association of state dam safety officials
79. Bergmann H, Maass V (2007) Sediment regulations and monitoring programmes in Europe. Sustainable management of sediment resources, vol 3, pp 207–231. https://doi.org/10.1016/S1872-1990(07)80067-5
80. Papanicolaou AN, Barkdoll BD (2011) Sediment dynamics upon dam removal (ASCE manuals and reports on engineering practice No. 122). https://doi.org/10.1061/9780784411360
81. Guénand Y, Perga ME, Chanudet V, Bouffard D (2020) Hydropower operations modulate sensitivity to meteorological forcing in a high altitude reservoir. Aquat Sci 82(3):1–13. https://doi.org/10.1007/s00027-020-00734-y
82. Facchini M, Vetsch DF, Boes RM, Siviglia A (2024) Modeling the morphological response of gravel-bed rivers subject to repeated sediment bypass tunnel operations. Front Earth Sci
83. Annandale GW, Morris GL, Karki P (2016) Extending the life of reservoirs: sustainable sediment management for dams and run-of-river hydropower. The World Bank. https://doi.org/10.1596/978-1-4648-0838-8
84. Ranzi R (2020) Climate change adaptation in water engineering (IAHR white paper series). International association for hydro-environment engineering and research
85. Balouchi B, Mirzaahmadi A, Bihs H, Rüther N (2022a) 3D CFD modeling of bed changes at a laboratory channel confluence. In: Proceedings of the 11th international conference on fluvial hydraulics (river flow 2022), Ottawa, Canada. https://doi.org/10.1201/9781003323037-46
86. Balouchi B, Rüther N, Mirzaahmadi A, Schwarzwälder K (2022b) Two-dimensional (2D) depth-averaged numerical modeling of a braided river morphodynamics upstream of a dam reservoir. In: Proceedings of the 7th IAHR Europe congress: innovative water management in a changing climate, Athens, Greece
87. Polvi LE, Lind L, Persson H, Miranda-Melo A, Pilotto F, Su X, Nilsson C (2020) Facets and scales in river restoration: Nestedness and interdependence of hydrological, geomorphic, ecological, and biogeochemical processes. J Environ Manage 265:110288. https://doi.org/10.1016/j.jenvman.2020.110288
88. Opperman JJ, Orr S, Baleta H, Garrick D, Goichot M, McCoy A, Morgan A, Schmitt R, Turley L, Vermeulen A (2020) Achieving water security's full goals through better integration of rivers' diverse and distinct values. Water Secur. https://doi.org/10.1016/j.wasec.2020.100063
89. Liermann CR, Nilsson C, Robertson J, Ng RY (2012) Implications of dam obstruction for global freshwater fish diversity. Bioscience. https://doi.org/10.1525/bio.2012.62.6.5
90. Anderson EP, Jenkins CN, Heilpern S, Maldonado-Ocampo JA, Carvajal-Vallejos FM, Encalada AC, Rivadeneira JF, Hidalgo M, Cañas CM, Ortega H, Salcedo N, Maldonado M, Tedesco PA (2018) Fragmentation of Andes-to-Amazon connectivity by hydropower dams. Sci Adv. https://doi.org/10.1126/sciadv.aao1642
91. Fearnside PM (2005) Brazil's Samuel dam: lessons for hydroelectric development policy and the environment in Amazonia. Environ Manage. https://doi.org/10.1007/s00267-004-0100-3

92. WWF (2024) Living planet report 2024—a system in peril. WWF, Gland, Switzerland
93. Sayer CA, Fernando E, Jimenez RR et al (2025) One-quarter of freshwater fauna threatened with extinction. Nature 638:138–145. https://doi.org/10.1038/s41586-024-08375-z
94. Harby A, Schäffer LE, Arnesen F, Bauhofer P, Beckitt A, Bockenhauer S, Botterud A, Christensen T, Middelton L, Nielsen N, Somani A, Gutierrez Tavarez E, Østlie C (2019) Flexible hydropower providing value to renewable energy integration. IEA hydropower annex IX white paper no. 1
95. Poff NL, Allan JD, Bain MB, Karr JR, Prestegaard KL, Richter BD, Sparks RE, Stromberg JC (1997) The natural flow regime. Bioscience 47(11):769–784
96. Mulligan M, van Soesbergen A, Sáenz L (2020) GOODD, a global dataset of more than 38,000 georeferenced dams. Sci Data 7:31. https://doi.org/10.1038/s41597-020-0362-5
97. Belletti B, Garcia de Leaniz C, Jones J et al (2020) More than one million barriers fragment Europe's rivers. Nature 588:436–441. https://doi.org/10.1038/s41586-020-3005-2
98. Krochta M, Anlauf-Dunn K, Bugni D, Chang H (2025) Effects of climate change on stream temperature and salmonid habitats in a Cascades river basin. J Environ Manage 387:125843. https://doi.org/10.1016/j.jenvman.2025.125843
99. He F, Zarfl C, Tockner K et al (2024) Hydropower impacts on riverine biodiversity. Nat Rev Earth & Environ 5:755–772. https://doi.org/10.1038/s43017-024-00596-0
100. Bozeman BB, Pracheil BM, Matson PG (2025) The environmental impact of hydropower: a systematic review of the ecological effects of sub-daily flow variability on riverine fish. Rev Fish Biol Fisheries 35:45–76. https://doi.org/10.1007/s11160-024-09909-4
101. Nielsen N, Szabo-Meszaros M (eds) (2022) Hydropower and fish–a roadmap for best practice management. Zenodo. https://doi.org/10.5281/zenodo.7805685
102. Hayes DS, Bruno MC, Alp M, Boavida I et al (2023) 100 key questions to guide hydropeaking research and policy. Renew Sustain Energy Rev 187:113729. https://doi.org/10.1016/j.rser.2023.113729
103. Feng M, Zolezzi G, Pusch M (2018) Effects of thermopeaking on the thermal response of alpine river systems to heatwaves. Sci Total Environ 612:1266–1275. https://doi.org/10.1016/j.scitotenv.2017.09.042
104. Auer S, Hayes DS, Führer S, Zeiringer B (2023) Effects of cold and warm thermopeaking on drift and stranding of juvenile European grayling (Thymallus thymallus). River Res Appl 39:401–411. https://doi.org/10.1002/rra.4077
105. Dorthe D, Pfister M, Lane SN (2025) The thermal future of a regulated river: spatiotemporal dynamics of stream temperature under climate change in a peri-Alpine catchment. Hydrol Earth Syst Sci 29:6309–6331. https://doi.org/10.5194/hess-29-6309-2025
106. Johnson MF, Albertson LK, Algar AC, Dugdale SJ, Edwards P et al (2024) Rising water temperature in rivers: ecological impacts and future resilience. WIREs Water. https://doi.org/10.1002/wat2.1724
107. Intergovernmental Panel on Climate Change (IPCC) (2023) Sixth assessment report. Working group I—the physical science basis, regional fact sheet—Europe
108. Hatry C, Binder TR, Hasler CT, Clarke KD, Katopodis C, Smokorowski KE, Cooke SJ (2011) Development of a national fish passage database for Canada (CanFishPass): rationale, approach, utility, and potential applicability to other regions. Can Water Resour J 36(3):219–228. https://doi.org/10.4296/cwrj3603869
109. Linnansaari T, Wallace B, Curry RA, Yamazaki G (2015) Fish passage in large rivers: a literature review (Mactaquac aquatic ecosystem study report series 2015–16). Canadian Rivers Institute, University of New Brunswick, p v + 55
110. O'Connor J, Hale R, Mallen-Cooper M, Cooke SJ, Stuart I (2022) Developing performance standards in fish passage: integrating ecology, engineering and socio-economics. Ecol Eng 182:106732. https://doi.org/10.1016/j.ecoleng.2022.106732
111. Silva AT, Lucas MC, Castro-Santos T, Katopodis C, Baumgartner LJ, Thiem JD, Aarestrup K, Pompeu PS, O'Brien GC, Braun DC, Burnett NJ, Zhu DZ, Fjeldstad H-P, Forseth T, Rajaratnam N, Williams JG, Cooke SJ (2017) The future of fish passage science, engineering, and practice. Fish Fish 00:1–23. https://doi.org/10.1111/faf.12258

112. Kampa E, Schmidt G (2023) Implementation of ecological flows in the EU. Developed under the framework contract 'Water for the green deal' (09020200/2022/869340/SFRA/ENV.C.1). European Commission
113. Arthington AH, Tickner D, McClain ME et al (2024) Accelerating environmental flow implementation to bend the curve of global freshwater biodiversity loss. Environ Rev 32:387–413. https://doi.org/10.1139/er-2022-0126
114. Arthington AH, Bhaduri A, Bunn SE, Jackson SE, Tharme RE, Tickner D, Young B, Acreman M, Baker N, Capon S, Horne AC, Kendy E, McClain ME, Poff NL, Richter BD, Ward S (2018) The Brisbane declaration and global action agenda on environmental flows. Front Environ Sci 6:45. https://doi.org/10.3389/fenvs.2018.00045
115. Nale J, Singh B, Rai N, Srivastava R, Johnson AJ, Sahoo A, McClain M, Schmutz S, Parasiewicz P, Hayes D, Schinegger R, Laïzé CLR (2020) Guidance document for environmental flows assessment and implementation in India. Deutsche Gesellschaft für Internationale Zusammenarbeit (GIZ) GmbH, New Delhi
116. Nautiyal JM, Gosselin M-P (2023) Optimal approach for upstream fish passage design: one -size-fits-all or made-to-order? River Res Appl 1–15. https://doi.org/10.1002/rra.4208
117. Apadula F, Bassini A, Elli A, Scapin S (2012) Relationships between meteorological variables and monthly electricity demand. Appl Energy 98(C):346–356
118. Gaudard L, Gilli M, Romerio F (2013) Climate change impacts on hydropower management. Water Resour Manage 27(15):5143–5156. https://doi.org/10.1007/s11269-013-0458-1
119. Rothstein B, Schroedter-Homscheidt M, Häfner C, Bernhardt S, Mimler S (2008) Impacts of climate change on the electricity sector and possible adaptation measures. In Hansjürgens B, Antes R (eds) Economics and management of climate change: risks, mitigation and adaptation. Springer, pp 231–241. https://doi.org/10.1007/978-0-387-77353-7_17
120. Hamududu B, Killingtveit A (2012) Assessing climate change impacts on global hydropower. Energies 5(2):305–322. https://doi.org/10.3390/en5020305
121. Harrison GP, Whittington HW (2001) The impact of climatic change on hydropower investment. In Honningsvag B (ed) Proceedings of the 4th international conference on hydropower in the new millennium (hydropower '01). A A Balkema Publishers, p 5
122. Cat Berro D, Acordon V, Di Napoli G (2008) Cambiamenti climatici sulla montagna piemontese
123. Giaccone E, Colombo N, Acquaotta F, Paro L, Fratianni S (2015) Climate variations in a high altitude alpine basin and their effects on a glacial environment (Italian Western Alps). Atmosfera 28(2):117–128. https://doi.org/10.1016/S0187-6236(15)30004-7
124. Peizhen Z, Molnar P, Downs WR (2001) Increased sedimentation rates and grain sizes 2–4 Myr ago due to the influence of climate change on erosion rates. Nature 410(6831):891–897. https://doi.org/10.1038/35073504
125. Lehner B, Czisch G, Vassolo S (2005) The impact of global change on the hydropower potential of Europe: a model-based analysis. Energy Policy 33(7):839–855. https://doi.org/10.1016/j.enpol.2003.10.018
126. Golombek R, Kittelsen SAC, Haddeland I (2012) Climate change: impacts on electricity markets in Western Europe. Clim Change 113(2):357–370. https://doi.org/10.1007/s10584-011-0348-6
127. Schwanghart W, Worni R, Huggel C, Stoffel M, Korup O (2016) Uncertainty in the Himalayan energy-water nexus: estimating regional exposure to glacial lake outburst floods. Environ Res Lett. https://doi.org/10.1088/1748-9326/11/7/074005
128. Beniston M (2003) Climatic change in mountain regions: a review of possible impacts. In: Climate variability and change in high elevation regions past present & future, vol 15, pp 5–31. https://doi.org/10.1007/978-94-015-1252-7_2
129. Harris C, Arenson LU, Christiansen HH, Etzelmüller B, Frauenfelder R, Gruber S, Haeberli W, Hauck C, Hölzle M, Humlum O, Isaksen K, Kääb A, Kern-Lütschg MA, Lehning M, Matsuoka N, Murton JB, Nötzli J, Phillips M, Ross N, Vonder Mühll D (2009) Permafrost and climate in Europe: monitoring and modelling thermal, geomorphological and geotechnical responses. Earth Sci Rev. https://doi.org/10.1016/j.earscirev.2008.12.002

130. Palmieri A, Shah F, Dinar A (2001) Economics of reservoir sedimentation and sustainable management of dams. J Environ Manage. https://doi.org/10.1006/jema.2000.0392
131. Wagner HJ, Mathur J (2011) Introduction to hydro energy systems: basics. technology and operation (green energy and technology). https://doi.org/10.1007/978-3-642-20709-9
132. Lettenmaier DP, Wood AW, Palmer RN, Wood EF, Stakhiv EZ (1999) Water resources implications of global warming: a U.S. regional perspective. Clim Change. https://doi.org/10.1023/A:1005448007910
133. Farinotti D, Usselmann S, Huss M, Bauder A, Funk M (2012) Runoff evolution in the Swiss Alps: projections for selected high-alpine catchments based on ENSEMBLES scenarios. Hydrol Process 26(13):1909–1924. https://doi.org/10.1002/hyp.8276
134. Gaudard L, Romerio F (2014) The future of hydropower in Europe: interconnecting climate, markets and policies. Environ Sci Policy 37:172–181. https://doi.org/10.1016/j.envsci.2014.05.005
135. Farinotti D, Pistocchi A, Huss M (2016) From dwindling ice to headwater lakes: could dams replace glaciers in the European Alps? Environ Res Lett 11(5):054022. https://doi.org/10.1088/1748-9326/11/5/054022
136. United Nations (2015) Transforming our world: the 2030 agenda for sustainable development (A/RES/70/1)
137. Branche E (2018) The multipurpose water uses of hydropower reservoirs: the SHARE concept. In: 26th international congress on large dams. https://doi.org/10.1016/j.crhy.2017.06.001
138. Chini CM, Schreiber KL, Barker ZA, Stillwell AS (2016) Quantifying energy and water savings in the U.S. residential sector. Environ Sci Technol. https://doi.org/10.1021/acs.est.6b01559
139. Hajkowicz S, Collins K (2007) A review of multiple criteria analysis for water resource planning and management. Water Resour Manage. https://doi.org/10.1007/s11269-006-9112-5
140. Markowska J, Szalińska W, Dąbrowska J, Brząkała M (2020) The concept of a participatory approach to water management on a reservoir in response to wicked problems. J Environ Manage. https://doi.org/10.1016/j.jenvman.2019.109626
141. Loch A, Adamson D, Mallawaarachchi T (2014) Role of hydrology and economics in water management policy under increasing uncertainty. J Hydrol. https://doi.org/10.1016/j.jhydrol.2013.10.049
142. Lu H, Huang G, He L (2009) Inexact rough-interval two-stage stochastic programming for conjunctive water allocation problems. J Environ Manage. https://doi.org/10.1016/j.jenvman.2009.08.011
143. Liu X, Chen L, Zhu Y, Singh VP, Qu G, Guo X (2017) Multi-objective reservoir operation during flood season considering spillway optimization. J Hydrol. https://doi.org/10.1016/j.jhydrol.2017.06.044
144. Lu S, Shang Y, Li W, Peng Y, Wu X (2018) Economic benefit analysis of joint operation of cascaded reservoirs. J Clean Prod. https://doi.org/10.1016/j.jclepro.2017.08.140
145. El Azhari M, Loudyi D (2019) Analysis of the water-energy nexus in central Oum Er-Rbia sub-basin—Morocco. Int J River Basin Manag. https://doi.org/10.1080/15715124.2018.1446966
146. Gourbesville P, Du M, Zavattero E, Ma Q (2016) DSS architecture for water uses management. Procedia Eng. https://doi.org/10.1016/j.proeng.2016.07.512
147. Ziogou I, Zachariadis T (2017) Quantifying the water-energy nexus in Greece. Int J Sustain Energ. https://doi.org/10.1080/14786451.2016.1138953
148. Liu D, Guo S, Liu P, Xiong L, Zou H, Tian J, Zeng Y, Shen Y, Zhang J (2019) Optimisation of water-energy nexus based on its diagram in cascade reservoir system. J Hydrol. https://doi.org/10.1016/j.jhydrol.2018.12.010
149. Ren L, He L, Lu H, Chen Y (2016) Monte Carlo-based interval transformation analysis for multi-criteria decision analysis of groundwater management strategies under uncertain naphthalene concentrations and health risks. J Hydrol. https://doi.org/10.1016/j.jhydrol.2016.05.063

150. Ren T, Liu X, Niu J, Lei X, Zhang Z (2020) Real-time water level prediction of cascaded channels based on multilayer perception and recurrent neural network. J Hydrol. https://doi.org/10.1016/j.jhydrol.2020.124783
151. Patro ER (2020) PhD Thesis. Climate, water, energy nexus: potential impacts on hydrology and hydropower production. Politecnico di Milano, Italy. http://hdl.handle.net/10589/166490

Clean Energy Transition and Policy Measures: Shaping the Future of Hydropower

Egidijus Kasiulis, József Kádár, Marija Lazarevikj, and Emanuele Quaranta

Technology enables the transition; policy shapes its direction.

Abstract This chapter examines the evolving role of hydropower within the European Clean Energy Transition, with a particular focus on the influence of policy frameworks at the EU and national levels. It begins by mapping the current position of hydropower within the European renewable energy landscape, highlighting its contribution to electricity generation and system stability. The analysis then explores the development of key policy instruments, including environmental regulation, renewable energy directives, financial mechanisms, and innovation policies, and their impact on hydropower development and modernisation. The chapter demonstrates that while hydropower remains a cornerstone of renewable electricity in Europe, its development is increasingly shaped by environmental constraints, market structures, and system integration needs. A shift is observed from capacity expansion towards optimisation of existing assets, increased flexibility, and integration with other renewable energy sources. The analysis further highlights the growing importance of multipurpose water infrastructure, energy storage, and the water-energy-society-ecosystem nexus. Finally, the chapter outlines the key dimensions shaping the future of hydropower, based on market and policy orientated prospects,

E. Kasiulis (✉)
Vytautas Magnus University, Kaunas, Lithuania
e-mail: egidijus.kasiulis@vdu.lt

J. Kádár
Arava Institute for Environmental Studies, Kibbutz Ketura, Israel
e-mail: Jozsef.Kadar@arava.org

M. Lazarevikj
Faculty of Mechanical Engineering, Ss. Cyril and Methodius University of Skopje, Skopje, North Macedonia
e-mail: marija.lazarevikj@mf.edu.mk

E. Quaranta
European Commission, Joint Research Centre (JRC), Ispra, Italy
e-mail: Emanuele.QUARANTA@ec.europa.eu

E. Doujak et al. (eds.), *Rethinking Hydropower: A Holistic Vision for a Sustainable Future*, Springer Tracts in Civil Engineering,
https://doi.org/10.1007/978-3-032-33328-5_10

and stakeholder engagement. It argues that the role of hydropower goes beyond electricity generation, positioning it as a critical component of resilient and integrated energy systems.

Keywords Clean energy transition · Energy policy · European Union · Future of hydropower

1 Introduction

Hydropower plays an important and evolving role in Europe's renewable energy transition. Traditionally, it has been considered a mature renewable energy source mainly used for electricity generation. However, its role today goes beyond simply producing power. As energy systems include more variable renewable sources, such as wind and solar, hydropower provides essential services like flexibility, storage, balancing, and grid stability. These functions are increasingly important for maintaining reliable and efficient energy systems. Therefore, hydropower supports the integration of renewable energy while also contributing to energy security, system resilience, and a reduced reliance on imported fossil fuels.

This chapter discusses how the European Union (EU) and national policy frameworks shape the status and future development of hydropower. It focuses on how regulatory frameworks at both the EU and national levels influence the planning, licensing, modernization, and operation of hydropower systems. In this context, hydropower is shaped by different policies, including renewable energy, water management, environmental protection, climate policy, and industrial development. This reflects the increasing complexity of the renewable energy transition, which requires not only expanding renewable generation capacity but also maintaining system stability and balancing environmental and societal considerations.

The chapter aims to present three interconnected objectives. First, the chapter focusing on its geography and connection to the rapid growth of variable renewables. Second, it reviews key EU and national policies, emphasising both synergies and tensions among energy, environmental, and industrial strategies. Third, it considers future paths for hydropower in the Clean Energy Transition, stressing its support for multi-use infrastructure through market and policy orientated prospects. Building on this, the chapter argues that hydropower should be understood not only in terms of electricity generation but also in its broader roles across policy, technological, environmental, and societal domains.

2 Mapping Hydropower in the Current Renewable Energy Sector in Europe

Following the strong focus of the EU on reducing greenhouse gas emissions from all industry sectors and achieving greater energy independence from imported energy sources, the clean energy transition is clearly reflected in statistics of electricity generation from the renewable energy sources. Between 2010 and 2023, the electricity generated from fossil fuels in the EU was reduced by approximately 45%, while generation from renewable energy sources increased by around 87%. Whereas in 2010 renewable electricity lagged behind nuclear and fossil fuels, by 2023 it became the leading source of electricity in the EU, reaching 1,218,484 GWh, or approximately 44% of total generation [1].

Hydropower, including pumped storage hydropower (PSH), generated 329,854 GWh of electricity in the EU in 2023, making it the number two source of renewable energy after wind, with a 12% share in the total EU electricity mix. As it is a resource dependent on hydrological and topographical conditions, hydropower is unevenly distributed across Europe. Therefore, this section considers the full European region, including non-EU countries, based on Eurostat data for 2023 [2].

The countries with the highest share of electricity generation from hydropower and PSH were Norway, Sweden, Türkiye, France and Austria (Fig. 1), while those with the smallest share were Hungary, the Netherlands, Moldova, Estonia and Denmark. These patterns reflect strong geographical dependence. Scandinavia, the Alpine

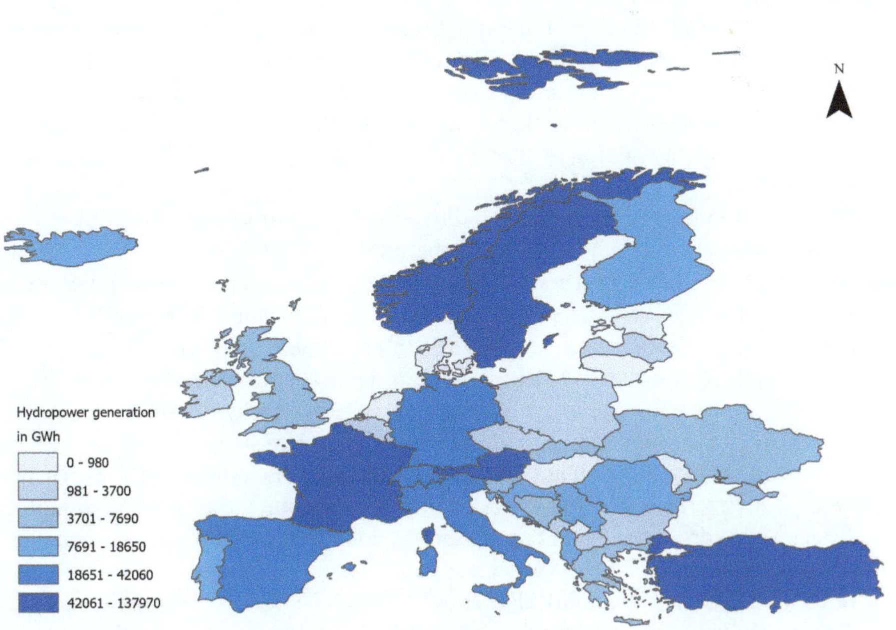

Fig. 1 Electricity generation from hydropower in European countries in 2023

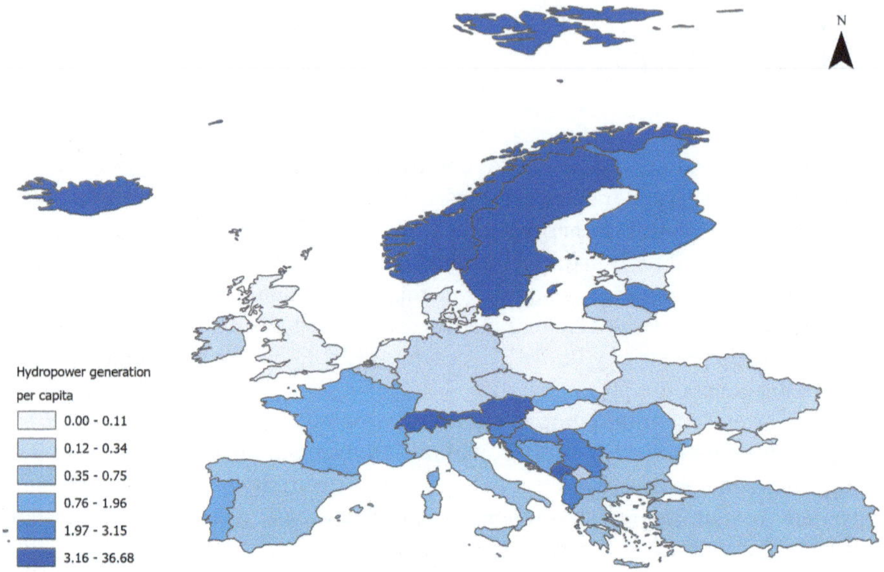

Fig. 2 Normalized value of hydropower generation in European countries per capita in 2023

region, and France stand out due to favourable hydrological and topographical conditions and a long-established infrastructure.

Much of Eastern Europe and the Baltic States also belong to the category of countries with a small generation share, indicating a more limited role of hydropower within national electricity systems. In these regions, predominant lowland areas restrict hydropower potential. Balkan countries generally occupy intermediate positions, reflecting a combination of existing infrastructure and remaining development potential [3].

The second map (Fig. 2) presents hydropower generation normalised by population. In the per capita statistics, the geographical pattern changes. While large countries such as France or Türkiye dominate in absolute terms, per capita values better reflect the role of hydropower within national systems. The Nordic region clearly dominates: Iceland reaches 36.7 MWh per capita, and Norway exceeds 25 MWh per capita, placing them far above the European average (2.7 MWh per capita). Sweden follows at 6.3 MWh per capita, closing a northern hydropower belt where water resources are deeply embedded in national energy systems.

The Alpine region also stands out in per capita terms. Austria (4.9 MWh per capita) and Switzerland (4.6 MWh per capita) demonstrate the continued structural importance of hydropower within mountainous countries. Several Balkan countries, including Montenegro (3.6 MWh per capita), Albania (3.2 MWh per capita), and Slovenia (2.5 MWh per capita) also show relatively strong performance despite smaller system sizes, indicating a significant role of hydropower within their energy mixes.

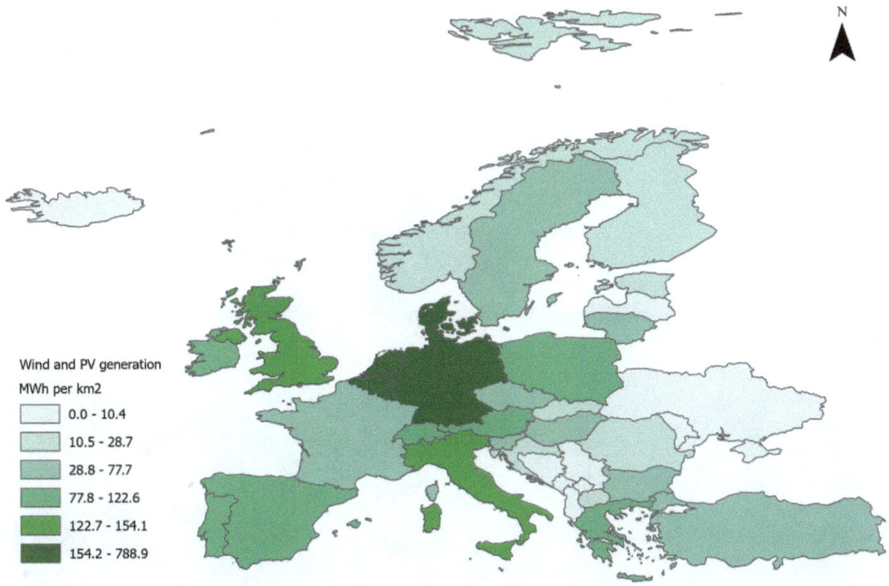

Fig. 3 Spatial distribution of electricity generation from wind energy and solar PV installations per km^2 in European countries in 2023

In contrast, much of Central and Eastern Europe remains in the lower per capita classes. Countries such as Poland (0.10 MWh per capita), Hungary (0.02 MWh per capita) and the Netherlands (0.004 MWh per capita) display structurally limited reliance on hydropower. This confirms that Europe's hydropower landscape is shaped not merely by scale, but fundamentally by geography, historical infrastructure development, and availability of natural resources.

However, geographical conditions alone do not fully explain national differences. The Baltic States provide an excellent example: despite similar physical conditions, Latvia reaches 2.0 MWh per capita, while Lithuania (0.3) and Estonia (0.02) remain much lower. This suggests that policy choices and historical development trajectories also play a significant role.

While hydropower reflects long-established geographical and infrastructural patterns, the rapid expansion of wind and solar power introduces a new spatial dynamic across Europe. Unlike hydropower, which is location-constrained, solar and wind installations are more widely distributed and often concentrated in areas with strong policy support and high land-use intensity. Analysing electricity generation from wind and solar per km^2 (Fig. 3) highlights the spatial pressure associated with the Clean Energy Transition.

When expressed per unit of land area, wind and solar generation show a markedly different pattern. The Netherlands reaches 789 MWh/km^2, followed closely by Malta at 716 MWh/km^2. Belgium exceeds 450 MWh/km^2, while Germany is above 400 MWh/km^2. These values indicate a high concentration of variable renewable energy

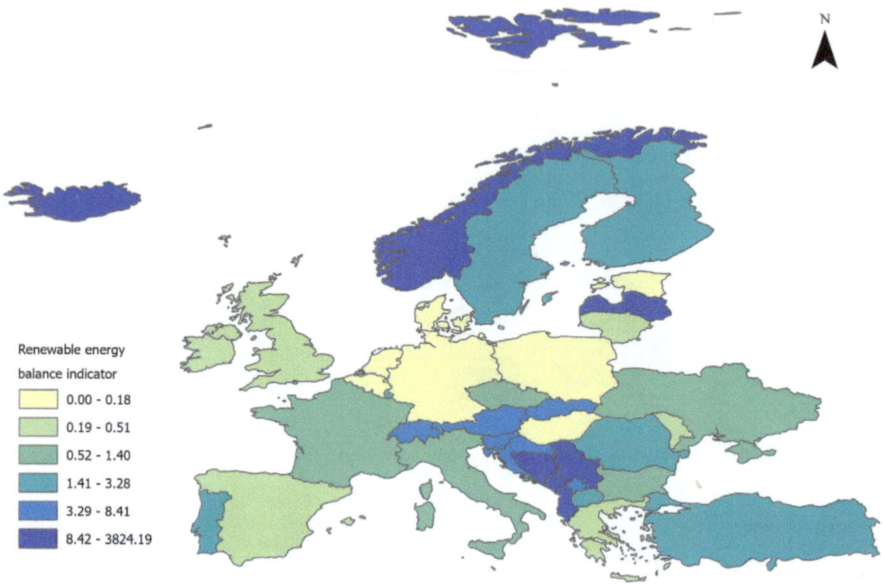

Fig. 4 Balance indicator as the ratio between hydropower and combined wind–solar electricity generation in European countries in 2023

in relatively small areas, particularly in Western and Central Europe. In contrast, Nordic countries, which dominate hydropower, show much lower densities of wind and solar deployment. Denmark also stands out with roughly 251 MWh/km^2, and Italy approaches 138 MWh/km^2. Most of the Balkan countries show even lower development patterns than the Nordic countries.

Germany has the largest installed capacity of both wind and solar in Europe, followed by Spain, France, and others. These patterns reflect strong policy-driven deployment of variable renewables, rather than purely geographical constraints. The rapid expansion of the electricity generation from wind and solar energy highlights the increasing importance of system flexibility. Variable renewable technologies depend on complementary sources capable of balancing fluctuations in output. In this chapter the ratio between hydropower and combined wind–solar generation is calculated as the renewable energy balance indicator (Fig. 4), to provide insight into the structural role of hydropower within national systems.

The balance indicator reveals clear differences. Countries such as Norway, Sweden, and several Balkan states show high ratios, indicating hydropower-dominated systems with strong internal balancing capacity. In contrast, countries such as Germany and Belgium exhibit low ratios, reflecting systems dominated by variable renewables. Between these extremes there are mixed systems, including France, Italy, Czechia, and Bulgaria, where hydropower and variable renewables coexist more evenly. These systems suggest a more balanced structure, although flexibility still depends on storage, grid integration, and interconnections. This is reflected in countries such as Norway, which generates over 90% of its electricity from hydropower and acts as a key balancing provider in the European electricity system through cross-border exchanges.

Although this indicator does not directly measure system stability, it provides insight into structural flexibility. Systems with low hydropower shares depend more heavily on interconnections, demand management or backup generation. At the same time, hydropower generation is increasingly affected by climate variability, with drought conditions in 2022 reducing output in several European countries [4]. Furthermore, a significant share of Europe's hydropower capacity was commissioned before the 1980s, indicating substantial potential for modernisation and efficiency improvements, which could increase generation by up to 10% without new large-scale developments [5].

The Clean Energy Transition in the EU and the whole Europe is therefore not only a question of capacity expansion, but also of maintaining a balance between variability and flexibility within interconnected energy systems. In this context, hydropower plays a key role not only as a source of renewable electricity, but as a provider of flexibility and system stability.

3 Evolution of Policy Framework and Its Impact on Hydropower Development

This section examines the evolution of the European policy framework and its influence on hydropower development across multiple governance levels—from the EU to national. While the technical potential of hydropower is largely determined by geographical and hydrological conditions, its actual deployment, modernisation, and integration into the energy system are strongly shaped by regulatory, strategic, and financial instruments. The following discussion outlines key directives, strategies, and funding mechanisms, and highlights how these policy instruments have progressively redefined the role of hydropower within the broader context of the Clean Energy Transition.

3.1 The EU Level

Hydropower development in Europe is shaped by a multi-level governance framework operating across the EU, national, and regional levels. While geographical conditions and existing infrastructure determine the physical potential of hydropower, policy frameworks play a decisive role in shaping how this potential is utilised, constrained, and integrated into evolving energy systems. In this context, hydropower should be understood not only as a source of renewable electricity, but also as a policy-mediated infrastructure situated at the intersection of energy, water, environmental, and societal domains.

Hydropower policies establish the strategic objectives including targets, indicators and timeframes, and the implementation instruments that shape the environmental performance of hydropower projects, making them central to how impact mitigation is defined and applied in practice. Across Europe, EU and national policy frameworks establish ecological and regulatory requirements that shape the planning, licensing, modernisation, and operation of hydropower systems, particularly in relation to water protection, biodiversity conservation, and renewable energy deployment [6].

The EU has introduced several directives and regulations to support RES integration in the European energy system. Adopted in 2000, the Water Framework Directive 2000/60/EC (WFD) [7] required that development projects ensure water bodies (both surface and groundwater) achieve or maintain good ecological status and prevent any deterioration or failure to reach planned improvements. Within this framework, Article 4(7) allows exemptions for new development, including hydropower projects, provided that any status deterioration resulting from river barriers, flow alterations and water abstractions, is justified and clearly documented in the River Basin Management Plans [8].

According to the European Environmental Agency (EEA) the environmental implications of hydropower under the WFD are substantial. Hydropower-related pressures affected around 9,000 surface European water bodies (6%), while around 6,000 water bodies have been designated as heavily modified, largely due to hydropower development. These impacts include river fragmentation, disruption of fish migration, altered sediment transport, and modified flow regimes. Consequently, compliance with ecological requirements, such as fish passages, environmental flows, and habitat restoration, has become a prerequisite for licensing, contributing to stricter permitting procedures and favouring refurbishment and modernisation over new large-scale developments. In addition, approximately 450,000 barriers have been found in European rivers, including around 225,000 weirs and dams (data from the AMBER atlas, EU funded project) [9], highlighting the cumulative impact of river regulation. The number of hydropower plants in Europe is estimated at around 25,000 [10].

Alongside environmental regulation, the renewable energy of the EU required each Member State, mainly through the Renewable Energy Directive (RED, 2001/77/EC), to adopt national indicative targets for renewable electricity by 2010, aiming for about a 21% RES share of the total electricity generation in the EU. Although it was not legally binding, the directive encouraged Member States to introduce support schemes, recognised hydropower as an eligible renewable source, and required progress to be monitored by the European Commission.

RED was replaced in 2009 with Renewable Energy Directive I (RED I, (2009/28/EC)) [11] which set a legally binding target of 20% renewable energy in gross final energy consumption by 2020. The EU countries were required to develop National Renewable Energy Action Plans (NREAPs) outlining their strategies for achieving the targets, for example through investment aid, tax exemptions or reductions, tax refunds, support schemes such as feed-in tariffs, premium payments, green certificates. The directive also clarified that electricity from PSH cannot be counted as renewable generation for the purpose of the targets if the water was pumped up using grid power. By 2020, the EU exceeded the 20% renewable energy target, reaching 22.1%, with hydropower continuing to provide a substantial share of renewable electricity, although growth in wind and solar dominated new capacity [12].

The renewable energy framework was further restructured under the Renewable Energy Directive II (RED II), adopted in 2018 as part of the Clean Energy for All Europeans Package. It set the more ambitious target of achieving at least 32% renewable energy in the EU's energy mix by 2030. Unlike RED I, this target is binding at the EU level since it is not divided into mandatory national targets but instead requires Member States to update their National Energy and Climate Plans (NECPs). RED II strengthened sustainability criteria (particularly for bioenergy), enhanced market integration of renewables, promoted self-consumption and renewable energy communities, and reinforced cooperation mechanisms among Member States. For hydropower, RED II did not introduce technology-specific provisions but maintained its eligibility as renewable energy source, leaving it up to Member States to continue supporting hydropower projects through refurbishment, modernisation, or new capacity additions, subject to compliance with environmental requirements under the WFD.

Adopted in June 2021, the EU Taxonomy Regulation 2020/852 and its Climate Delegated Act (Commission Delegated Regulation (EU) 2021/2139) [13] recognises hydropower as an eligible economic activity contributing to climate change mitigation, but only under clearly defined and stringent technical screening criteria (TSC). Electricity generation from hydropower must demonstrate a substantial contribution to mitigation by fulfilling at least one of the following conditions: operation as a run-of-river system without artificial reservoir, achievement of a power density above 5 W/m^2, or limitation of lifecycle greenhouse gas emissions to less than 100 gCO2e/kWh, as calculated by recognized LCA methodologies and verified by an independent third party. Beyond climate mitigation, the act imposes strict "Do No Significant Harm" (DNSH) [14] requirements which constitute the most decisive constraints for hydropower development. Central to these is full compliance with the Water Framework Directive, particularly the obligation to prevent deterioration and to ensure or achieve good ecological status or potential of affected water bodies. For new projects, requirements are particularly stringent, whereas for existing installations, the focus is on upgrading and improving environmental performance. As such, the taxonomy acts as both a regulatory and financial filter, restricting new developments while incentivising modernisation and low-impact designs.

PSH is addressed within the electricity storage activities, but it remains subject to similar environmental rules. Where it is connected to a river system, it must comply with the same DNSH criteria as conventional hydropower, particularly those related to water protection and ecological status, while off-river (closed-loop) systems are assessed under separate but still strict requirements for water protection. In addition, storage projects are required to demonstrate alignment with circular economy principles, including the implementation of waste management plans ensuring reuse or recycling at end of life.

In the context of the 2022 energy crisis, the EU launched the REPowerEU [15] plan to rapidly reduce reliance on Russian fossil fuels, drive energy savings and accelerate the Clean Energy Transition. The plan encourages energy storage through investment incentives, facilitated permitting and integration into building renovations, emphasizing its significance for ensuring security of supply and supporting RES. PSH, as the most mature large-scale storage technology, was thus effectively reinforced by accelerating PSH proposals and increasing funding for energy storage (for example, through EU Recovery and Resilience Facility funds under REPowerEU chapters).

Under the European Green Deal and the "Fit for 55" package, the EU renewable energy framework was further strengthened, with RED III (Directive (EU) 2023/2413) setting a binding EU-level target of at least 42.5% renewable energy by 2030. Rather than reintroducing binding national targets, the revised framework increases ambition through additional sectoral requirements, including binding targets for the transport sector and heating sectors, while maintaining flexibility for Member States in other areas. In this context, hydropower development remains subject to existing environmental obligations, as the requirements of the WFD continue to apply, including in areas designated for accelerated renewable deployment, ensuring that any potential impacts on water bodies are justified and that appropriate mitigation measures are implemented.

In parallel, several sectoral frameworks create new opportunities. Even if not specifically developed for hydropower, they may trigger investments, especially on "hidden hydro" or on multi-purpose schemes within the water-energy nexus.

The EU Water Resilience Strategy (WRS) [16] helps to improve water resources management, competitiveness and innovation of the European water industry. In this context, saving water and energy is of utmost importance. The WRS also launched the European Water Academy to address capacity needs, in Europe's water sector and to bridge research and real-world application of technologies. Through advanced training, support for innovation, and fostering collaboration between public and private actors across Europe, the Academy strengthens the social dimension of water governance—exemplifying the Water-Energy-Society-Ecosystem (WESE) nexus in practice. The WRS mentions water storage for integrated water resources management as an important option for adaptation to climate change, including artificial water storage using dams. Within this framework, dams could be equipped for hydropower generation even if they primarily serve other purposes.

The recast of the EU Wastewater Treatment Directive (European Commission, EC, 2025b) requires Member States (MSs) to ensure that the total annual energy used by all urban wastewater treatment plants (WWTPs) on their territory, treating an effluent load equivalent to 10,000 persons or more, does not exceed the production of energy from renewable sources. This objective must be met at national level. This also includes hydropower installations, for example on the wastewater effluent. The recast Drinking Water Directive (DWD), the EU's main law on drinking water, was adopted in 2020. In order to improve the efficiency of the water infrastructure, including the avoidance of over-exploitation of scarce water resources, water leakage levels should be assessed by all Member States and reduced if they are above a certain threshold. This would reduce the energy consumption of the sector. Small hydropower devices could be installed to exploit hidden hydro potential in the water networks. The estimated potential in the EU is slightly below 1 TWh per year [10, 17].

The Social Climate Fund (SCF) [18] supports the EU's population with structural measures and investments in energy efficiency and renovation of buildings, clean heating and cooling and integration of renewable energy, and it provides for investments in zero- and low-emission mobility solutions. The refurbishment and modernisation of small hydropower plants is eligible under the SCF. Small hydropower plants shall not be located either in areas earmarked as restoration areas in national restoration plans, or in structures that need to be removed. It was estimated that efficiency and operation improvement can increase energy generation by up to 10% [19].

Industrial and innovation policies are increasingly shaping the future of hydropower. The Net-Zero Industry Act (NZIA) aims to strengthen European manufacturing capacity for clean technologies, including hydropower components such as turbines. The Regulation is expected to increase the competitiveness of the net-zero technology sector, attract investments and improve market access for clean technologies in the EU. This supports the clean energy transition and improves the EU's energy resilience. Under the Act, net-zero manufacturing capacity must meet at least 40% of the EU's annual deployment needs by 2030, providing predictability, certainty and long-term orientation to manufacturers and investors. NZIA will facilitate investments by shortening and streamlining permitting processes. The EU is already the global leader in turbine production and trade, and it already comfortably exceeds the identified targets.

At the same time, research and innovation programmes, particularly Horizon Europe, play a key role in advancing hydropower technologies. Between 2015 and 2024, approximately 40 hydropower-related projects were funded, in particular projects with focus on sustainable refurbishment, flexibility and digitalisation [20]. Other funding schemes for research, such as the LIFE, Interreg and the Innovation Fund programmes, are also relevant to hydropower. These developments indicate a shift from capacity expansion towards system optimisation and integration.

The influence of EU policies extends well beyond its geographic boundaries. Through initiatives such as Global Gateway, the EU supports international energy infrastructure investments. Hydropower accounts for a notable share of these projects, highlighting its continued global relevance as part of energy transition strategies. More than 20% of all Global Gateway Flagship projects between 2023–2025, were energy projects, and around 7% were hydropower projects. Most hydropower investments have focused on large-scale projects, for example, Nachtigal (420 MW) in Cameroon or Bac Ai PSH (1200 MW) in Vietnam. The Global Gateway has also financed SP Nare (20 MW) in Colombia and Achwa I (42 MW) in Uganda.

Table 1 summarises the analysed policy instruments and highlights their respective implications for hydropower.

Table 1 The EU policy frameworks and their impact on hydropower development in Europe

Policy framework	Type of instrument	Specific instrument	Impact on hydropower
Water Framework Directive (WFD)	Regulatory	Ecological status requirements; "no deterioration" principle; Article 4(7) exemptions	Strong constraint on new hydropower projects; prioritises ecological protection; requires mitigation measures; favours modernisation over new development
Renewable Energy Directive I (RED I)	Regulatory	Binding renewable energy targets; national support schemes (FiTs, FiPs, certificates)	Supports hydropower as part of renewable energy mix; incentivises deployment, particularly small-scale projects
Renewable Energy Directive II (RED II)	Regulatory	EU-level renewable target (32% by 2030); NECPs; market integration measures	Strengthens deployment and integration of renewable energy, including hydropower; reinforces environmental compliance requirements
Renewable Energy Directive III (RED III)	Regulatory	Binding EU-level renewable energy target (minimum 42.5% by 2030); sectoral targets; accelerated permitting	Supports hydropower within renewable framework while maintaining environmental requirements; may facilitate permitting in designated areas
REPowerEU	Strategic/Financial	Energy security measures; storage investments; accelerated deployment of renewables	Reinforces role of PSH for flexibility and system balancing; supports investment in storage capacity

(continued)

Table 1 (continued)

Policy framework	Type of instrument	Specific instrument	Impact on hydropower
EU Taxonomy Regulation	Regulatory/Financial	Technical Screening Criteria; DNSH principle; WFD compliance	Restricts new hydropower developments; prioritises low-impact projects; incentivises refurbishment and environmentally compliant investments
Water Resilience Strategy (WRS)	Strategic	Integrated water resource management; promotion of water storage; capacity-building initiatives	Indirect support for multipurpose water infrastructure and water–energy nexus approaches
Wastewater Treatment Directive	Regulatory	Energy neutrality requirement for WWTPs; renewable energy obligations	Creates direct opportunity for small hydropower in wastewater systems
Social Climate Fund (SCF)	Financial	Funding for energy efficiency and renewables; eligibility for refurbishment	Supports modernisation of existing hydropower; excludes environmentally sensitive areas
Net-Zero Industry Act (NZIA)	Enabling/Innovation	Streamlined permitting; manufacturing targets; investment facilitation	Indirect support through improved supply chains; faster project development; increased investment certainty
Horizon Europe/EU R&I programmes	Financial/Innovation	Research funding; innovation support; cross-sector collaboration	Promotes innovation in hydropower (digitalisation, flexibility, refurbishment); supports system integration

3.2 National Implementation and Support Mechanisms

The evolution of the Renewable Energy Directives has also led to important changes in governance. While earlier frameworks, particularly RED I, introduced binding renewable energy targets at the Member State level for 2020, the policy approach has shifted since RED II and is maintained under RED III. From 2030 onwards, renewable energy targets are binding at EU level, while national targets are no longer legally binding.

This governance shift provides greater flexibility to the Member States in designing their national energy strategies, while still operating within a common EU framework. As a result, implementation varies across countries, leading to differences in regulatory approaches, support schemes, and the role assigned to hydropower. National policies are shaped by several factors, including the need for renewable energy, compliance with EU directives, environmental considerations and broader economic priorities.

Across Europe, national policies for hydropower increasingly prioritise small-scale development and strict environmental compliance, rather than the expansion of new large-scale capacity. Renewable energy support is implemented through a range of mechanisms, including feed-in tariffs (FiTs), feed-in premiums (FiPs), quota systems, and competitive tendering schemes. These are often complemented by financial instruments such as investment grants, soft loans, loan guarantees, and tax incentives. For hydropower specifically, FiTs have been widely used to provide stable revenue for small-scale projects, while in some countries they have been replaced or complemented by FiPs and market-based mechanisms. In addition, governments frequently support hydropower through investment grants and low-interest financing, particularly for refurbishment and modernization.

Figure 5 presents the share of hydropower in supported renewable electricity across selected European countries in 2022 and 2023, based on data from Council of European Energy Regulators (CEER) report [21]. It is important to note that this indicates only the share of supported renewable electricity and does not reflect the total contribution of renewables in a country, as existing installations may no longer receive support and new capacity may be deployed without support, while still contributing to overall generation.

Significant variation can be observed across countries, reflecting differences in national policy frameworks and energy system characteristics. In countries such as Slovenia and North Macedonia, hydropower accounts for a dominant share of supported renewable electricity, exceeding 50% in 2023. Both countries show high values in 2022 as well, with a slight decrease observed in North Macedonia between 2022 and 2023. This indicates that hydropower remains a central component of supported renewable energy in these systems.

A second group of countries, including Austria, Czechia, Slovakia, and Italy, exhibits moderate hydropower shares, generally ranging from 10% to 20%. Within this group, different developments can be observed. Austria and Czechia show an increase in hydropower share between 2022 and 2023, while Slovakia shows a noticeable decrease. Italy remains relatively stable, with only minor variation between the two years.

In contrast, countries such as Germany, Spain, the United Kingdom, and Sweden display very low shares of hydropower within supported renewable electricity, generally below 3% in both years. The values remain consistently low between 2022 and 2023, with only small changes observed. Other countries such as France, Greece, and Romania show only small changes in hydropower share.

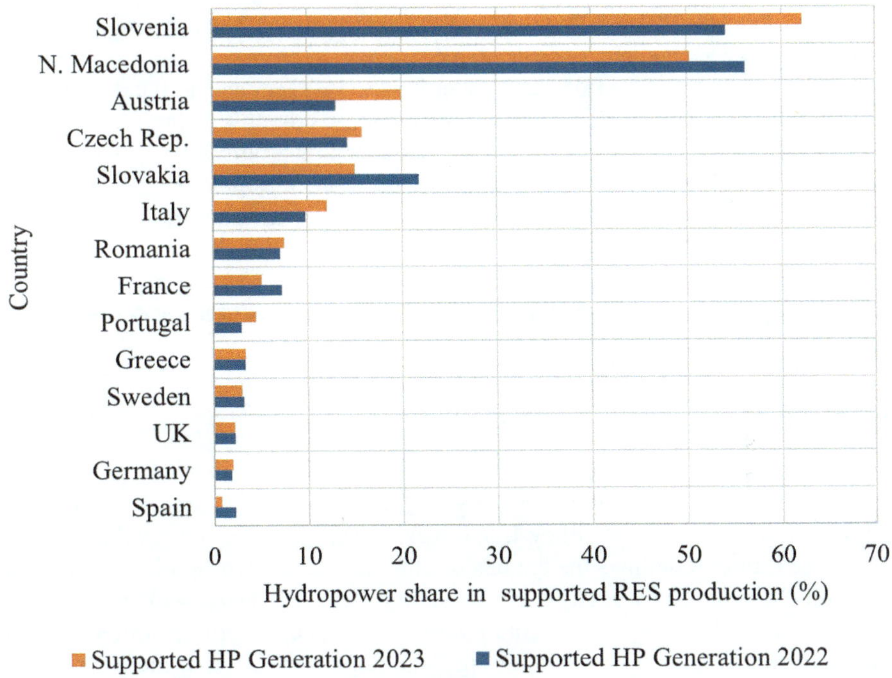

Fig. 5 Share of hydropower in supported renewable electricity production across selected European countries in 2022 and 2023 (adapted from CEER report [21])

Recent developments across Europe indicate that hydropower, and particularly PSH, is supported through a combination of financial, regulatory, and market-based instruments. Countries such as Austria, Spain, Italy, and the United Kingdom demonstrate active policy measures alongside multiple hydropower projects under development or construction. In these contexts, support mechanisms include capacity market developments (Italy, 2024), revenue stabilisation models (e.g. the United Kingdom's Cap and Floor scheme, 2024–2025), and national energy and climate plans with storage targets (Spain, updated in 2023–2024). These policy frameworks are closely linked to ongoing and planned PSH investments [22].

In countries such as Switzerland and Greece, policy measures focus on existing infrastructure and planned expansion. Switzerland requires the assessment and potential modernisation of existing hydropower plants, while Greece has set national storage targets and is developing a pipeline of PSH projects. Planning instruments are also used in countries such as Czechia, where potential PSH sites were identified in 2024, supporting future project development.

Financial instruments introduced or expanded in 2024 are observed in several countries. Loans and grants from institutions such as the European Investment Bank and national development banks support both expansion and modernisation projects, including in Latvia, Lithuania, Serbia, and Spain. Procurement mechanisms, such as tenders (Estonia, 2024–2025), and targeted investments (Norway, 2024) further contribute to advancing hydropower projects.

In Northern and Central European countries such as Germany, Finland, and Sweden, recent developments focus on refurbishment, recommissioning, and upgrading of existing assets, including projects announced or progressing between 2024 and 2025. These include investment-driven projects and feasibility studies assessing additional hydropower potential, with several initiatives announced or progressing between 2024 and 2025.

Table 2 groups recent national policy instruments according to their type and illustrates their application across countries and their effects on hydropower development.

4 Shaping the Future

Building on the analysis of the current state of hydropower in the renewable energy sector and the evolution of policy frameworks across Europe, this section takes a look at the future. It outlines the dimensions through which hydropower is expected to contribute to the Clean Energy Transition in the coming decades. Beyond its traditional role in electricity generation, hydropower is increasingly positioned as an important multifaceted asset—supporting technological innovation, energy security, grid stability, and climate adaptation. The following sections highlight these interconnected roles and illustrate how policy measures can shape the future development of hydropower in the evolving energy systems and why stakeholder engagement is important.

4.1 Market-Orientated Prospects

Manufacturing of hydropower equipment is a key pillar of the EU's industrial and innovation capacity. EU-based manufacturers account for more than 80% of high-value inventions (inventions on hydropower patented in more than one patenting office) demonstrating strong technological leadership and global relevance. Leading companies such as Voith Hydro, GE and Andritz Hydro together share more than 50% of high-value inventions among the top 10 of leading companies worldwide. In addition, EU manufacturers account for nearly 50% of global hydropower turbine and related component exports, reinforcing their competitiveness in international markets [20].

Beyond conventional turbine technologies such as Pelton, Francis and Kaplan, European companies are at the forefront of developing advanced and specialised solutions with strong export potential. These include innovative turbine designs such as VLH and Ossberger turbines, as well as emerging technologies such as the DIVE, Streamdiver and Hydromatrix systems. In parallel, advancements in variable speed turbines, as well as variable speed generators, which have been studied extensively

Table 2 National policy instruments for hydropower in Europe and associated project examples

Policy instrument	Countries (examples with projects)	Effect on hydropower
Capacity remuneration mechanisms (market-based)	United Kingdom: Coire Glas PSH (1.3 GW), Cruachan 2 (600 MW), supported under Cap & Floor scheme (2024–2025) Italy: Campolattaro PSH (600 MW), Pescopagano & Villarosa PSH (~500 MW), linked to MACSE capacity market (2024)	Improves revenue certainty for large PSH projects, enabling investment in long-duration storage
Public financing (loans, grants, EU funding)	Spain: Alcántara PSH (440 MW, €44.9 million grant), Valdecañas PSH (275 MW) Lithuania: Kruonis PSH expansion (+110 MW, EIB loan) Latvia: Daugava River HPP modernisation (€230 million NIB loan) Serbia: Bistrica PSH (~2.4 GW planned), EU/EIB funding (2024)	Supports modernisation and expansion, reduces investment risk, and accelerates implementation
Tendering/procurement mechanisms	Estonia: Paldiski PSH (500 MW, tender issued 2024) Iceland: Hvammur HPP (95 MW), Sigalda expansion (+65 MW turbine), ongoing tenders	Facilitates project realisation through competitive processes and structured development phases
Regulatory streamlining (permitting laws)	Turkey: 2023 law simplifying approval procedures; ongoing projects (~600 MW under construction) and expansion of existing hydropower facilities	Reduces administrative delays and enables faster deployment of new and upgraded capacity
National energy and climate planning (targets)	Spain: NECP target ~22.5 GW storage by 2030, supporting PSH (Tamega complex: Gouvaes 880 MW, Alto Tamega 160 MW) Greece: ~3 GW storage target, multiple PSH projects under development	Provides direction for investment in hydropower, especially storage, and its integration into the energy system

in the EU-funded projects such as XFLEX Hydro, enhance operational flexibility and support the integration of variable energy sources.

The EU also demonstrates leadership in low-head and small-scale hydropower technologies, including water wheels and Archimedes screws, which are particularly relevant for decentralised and embedded applications within existing water infrastructure [23].

Over 2022–2024, the EU accounted for 47% of global exports (including intra-EU trade), while extra-EU exports (excluding intra-EU trade) represented 38% of global transactions. The EU met 80% of its import needs of turbines and parts through intra-EU trade. However, the real dimension of the hydropower sector would be much larger if all the technologies associated with the hydropower value chain were considered [20].

Hydropower does not need critical materials and therefore is avoiding dependency on third countries, especially China. If in the EU, the growth of large hydropower is limited to the expansion of PSH and refurbishment of existing hydropower fleet, the market is growing elsewhere with extensive involvement of the EU-based hydropower companies. However, this technological leadership is increasingly challenged by growing competition from non-EU manufacturers. Lower-cost technologies may undermine the global market position of European firms and are often associated with significantly weaker environmental and social standards, raising concerns regarding long-term sustainability [24].

The strong technological and industrial base of the EU is closely linked to its ability to shape regulatory frameworks and sustainability standards. In this context, technological leadership does not operate in isolation but co-evolves with policy development, enabling the EU to influence both domestic governance and international standards in hydropower.

4.2 *Policy-Orientated Prospects*

The peculiarity of hydropower, that separates it from other renewable energy sources and especially solar PV and wind, is that the created reservoir can be used for multiple purposes. In the EU, there are almost 4,500 large dams, of which 67% are officially registered as single-purpose dams. These account for approximately 50% of the water volume stored by large dams, and excluding dams that, even if not officially registered as multipurpose, provide more than one service and are, practically, multipurpose. In the EU, 46% of single purpose reservoirs are for hydropower (which store 83% of the volume of single-purpose dams) and 23% of multiple purpose reservoirs have hydropower as a first use (34% have water supply as first use). Furthermore, 6% of all reservoirs have hydropower as second use and 3.5% as third or fourth use. Overall, 48% of EU's reservoirs (associated to large dams) utilize hydropower. EU large-dam's reservoirs store 285 km^3 of water: 124 km^3 are in hydropower reservoirs as single use, while 111 k^3 are in multipurpose reservoirs equipped with hydropower [25].

The role of multipurpose water storage is key in the EU according to the Water Resilience Strategy, that says: "Action on water management should prioritise nature-based solutions but also needs to rely on man-made structures or a combination of both. The planning of new dams and reservoirs should carefully evaluate their environmental impacts, involving all relevant actors, and ensure that such actions are part of an integrated and sustainable water management strategy, fully reflecting long-term climate reference scenarios and projections, to avoid stranded investments".

This also means that during all stages of hydropower project's lifespan a variety of different skilled professionals are involved, from engineers and geologists to economists and ecologists. It was estimated that every 1 MW of community-owned hydropower installed generates ten full-time equivalent jobs in every year of its operation. This is significantly more than any other renewable energy technology [20].

Finally, the importance of hydropower for reaching the Clean Energy Transition targets, for grid stability and system resilience, for energy security and energy justice, for protection from floods and droughts, and for basically providing water for drinking, agricultural, recreational and other purposes are well documented.

Still, today the benefits of hydropower are challenged by financial, regulatory and environmental aspects and multiple water supply, grid stability and other services are usually not officially recognized. Investment risks, especially, for large hydropower schemes are higher, requiring specific policy instruments and incentives, as well as a longer-term policy perspective and vision. Even development of small or micro hydropower, for example, repowering of small historical watermills, are challenged with the same intensity, although their benefits for local communities can be proven.

The holistic, all-encompassing view of hydropower needs to be recognized in the policy instruments. If hydropower will be understood as pure electricity source, or in some cases grid stability provider, there will be a legislative gap between the renewable energy development orientated policy instruments such as RED III or REPowerEU, and environmental protection regulations such as the Water Framework Directive (WFD).

These challenges should serve as a catalyst for a more comprehensive dialogue among stakeholders. To gain acceptance and to obtain a win-win solutions among all stakeholders' needs, new hydropower or even modernization projects must be designed as multipurpose ones by multidisciplinary teams. Today exists situation when some stakeholder groups such as hydropower suppliers, local civil society, and academics are under-engaged in fixing regulatory uncertainty [26].

Stakeholder engagement can become a central component of hydropower governance. Conflicts between energy production and environmental objectives require negotiation and adaptive management approaches that balance competing interests. At the same time, participatory processes can enhance legitimacy, transparency, and the long-term sustainability of hydropower projects.

Addressing these challenges will require integrated planning approaches, stronger coordination across policy domains, and continued innovation in both technology and governance. In this context, hydropower is not only part of the energy transition, but also a critical component of broader efforts to build resilient and sustainable socio-technical systems.

Acknowledgements This chapter is based upon work from the WG4 from the COST Action PEN@Hydropower (CA21104), supported by COST (European Cooperation in Science and Technology).

Competing Interests The authors have no conflicts of interest to declare that are relevant to the content of this chapter.

References

1. Eurostat (2026) Electricity and heat statistics. https://ec.europa.eu/eurostat/statistics-explained/index.php?title=Electricity_and_heat_statistics#Consumption_of_electricity_and_derived_heat. Accessed: 2026-04-23
2. Eurostat (2026) Production of electricity and derived heat by type of fuel. https://ec.europa.eu/eurostat/databrowser/view/nrg_bal_peh__custom_20205725/default/table?lang=en. Accessed: 2026-04-23
3. Rajić M, Quaranta E, Kuriqi A, Kasiulis E (2025) Renewable energy in the Western Balkans and EU policy: pathway to regional Net-Zero. In: IOP conference series: earth and environmental science, vol. 1442, no. 1, pp. 012002. IOP Publishing
4. European Environment Agency (2024) Environmental statement 2023. https://www.eea.europa.eu/en/analysis/publications/environmental-statement-2023. Accessed: 2026-04-24
5. Quaranta E (2024) Hydropower in the European Union: state-of-the art, future opportunities and main issues within the current social, energy, geopolitical, and climate situation. In: Sustainable development of renewable energy. Elsevier, pp. 281–302
6. Rutschmann P, Kampa E, Wolter C, Albayrak I, David L, Stoltz U, Schletterer M (2022) Novel developments for sustainable hydropower. Springer
7. European Commission (2000) Water framework directive. https://environment.ec.europa.eu/topics/water/water-framework-directive_en. Accessed: 2026-04-24
8. European Investment Bank (2019) Environmental, climate and social guidelines on hydropower development. https://www.eib.org/en/publications/environmental-climate-and-social-guidelines-on-hydropower-development. Accessed: 2026-04-24
9. Belletti B, Garcia de Leaniz C, Jones J, Bizzi S, Börger L, Segura G, Castelletti A, Van de Bund W, Aarestrup K, Barry J et al (2020) More than one million barriers fragment Europe's rivers. Nature 588(7838):436–441
10. Quaranta E (2026) Policy, science and industry synergies to unlock hidden small hydropower in the European Union under the water-energy-society-ecosystem nexus. Energy Strat Rev 64:102174. ISSN: 2211-467X, https://doi.org/10.1016/j.esr.2026.102174, https://www.sciencedirect.com/science/article/pii/S2211467X26001409
11. European Commission (2009) Renewable energy directive. https://energy.ec.europa.eu/topics/renewable-energy/renewable-energy-directive-targets-and-rules/renewable-energy-directive_en. Accessed: 2026-04-24
12. European Commission (2022) 2022 Report on the Achievement of the 2020 Renewable Energy Targets. https://eur-lex.europa.eu/legal-content/EN/TXT/PDF/?uri=CELEX:52022DC0639. Accessed: 2026-04-24
13. European Commision (2020) Regulation (EU) 2020/852 of the European Parliament and of the Council of 18 June 2020 on the establishment of a framework to facilitate sustainable investment, and amending Regulation (EU) 2019/2088. https://eur-lex.europa.eu/eli/reg/2020/852/oj/eng. Accessed: 2026-04-24
14. Achterfeldt S, Driescher M, Alessi M, Stogl O, Valentová M, Traverso M (2025) Enabling voluntary EU Taxonomy disclosure for the clean energy sector: introducing the Clean Leap Tool. Environ Sci Eur 37(1):1–24

15. Vezzoni R (2023) Green growth for whom, how and why? The REPowerEU Plan and the inconsistencies of European Union energy policy. Energy Res. Soc. Sci. 101:103134. ISSN: 2214-6296. https://doi.org/10.1016/j.erss.2023.103134, https://www.sciencedirect.com/science/article/pii/S2214629623001949
16. European Commision (2025) Communication from the Commission to the European Parliament, the Council, the European Economic and Social Committee and the Committee of the Regions. European Water Resilience Strategy. https://eur-lex.europa.eu/legal-content/EN/TXT/?uri=CELEX%3A52025DC0280&qid=1750857768458. Accessed: 2026-04-24
17. Quaranta E, Bódis K, Kasiulis E, McNabola A, Pistocchi A (2022) Is there a residual and hidden potential for small and micro hydropower in Europe? A screening-level regional assessment. Water Resour Manage 36(6):1745–1762
18. European Union (2024) Regulation (EU) 2023/955 of the European Parliament and of the Council of 10 May 2023 establishing a Social Climate Fund and amending Regulation (EU) 2021/1060. https://eur-lex.europa.eu/eli/reg/2023/955/oj/eng. Accessed: 2026-04-24
19. Quaranta E, Aggidis G, Boes RM, Comoglio C, De Michele C, Patro ER, Georgievskaia E, Harby A, Kougias I, Muntean S et al (2021) Assessing the energy potential of modernizing the European hydropower fleet. Energy Convers Manage 246:114655
20. Quaranta E, Aggidis G, Boes RM, Comoglio C, De Michele C, Patro ER, Georgievskaia E, Harby A, Kougias I, Muntean S, et al (2025) Clean energy technology observatory: hydropower and pumped storage hydropower in the European Union - 2025 - Status report on technology development, trends, value chains and markets
21. Council of European Energy Regulators (2025) Status review of renewable support schemes in Europe for 2022 and 2023. https://www.ceer.eu/publication/status-review-of-renewable-support-schemes-in-europe-for-2022-and-2023/. Accessed: 2026-04-24
22. International Hydropower Association (2025) 2025 world hydropower outlook. https://www.hydropower.org/publications/2025-world-hydropower-outlook. Accessed: 2026-04-24
23. Pauwels IS, Tuhtan J, Coeck J, Buysse D, Baeyens R (2022) Archimedes screw–an alternative for safe migration through turbines?. In: Novel developments for sustainable hydropower. Springer, pp 125–133
24. Technology transfer in the hydropower industry: An analysis of Chinese dam developers' undertakings in Europe and Latin America. Energy Policy 113:546–558 (2018). ISSN: 0301-4215
25. International Commission on Large Dams (2023) World register of dams. https://www.icold-cigb.org/GB/publications/world_register_of_dams.asp. Accessed: 2026-04-24
26. Kasiulis E, Cerpinska M, Pummer E, Finger D, Malicka E, Fischer-Aupperle B, Doujak E, Anagnostopoulos I, Hocevar M, Cavazzini G (2026) Action towards increase in sustainable hydropower: a Pan-European assessment and strategy. Renew Energy 263:125503. ISSN: 0960-1481. https://doi.org/10.1016/j.renene.2026.125503, https://www.sciencedirect.com/science/article/pii/S0960148126003289

Holistic Assessment and Stakeholder Interaction

Marija Lazarevikj, Valbona Karapici, Iraia Nuñez Aguirre, and Shaua Fui Chen

Integrating technical, economic, environmental, and social dimensions for sustainable hydropower.

Abstract Hydropower systems are increasingly assessed within complex socio-technical contexts that extend beyond technical performance alone. Their development and operation involve interconnected technical, environmental, economic, regulatory, and social dimensions, creating trade-offs between energy production, ecosystem protection, market value, public acceptance, and long-term sustainability. This chapter addresses the need for a holistic assessment framework that integrates these dimensions while explicitly incorporating stakeholder engagement. It examines technical and environmental performance alongside economic aspects such as cost structures, market participation, value creation, and policy influences on long-term viability. At the same time, it highlights the social dimension of hydropower, including public acceptance, stakeholder perceptions, fairness in decision-making, communication practices, and the role of trust in shaping project outcomes. Particular attention is given to the interactions among operators, regulators, transmission system operators, local communities, and other affected actors, whose perspectives influence how impacts and benefits are interpreted and how trade-offs are negoti-

M. Lazarevikj (✉)
Faculty of Mechanical Engineering, Ss. Cyril and Methodius University of Skopje, Skopje, North Macedonia
e-mail: marija.lazarevikj@mf.edu.mk

V. Karapici
Faculty of Economy, University of Tirana, Tirana, Albania
e-mail: valbona.karapici@unitir.edu.al

I. Nuñez Aguirre
Zabala Innovation Consulting, Brussels, Belgium
e-mail: inunez@zabala.eu

S. F. Chen
Department of Interdisciplinary Studies of Culture, Norwegian University of Science and Technology, Trondheim, Norway
e-mail: shaua.f.chen@ntnu.no

E. Doujak et al. (eds.), *Rethinking Hydropower: A Holistic Vision for a Sustainable Future*, Springer Tracts in Civil Engineering,
https://doi.org/10.1007/978-3-032-33328-5_11

ated. The chapter proposes a stakeholder-centred framework for holistic hydropower assessment, supporting more transparent, inclusive, and adaptive decision-making in the context of evolving energy systems.

Keywords Stakeholders · Social acceptance · Economic performance · Interlinkages

1 Introduction

Hydropower occupies a distinctive and strategic position within modern energy systems, combining renewable electricity generation with storage capacity, operational flexibility, and water resource management functions. As the penetration of variable renewable energy sources (VRES) increases, hydropower is increasingly required to balance fluctuations in supply and demand, thereby contributing to system reliability, grid stability, and decarbonization pathways. Despite these advantages, hydropower systems cannot be understood solely in terms of their technical performance. They are embedded within complex socio-technical systems that integrate ecological processes, river basin dynamics, electricity markets, regulatory frameworks, and societal expectations. These interconnected dimensions create interdependencies that influence project design, operation, and policy decisions, often giving rise to trade-offs between energy production, environmental protection, and social acceptance. In this context, there is an urgent need for a holistic assessment of hydropower that considers not only technical aspects such as efficiency and lifetime extension, but also environmental impacts on biodiversity and ecosystems, social dimensions including public acceptance and stakeholder perceptions, and economic factors related to investment and long-term profitability. Traditional approaches that address these dimensions in isolation are no longer sufficient to capture the complexity of modern hydropower systems. Stakeholder engagement plays a central role within such a holistic perspective. The development and operation of hydropower projects involve multiple stakeholder groups, including operators, policymakers, local communities, and environmental organizations, each with different interests, knowledge, and expectations. Their interactions shape how impacts are perceived, how trade-offs are evaluated, and ultimately how decisions are made. As a result, effective engagement processes are essential to ensure transparency, build trust, and support socially acceptable and sustainable outcomes. This chapter explores approaches for integrating stakeholder engagement into holistic hydropower assessment. Examines the interconnections between technical, environmental, social and economic dimensions and it discusses how stakeholder-centered frameworks can support more inclusive, adaptive, and robust decision-making processes in the context of the energy transition.

2 Economic Performance and Market Integration of Hydropower

Assessing the economic performance of hydropower requires a perspective that goes beyond conventional cost-based indicators to capture the full economic value.

2.1 Cost Competitiveness and Economic Performance of Hydropower

The economic competitiveness and performance of hydropower is determined primarily by its cost structure, which is characterized by high capital expenditures and relatively low operational and maintenance costs throughout the long lifetime of the asset. The main indicator for the comparison of hydropower with alternative generation technologies is the levelized cost of electricity (LCOE). This, however, often underestimates the full economic value of hydropower generation within the energy system. Unlike renewable sources, hydropower offers dispatchability, storage capability, particularly in pumped storage hydropower (PSH), and ancillary services such as frequency regulation and reserve provision. These qualities contribute significantly to security of energy supply, grid stability, and the overall efficiency of the power system. According to [1] the LCOE provides a practical decision-making framework under traditional cost-of-service regulation, even though in the context of liberalised and integrated electricity markets its importance has weakened. In particular, LCOE fails to capture the temporal and spatial dimensions of electricity value, as well as the system-level contributions of flexible and dispatchable resources, which are increasingly critical in modern power systems [1–3]. It represents an average cost metric that assumes constant output over time, therefore neglecting the variability of electricity demand and price fluctuations across hours, seasons, and locations. However, in modern electricity markets, the economic value of electricity is highly time-dependent, with significant price differentials between peak and off-peak periods, as well as across geographically interconnected markets [1, 2]. While LCOE provides a useful benchmark for comparing generation costs, it does not fully reflect the economic performance of hydropower in modern electricity systems, where value creation also depends on flexibility, timing of production, and contribution to system reliability [1]. In the European context, the transition toward market-based dispatch is embodied in the Single Day-Ahead Coupling (SDAC), where electricity prices are determined through the EUPHEMIA algorithm. This algorithm simultaneously clears supply and demand across multiple bidding zones by maximising social welfare, subject to network constraints and cross-border transmission capacities. As a result, electricity prices are no longer determined by local cost structures alone, but by system-wide optimisation across interconnected markets [4]. Within this framework, the economic performance of hydropower is increasingly shaped by its ability to respond to price signals over time and across regions. Unlike inflexible generation technologies, hydropower can optimise its production by shifting output to periods of higher prices, engaging in temporal arbitrage, and contributing to congestion management through cross-border flows. These characteristics are not reflected in LCOE metrics but are central to value creation in market coupling environments. Consequently, the economic evaluation of hydropower requires a transition from a cost-based model to a value-based framework, where flexibility, timing, and system integration become the primary drivers of economic performance. The ability of Hydropower to provide reliable and dispatchable generation, combined with storage capabilities, makes it a fundamental asset in managing supply-demand imbalances and responding to system shocks, generating broader economic benefits that are not reflected in traditional cost-based indicators. In this context, the competitive advantage of hydropower lies not only in its cost structure, but in its strategic role within a coupled and increasingly complex electricity market [5].

2.2 *Market Integration and Value Creation*

The participation of hydropower in electricity markets is shaped by its operational flexibility and ability to respond to price signals. Hydropower operators participate in multiple market segments, including day-ahead, intraday, and balancing markets, by optimizing generation across electricity markets to maximize revenues. Therefore, the economic value of hydropower is strongly influenced by price variability, peak demand periods, and cross-border electricity flows. Moreover, value creation is influenced by multiple stakeholders, including generators, transmission system operators (TSOs), regulators, and water resource users. The interaction between these stakeholders reflects the complexity of the water–energy–environment nexus, where economic optimization must be balanced against hydrological constraints, environmental regulations, and competing water uses such as irrigation and industrial demand. In this framework, hydropower plays a strategic role due to its ability to shift production to a high-price period, and to support congestion management through flexible dispatch. This operational flexibility enables hydropower to capture price differentials across hours and markets, thus enhancing its overall market value. In interconnected electricity systems, particularly within the European market coupling framework, such flexibility contributes to the efficient allocation of resources across bidding zones and facilitates cross-border electricity trade [2]. The contribution of hydropower to value creation is therefore linked not only to energy generation but also to its role in supporting system integration and reliability. By providing balancing and reserve capacity, hydropower enables the integration of variable renewable energy sources such as wind and solar, reducing curtailment and mitigating price volatility [6, 7]. In this context, its value is increasingly determined by its responsiveness to system conditions rather than by its average production cost alone [7].

2.3 Policy and Regulatory Influences on Economic Viability

The economic viability of hydropower projects is strongly influenced by policy and regulatory frameworks. Market design, including pricing mechanisms and access to ancillary service markets, plays a critical role in determining revenue streams. At the same time, environmental legislation, permitting procedures, and water management policies impose constraints that can affect both investment decisions and operational flexibility. Support mechanisms, such as capacity remuneration schemes or incentives for renewable generation, may enhance the financial attractiveness of hydropower, particularly for capital-intensive projects with long payback periods. However, regulatory uncertainty and evolving environmental standards can introduce investment risks, underscoring the importance of stable and transparent policy frameworks. Overall, the integration of hydropower into modern electricity markets requires a holistic approach that accounts for its multifunctional role within the energy system and its interaction with broader environmental and economic objectives [7].

3 Social Acceptance of Hydropower

New energy projects are often met with opposition from local stakeholders, making social acceptance a critical factor in their successful implementation. Limited acceptance of renewable energy projects can hinder the effectiveness of energy transition policies and create gaps between strategic objectives and practical outcomes. Consequently, understanding the different dimensions of social acceptance is essential for ensuring the effective implementation of energy policies [8].

3.1 Conceptual Framework of Social Acceptance in the Field of Renewable Energy

Social acceptance of renewable energy technologies is a multi-dimensional and context-dependent concept, encompassing a range of attitudes and behaviors of individuals and social groups towards energy systems. It has been conceptualised as a continuum of responses, ranging from active opposition to active support [9], as well as a combination of attitudes and actions reflecting different levels of engagement [10]. A widely adopted framework in the field of renewable energy is proposed by [11], who distinguishes three interrelated dimensions of social acceptance: socio-political, community, and market acceptance, highlighting that acceptance operates at multiple levels within society. Factors influencing socio-political and community acceptance help explain the gap between broad public support for renewable energy and challenges in implementing specific projects, while market acceptance remains less explored and offers scope for further research. Socio-political acceptance refers to general public support and policy endorsement of renewable energy technologies, shaped by broader societal values, national priorities and trust in institutions, with governance processes and perceived fairness playing an important role [12]. Community acceptance captures local attitudes toward specific projects, influenced by place-based perceptions, environmental and social impacts, and the quality of stakeholder engagement, including aspects such as procedural and distributive justice, trust, and local identity [13, 14]. Market acceptance relates to the willingness of investors, developers, and consumers to adopt and support renewable energy technologies, and is shaped by economic viability, regulatory stability, and market conditions, including perceived risks, costs, and expected returns [11]. Furthermore, social acceptance is increasingly understood as a dynamic and evolving process, influenced by psychological, contextual, and institutional factors [15], and closely linked to perceptions of fairness and legitimacy. In this context, procedural justice refers to the fairness and transparency of decision-making processes, while distributive justice concerns the perceived equity in the sharing of benefits and burdens [16, 17]. Early work by [14] demonstrated the importance of fairness in shaping community responses to energy projects, while [13] emphasised the role of public engagement and place-based perceptions. These perspectives are commonly brought together within the energy justice framework [18], which also introduces recognition justice, emphasising the importance of acknowledging diverse local values and concerns.

3.2 Hydropower Flexibility and Social Acceptance

In the context of hydropower, social acceptance is shaped by a combination of environmental, socio-economic, and institutional factors, reflecting how communities experience and interpret project impacts [19]. Perceived environmental effects, such as alterations in flow regimes, impacts on biodiversity and fish migration, sediment transport disruption, and landscape modification, are among the most influential drivers of acceptance and sources of concern. These impacts are often evaluated through place-based and cultural values, leading to different interpretations of sustainability and fairness among stakeholders [17, 20, 21]. Acceptance therefore depends not only on the technical mitigation of impacts, but also on the relationships and trust established between operators, institutions and local communities [22]. In addition, the distribution of local benefits and burdens, as well as the quality of stakeholder engagement, play a critical role in shaping public responses. Acceptance tends to be higher where projects provide local benefits and address environmental concerns, reflecting the importance of fairness in the distribution of impacts and benefits [16], and where decision-making processes are perceived as fair and inclusive [14].

Flexible hydropower and controversy triggers

Particular controversy arises in the context of flexible hydropower operation, where practices such as hydropeaking and rapid flow fluctuations introduce visible, frequent, and sometimes unpredictable changes in river conditions. These operational changes can intensify ecological disturbances, affect biodiversity and riverbank stability, and they can also alter the usability and safety of river areas, thereby directly influencing how local communities experience and evaluate hydropower projects [17, 21, 23]. Such impacts have also driven increasing efforts in river restoration, aimed at mitigating hydromorphological alterations and restoring ecological continuity, all of which highlights the extent to which regulated flow regimes can affect river systems [24]. Importantly, social acceptance is not determined only by the presence of these impacts, but by their frequency, intensity, and visibility, with frequent and abrupt changes typically generating stronger negative responses than gradual or less noticeable variations [21]. As a result, flexible operation is often perceived as more disruptive than conventional, steady operation. This further exacerbates tensions between energy system needs and local environmental and social priorities.

3.3 Stakeholder Engagement in Shaping Public Perception and Social Response

The case of flexible hydropower highlights the importance of interaction processes through which operational practices are translated into social responses. Flexible operation introduces not only changes in river conditions, but also decision-making dynamics that involve multiple stakeholders with differing roles, expectations, and levels of influence. In this context, social acceptance is shaped less by the existence of impacts alone, and more by how these changes are communicated, managed, and negotiated among stakeholders, ultimately influencing both public perception and decision-making processes [13, 15]. This is supported by empirical evidence showing that impact management actions are significantly correlated with how environmental and social impacts are perceived by stakeholders [25]. A central element of this process is the operational profile of hydropower plants, which shapes how changes are experienced over time and interpreted by stakeholders. Highly variable and rapidly changing operational regimes can create practical challenges and increase stakeholder concerns, particularly when changes are difficult to anticipate, highlighting the need for effective communication and stakeholder engagement [21, 23]. Communication acts as the primary translation mechanism between technical operation and public understanding. In the context of flexible hydropower, it should be understood as a two-way process that forms the basis of sustained stakeholder relationships throughout the project lifecycle. This involves dialogue between hydropower developers, operators, and contractors, as well as affected communities, local and national authorities, and other relevant actors. The effectiveness of communication depends on its clarity, timing, transparency and accessibility, but also on its ability to facilitate feedback and mutual understanding. Information provided in a timely and accessible manner enables stakeholders to better interpret operational practices and associated trade-offs, including the relationship between local impacts and broader system benefits such as energy reliability, whereas insufficient communication can lead to greater uncertainty and mistrust [21, 22]. The interpretation of operational changes involves multiple stakeholder groups, each contributing through distinct roles, forms of knowledge, and levels of influence. Perceptions of hydropower impacts may differ significantly across stakeholder groups, with local communities often expressing more critical views than industry or policymakers [25] Local communities are directly exposed to operational changes and provide experiential, place-based knowledge, which strongly influences how changes are perceived and evaluated in everyday contexts. This reflects the broader observation that support for renewable energy at a general level does not necessarily translate into acceptance of specific local projects, where impacts are directly experienced [26]. Here the responses are often linked to local priorities, such as safety, livelihoods, and the use of river spaces, which consequently become central factors in shaping acceptance at project level. Non-governmental organisations (NGOs) typically operate at the interface between local concerns and broader environmental or societal objectives. Through advocacy, dissemination of knowledge, and participation in formal decision-making processes, they can amplify specific issues, influence public narratives, and contribute to shaping policy debates. Their role is particularly significant in framing how operational practices are interpreted within wider sustainability and environmental discourses. Hydropower operators and industry actors are responsible for defining operational strategies and implementing communication approaches, and therefore play a key role in shaping both the technical and social dimensions of system performance. Their decisions influence not only operational outcomes, but also the transparency of information, responsiveness to stakeholder concerns, and overall credibility of the project. At the same time, operators are often required to balance competing objectives, including system flexibility, regulatory compliance, and social acceptance. Regulators and public authorities establish the legal, institutional and procedural framework within which both operation and stakeholder participation occur. Through licensing procedures, environmental requirements, and consultation obligations, they define the conditions under which stakeholders can engage and influence decisions. Their role is therefore critical in ensuring that engagement processes are not only formal, but also meaningful and effective. These agents do not operate independently, but interact with each other, but interact through continuous exchanges of information, mutual influence and feedback. Relationships between them can amplify or mitigate project-related risks and influence outcomes. Community responses can shape regulatory decisions and operational adjustments, NGOs influence public narratives and policy priorities, while operators adapt communication and management strategies in response to stakeholder concerns and regulatory requirements. Engagement processes, based on stakeholder roles, constitute the institutional framework through which interaction is structured. These processes, ranging from formal consultations to participatory planning and co-creation, determine the extent to which stakeholders can access decision-making processes and influence outcomes. Effective engagement is characterised by inclusiveness, continuity, and responsiveness, and should account for diverse stakeholder needs, including vulnerable groups, cultural contexts, gender considerations, and differences in access to information. Such approaches are reflected in established sustainability frameworks, including the Hydropower Sustainability Tools developed by the International Hydropower Association, which emphasise the role of structured communication and consultation in improving project performance across the planning, construction, and operational phases.

Best practices and recommendations for improving stakeholder engagement strategies

Building on the interaction mechanisms described above, improving social acceptance of flexible hydropower requires targeted strategies that strengthen communication, stakeholder involvement, and governance processes. In this context, stakeholder engagement should be seen as a continuous and adaptive process, rather than a one-time procedural requirement. These recommendations are derived from a synthesis of existing literature on social acceptance and stakeholder engagement, and are consistent with established hydropower sustainability frameworks, including the Hydropower Sustainability Tools of the International Hydropower Association. These strategies can be grouped into key areas related to the timing of engagement, communication practices, integration of stakeholder knowledge and benefits, and adaptive governance and monitoring. A key principle is the early and continuous involvement of stakeholders throughout the project lifecycle. Engaging stakeholders from the initial planning stages enables the identification of local concerns, expectations, and potential conflicts before they escalate. Continuous engagement during operation is particularly important, as evolving operational patterns may influence stakeholder perceptions over time. Effective engagement depends on two-way communication, allowing not only the provision of information but also the incorporation of stakeholder feedback into decision-making processes. Transparent, timely, and accessible communication supports an informed understanding of operational practices and associated trade-offs, while also reducing uncertainty and strengthening trust between stakeholders and project developers [16, 19]. Ensuring procedural fairness, including transparent decision-making and equal opportunities for stakeholder participation, is essential for enhancing perceived legitimacy and acceptance. Clear communication of project limitations and realistic expectations is important to avoid misunderstandings and potential conflicts over time. Another critical aspect is the integration of local knowledge and stakeholder concerns into planning and operational decisions. Local communities possess context-specific knowledge related to river use, safety and environmental conditions, which can improve the design and implementation of operational strategies. Incorporating this knowledge enhances the relevance, legitimacy and acceptability of decisions. To further strengthen acceptance, benefit-sharing mechanisms should be implemented to ensure that local communities receive tangible and locally relevant benefits from hydropower projects. These may include economic compensation, local infrastructure development, or support for community initiatives. Equitable distribution of benefits can help address perceived imbalances between local impacts and broader system-level

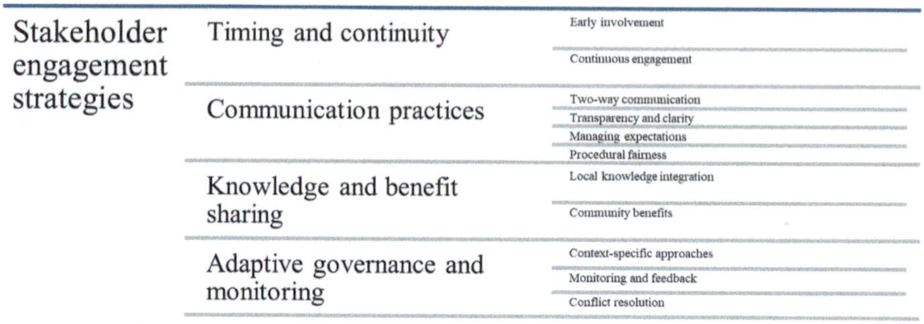

Fig. 1 Framework of stakeholder engagement strategies for social acceptance of hydropower

benefits. Engagement strategies should be context-specific and adaptive, reflecting differences in social, cultural, and institutional conditions. Since flexible hydropower projects often operate in diverse environments, they require tailored approaches that account for stakeholder diversity, including vulnerable groups and differing capabilities to participate. Adaptive governance approaches, which enable iterative learning and adjustment over time, are particularly important in addressing evolving stakeholder expectations and operational conditions. Finally, the establishment of transparent and participatory monitoring and feedback mechanisms is essential for maintaining long-term trust and accountability. Involving stakeholders in monitoring project impacts and performance creates opportunities for continuous dialogue, supports timely identification of emerging concerns, and can inform adjustments to both operational and engagement strategies. Establishing mechanisms for conflict resolution and grievance management provides stakeholders with formal channels to express concerns and supports timely resolution of disputes. These practices, as illustrated in Fig. 1, collectively support the development of stakeholder engagement strategies that are inclusive, responsive, and capable of enhancing the long-term social acceptance of flexible hydropower.

Stakeholder engagement and social acceptance: The case of the Bakun Dam

The Bakun dam project spanned several decades, from its inception in the 1970s to its commissioning in 2011. Construction of the Bakun dam began in 1994. It was suspended following the Asian financial crisis in 1997 and resumed again in 2000. The project, initiated by the Malaysian federal government, was fully taken over by the Sarawak state government in 2017. The project was included in the Sarawak Corridor of Renewable Energy (SCORE) in 2008 to supply electricity to energy-intensive industries. The project experienced several critical periods, during which stakeholder combinations included foreign experts, national and state government agencies, and environmental and indigenous communities affected by the project, among others. It weathered the peak period of the anti-dam movement, hen the WCD put the spotlight on the social and environmental impacts of dams, but then benefited from the renewed interest in hydropower as the largest renewable energy source, enabling the global energy transition. It also saw a shift from western-focused to China-led hydropower expertise in global hydropower development. The feasibility studies were carried out in the 1970s and 1980s by Western experts from Australia, German, and Swiss enterprises, but the project was ultimately constructed by China's Sinohydro upon its resumption in 2000. This project serves as an instructive case for understanding how IHA's Sustainability Standards were able to help develop a stakeholder engagement plan, which went through at least three stakeholder group combinations at different stages of the project. The project was initiated by the Malaysian federal government in the 1990s, and planning was carried out at the federal level. It was a privatized project that was awarded to Ekran Berhad without competitive bidding. During the planning stage of the project, the Coalition of Concerned Non-governmental Organizations (Gabungan), the indigenous communities affected by the dam project and the environmental group Sahabat Alam Malaysia (SAM) all criticized the lack of transparency in the planning process, noting that there had not been any public consultation on the project and that the directly affected indigenous communities had not been engaged in the planning process. Many felt that they had been pushed into accepting the resettlement scheme. The resettlement of 10,000 indigenous communities, comprising 1,693 families living in 15 longhouses, was carried out between 1998 and 1999. However, the project was then shelved. The resumption of the project, now with a capacity reduced by 500MW, involved restructuring. China's Sinohydro joined as the lead contractor under the EPC/turnkey contract and established a joint venture with Sime Engineering Berhad. China's EXIM Bank provided the financing for the project. At this stage, Gabungan questioned the justification of the large-scale resettlement scheme in view of the reduced capacity. The indigenous-led NGO, Save Rivers Network, was officially established in 2011 and highlighted the government's unfulfilled promises as well as the hardships faced by resettled indigenous people. The network became a crucial stakeholder in the hydropower dam development in Sarawak and also opposed a new wave of mega dams following the launch of the SCORE. In 2017, the Sarawak state-owned utility company, Sarawak Energy Berhad (SEB), acquired the Bakun dam from the federal government. As the project progressed to the operational stage, the stakeholders included state-level government agencies as well as the directly affected indigenous communities. The implementation and monitoring of the project focused on mitigating its social and environmental impacts, especially the consequences of resettlement on the affected indigenous communities. One of the criticisms was the lack of a mechanism to express their grievances to the relevant authority. SEB then upgraded its corporate sustainability practices and applied the Hydropower Sustainability Standard (HSS) to guide its response to sustainability shortcomings. The assessment of the Bakun dam for certification was performed in 2024, and the project received "silver" certification in August 2025. Based on the assessment report, SEB has improved engagement with communities across all areas, establishing a grievance mechanism and a stakeholder management plan (SMP) in February 2024. The SMP included mapping of over 30 stakeholder groups by level of interest and influence, identifying issues of concern for about 50 stakeholder groups, and determining engagement methods. The SMP helped SEB identify the frequency, timing, scope, and level of engagement with stakeholders, including institutional stakeholders and community-level stakeholders such as the long house committee, the village headman, and women in the communities, in order to discuss corporate social responsibility initiatives.

4 Interconnections Across Hydropower Dimensions

Hydropower systems cannot be understood solely in terms of their technical capabilities. They are embedded within a broader socio-technical system that encompasses ecological dynamics, water resource management, electricity markets, regulatory frameworks, and societal expectations. These dimensions are deeply interconnected, creating complex interdependencies that shape decision-making processes and give rise to trade-offs among competing objectives. This chapter builds on research conducted under the COST Action PEN@Hydropower, particularly within Working Group 5, and examines how technical, environmental, economic, social, and policy factors interact in hydropower planning and operation. The aim is to translate scientific and technical evidence into structured, policy-relevant insights that support the development of integrated strategies and implementation roadmaps. The analysis follows a clustering approach in which scientific outputs are synthesized into macro-level policy areas. This enables the identification of systemic challenges that emerge across different hydropower dimensions, highlighting the interdependencies that constrain or enable decision-making. As illustrated in Fig. 2, the process consists of several interconnected steps, including the review and screening of content, identification of cross-domain linkages, clustering of findings into policy-relevant domains, extraction of key policy issues, and the development of a structured data matrix, in order to support the transformation of complex, multidisciplinary evidence into actionable insights, facilitating the development of coherent and integrated policy responses.

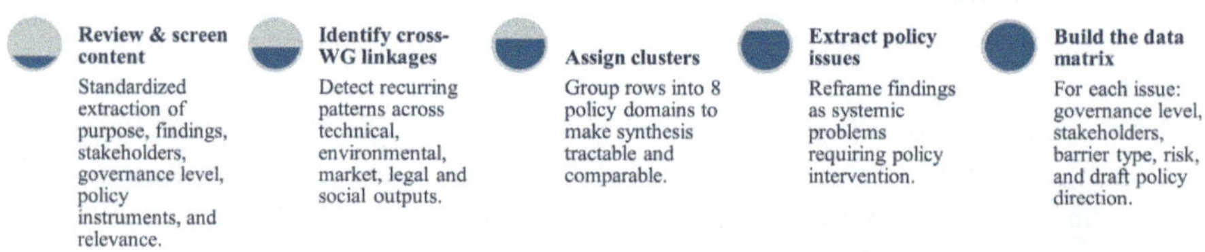

Fig. 2 Methodological framework for translating scientific evidence into policy-relevant insights through clustering and synthesis across hydropower dimensions

Fig. 3 Framework for structuring hydropower policy issues into actionable policy directions

4.1 Systemic Challenges Across Interconnected Domains

The identified issues are then further structured into a policy-oriented analytical framework, as shown in Fig. 3, where each issue is systematically characterized in terms of its cluster, governance level, involved stakeholders, barrier types and associated risks, with the aim of shaping initial policy directions. This step enables the translation of complex findings into a consistent and comparable structure.

Market design, flexibility and system adequacy

Hydropower is increasingly required to provide flexibility services including ramping, frequency regulation, and balancing to support power systems with high shares of VRES. However, existing electricity market designs remain largely oriented toward energy production rather than the provision of system services. Evidence from the PEN@Hydropower COST Action indicates that the flexibility of hydropower is often undervalued, while intensified operational regimes increase wear and maintenance costs. Stakeholder discussions within WG5 further highlighted that remuneration mechanisms for flexibility services, including balancing, ramping, and storage, remain highly dependent on national contexts. In addition, the costs associated with flexible operation, including operational risks and reduced asset lifetime, are difficult to quantify and are often not adequately reflected in revenue streams. This misalignment between system needs and economic incentives discourages investment in flexibility assets and undermines efficient system integration [27–29]. Participants in WG5 activities also noted that unclear economic signals and regulatory uncertainty, particularly in contexts where concession frameworks are approaching renewal, may delay or discourage investments in assets essential for long-term system adequacy and resilience. PSH is a critical enabler of Europe's energy transition, but its optimal deployment must be tailored to national system characteristics, realistic development potential, and VRES profiles. The findings provide robust technical evidence to support differentiated, system-specific storage strategies in future European energy policies and planning frameworks [30]. In parallel, the integration of VRES requires not only short-term balancing but also long-duration storage to ensure system adequacy during prolonged periods of low generation. While battery technologies are well suited for short-term flexibility, pumped storage hydropower (PSH) remains the only mature large-scale solution for long-duration storage. However, current planning frameworks tend to underestimate this requirement, often favoring short-duration technologies. This creates a structural bias that may compromise system resilience under extended supply deficits [28].

Environmental governance and climate legitimacy

Hydropower development and operation significantly influence river ecosystems, affecting fish migration, sediment transport and flow regimes. Research highlights both the scale of these impacts and the importance of mitigation and restoration measures. At the same time, environmental governance frameworks—most notably the EU Water Framework Directive—introduce stringent constraints that may limit the deployment or optimization of hydropower assets. This creates a persistent tension between renewable energy objectives and biodiversity protection. Stakeholder discussions within WG5 further emphasized that environmental challenges in hydropower are increasingly driven by climate-induced changes in water availability and competing water uses. Increasing aridity, changing precipitation patterns, decreased runoff, and reduced reservoir inflows are identified as growing concerns directly affecting hydropower generation potential. In parallel, rising water demand, especially from agricultural irrigation, intensifies competition for water resources and places hydropower within a broader water–energy–food nexus. In addition, stakeholders highlighted the increasing frequency and intensity of extreme hydrological events, including floods and prolonged droughts, which may compromise electricity generation reliability and system resilience. From a governance perspective, these challenges require stronger coordination across energy, water, agricultural, and land-use policies. The challenge is further compounded by uncertainties associated with greenhouse gas emissions from reservoirs due to the absence of harmonized methodologies for estimating GHGs emissions which introduces ambiguity into carbon accounting practices [31]. Such inconsistencies complicate the inclusion of hydropower in sustainable finance frameworks and EU taxonomy classifications, ultimately affecting its climate legitimacy [32, 33].

Digitalization, modernization and asset resilience

The European hydropower fleet is aging [27], but at the same time operational demands are increasing due to growing flexibility requirements. Stakeholder discussions within WG5 strongly confirmed that many hydropower assets across Europe have been operating for several decades, limiting their ability to respond to increasingly dynamic and variable power system conditions. Participants in WG5 activities emphasized that hydropower must adapt to evolving grid demands not only through physical refurbishment, but also through the integration of digital technologies and advanced operational strategies. Digitalization, modernization, and asset resilience play a central role in addressing these challenges, offering the potential to enhance efficiency, enable predictive maintenance, and improve system reliability [32]. Stakeholders further highlighted that predictive maintenance, real-time monitoring, and optimized dispatch are not feasible without adequate digitalisation, creating a strong interdependency between modernization and digital transformation. However, modernization efforts are often constrained by regulatory uncertainty and insufficient financial incentives. Additional barriers identified by participants include lengthy consultation and permitting procedures, regulatory gaps affecting smaller hydropower installations, and cultural resistance to change linked to legacy technologies and established operational practices. Stakeholders also emphasized the complexity of the institutional landscape, where broad involvement of public authorities, private operators, and emerging actors such as energy communities may create coordination challenges and slow decision-making processes. This creates a gap between the technical potential for upgrading hydropower assets and the institutional conditions required to realize it [34, 35]. To address these challenges, participants emphasized the need for stronger alignment between policy frameworks and funding mechanisms, improved technical skills and human capital for digital deployment, and better communication regarding the role of hydropower in the energy transition.

Social acceptance, energy justice and transboundary harmonisation

Hydropower development is shaped not only by technical and environmental considerations but also by social acceptance and energy justice dynamics [36]. While hydropower is broadly supported as a renewable energy source, local opposition to specific projects often arises due to environmental concerns and perceived inequities in the distribution of costs and benefits. Stakeholder discussions within WG5 further confirmed that local concerns frequently include displacement, loss of cultural heritage, reduced recreational opportunities, and impacts on agricultural activities, with affected communities often perceiving that they bear a disproportionate share of negative impacts while benefits are distributed more broadly. Energy justice perspectives highlight the importance of procedural fairness, equitable distribution, and recognition of affected communities [37]. Participants in WG5 activities also emphasized that social acceptance is strongly influenced not only by material impacts, but also by trust, transparency, governance quality, and public perception. Lack of transparency and perceived unfairness in decision-making processes were identified as major drivers of opposition, while negative narratives in public discourse and social media may further reinforce resistance to hydropower projects. At the same time, investment in hydropower is increasingly influenced by sustainable finance and ESG frameworks. Although instruments such as sustainability standards and climate disclosure mechanisms are gaining prominence, inconsistencies across regulatory and financial systems create uncertainty for investors and limit access to green finance. Stakeholders further highlighted the importance of early recognition of social concerns, stronger communication and engagement strategies, transparent regulatory frameworks, and benefit-sharing mechanisms to improve procedural justice and strengthen public trust. Investment in education and awareness initiatives was also identified as important for improving understanding of hydropower role in the energy transition. These challenges are further intensified in transboundary river basins, where coordination across jurisdictions is required. In this context, transboundary governance and legal harmonization become a critical issue, as differing national priorities and regulatory frameworks may generate conflicts or inefficiencies. While international water law provides guiding principles, its incomplete integration into operational and sustainability frameworks results in persistent governance gaps [38].

Workforce and capacity development

Workforce and capacity development are critical but often overlooked in the context of hydropower transition. The sector has traditionally relied on specialized engineering expertise, but the energy transition is shifting demand toward interdisciplinary skills combining engineering, environmental science, digital systems, policy, and stakeholder engagement, something not yet fully reflected in academic curricula. Addressing this gap requires coordinated action across educational, industrial and public institutions. Early engagement at the secondary level is important, alongside stronger university–industry collaboration through internships, joint programs, and targeted training. Developing interdisciplinary educational pathways is key to preparing professionals for complex hydropower challenges, while improved working conditions, career prospects, and remuneration are essential to attract and retain talent [39].

4.2 Toward Integrated Policy Strategies

The interconnected nature of hydropower systems gives rise to structural trade-offs that cannot be resolved within a single domain. Flexibility provision may conflict with ecological integrity, while efforts to enhance energy security can generate local environmental and social impacts. Similarly, prioritizing short-term efficiency may undermine long-term system resilience. Climate-induced changes in water availability, increasing aridity, competing water uses, and more frequent extreme hydrological events further intensify these trade-offs, placing hydropower within a broader water-energy-food nexus and requiring stronger coordination across energy, water, agricultural, and land-use policies. At the same time, important synergies emerge across domains. Hydropower plays a central role in enabling the integration of variable renewables, while digitalization can simultaneously improve operational performance and environmental monitoring. Advanced digital solutions, including predictive maintenance, real-time monitoring, and optimized dispatch, may also strengthen asset resilience and improve the capacity of aging hydropower infrastructure to respond to increasingly dynamic power system conditions. In addition, sustainability standards have the potential to enhance investor confidence and support access to finance. Despite these synergies, significant policy gaps persist. Market frameworks often fail to adequately value flexibility and long-duration storage. Stakeholder discussions further highlighted that remuneration mechanisms for flexibility services remain fragmented across national contexts, while the costs associated with flexible operation, including operational risks and long-term asset degradation, are often insufficiently reflected in market incentives. Environmental, climate, and energy policies remain fragmented, limiting the effectiveness of integrated solutions. Social acceptance considerations are not systematically embedded in planning processes, and alignment between policy objectives and financial mechanisms remains weak. In addition, lack of transparency, limited stakeholder engagement, and perceived inequities in the distribution of project impacts and benefits may further undermine legitimacy and delay project implementation. These gaps highlight the need for more coherent and coordinated governance approaches. Addressing the identified challenges requires policy approaches that explicitly recognize the interconnected nature of hydropower systems. This entails moving beyond sectoral thinking and adopting integrated strategies that align market design, environmental regulation, technological innovation, and social considerations. Key priorities include reforming electricity market design to properly value flexibility and system services, as well as integrating long-duration storage into energy planning frameworks. There is also a need for improved methodologies to quantify the full costs of flexibility provision, including maintenance impacts, operational risks, and asset lifetime reduction. In parallel, there is a need to develop adaptive environmental performance standards and harmonize carbon accounting methodologies to ensure regulatory consistency and climate credibility. Supporting the modernization and digitalization of existing hydropower assets is equally critical, particularly in consideration of the aging infrastructure and evolving system demands. This should be accompanied by stronger alignment between policy frameworks, funding mechanisms, and technical capacity-building efforts necessary for the deployment and operation of digital solutions. At the same time, embedding social engagement processes and energy justice principles can enhance legitimacy and reduce conflict. Early recognition of social concerns, transparent governance frameworks, benefit-sharing mechanisms, and improved communication and awareness strategies may play a key role in strengthening trust and procedural justice. Finally, a better alignment between sustainable finance frameworks and public policy objectives, together with improved transboundary governance mechanisms, will be essential to enable coordinated and long-term investment in hydropower systems.

5 Conclusion

Hydropower should be understood as a complex socio-technical system shaped by the interaction of technical, environmental, economic, and social dimensions. These interdependencies are not peripheral but central to how decisions are made and how trade-offs are managed. This chapter has shown that the challenges facing hydropower are systemic in nature and cannot be addressed through isolated interventions. Instead, they require integrated policy frameworks capable of aligning multiple objectives and governance levels. The findings further demonstrate that hydropower is increasingly shaped by evolving flexibility requirements, climate-induced pressures on water availability, digital transformation needs, and growing societal expectations regarding transparency, participation, and fairness. These dynamics place hydropower within broader interconnected systems, including the water-energy-food nexus and the wider energy transition. By structuring diverse evidence into interconnected domains, the analysis provides a foundation for the development of coherent strategies that support the role of hydropower in the energy transition while ensuring environmental sustainability and social legitimacy.

Competing Interests The authors have no conflicts of interest to declare that are relevant to the content of this chapter.

References

1. Joskow PL (2011) Comparing the costs of intermittent and dispatchable electricity generating technologies. Am Econ Rev 101(3):238–241. https://doi.org/10.1257/aer.101.3.238, ISSN: 0002-8282
2. Hirth L (2013) The market value of variable renewables. Energy Econ 38:218–236. https://doi.org/10.1016/j.eneco.2013.02.004, ISSN: 01409883
3. International Energy Agency (IEA) (2020) World Energy Outlook 2020
4. All NEMO Committee (2025) EUPHEMIA Public Description Single Price Coupling Algorithm
5. Milligan M, Ela E, Hodge B-M, Kirby B, Lew D, Clark C, DeCesaro J, Lynn K (2011) Integration of variable generation, cost-causation, and integration costs. Electr J 24(9):51–63. https://doi.org/10.1016/j.tej.2011.10.011
6. ENTSO-E (2022) ENTSO-E Market report 2021
7. European Commission (2020) Powering a climate-neutral economy: an EU strategy for energy system integration. Brussels
8. Bragolusi P, Righettini MS (2022) Dimensions of social acceptance in energy transition. Springer, Cham, pp 1780–1789. https://doi.org/10.1007/978-3-031-06825-6_172
9. Wolsink M (2007) Wind power implementation: the nature of public attitudes: equity and fairness instead of 'backyard motives'. Renew Sustain Energy Rev 11(6):1188–1207. https://doi.org/10.1016/j.rser.2005.10.005, ISSN: 13640321
10. Schweizer-Ries P (2008) Energy sustainable communities: environmental psychological investigations. Energy Policy 36(11):4126–4135. https://doi.org/10.1016/j.enpol.2008.06.021, ISSN: 03014215
11. Wüstenhagen R, Wolsink M, Bürer MJ (2007) Social acceptance of renewable energy innovation: an introduction to the concept. Energy Policy 35(5):2683–2691. https://doi.org/10.1016/j.enpol.2006.12.001, ISSN: 03014215
12. Aitken M (2010) Why we still don't understand the social aspects of wind power: a critique of key assumptions within the literature. Energy Policy 38(4):1834–1841. https://doi.org/10.1016/j.enpol.2009.11.060, ISSN: 03014215
13. Devine-Wright P (2009) Rethinking NIMBYism: the role of place attachment and place identity in explaining place-protective action. J Community Appl Soc Psychol 19(6):426–441. https://doi.org/10.1002/casp.1004, ISSN: 1052-9284
14. Gross C (2007) Community perspectives of wind energy in Australia: the application of a justice and community fairness framework to increase social acceptance. Energy Policy 35(5):2727–2736. https://doi.org/10.1016/j.enpol.2006.12.013, ISSN: 03014215
15. Batel S, Devine-Wright P (2015) A critical and empirical analysis of the national-local 'gap' in public responses to large-scale energy infrastructures. J Environ Plan Manag 58(6):1076–1095. https://doi.org/10.1080/09640568.2014.914020, ISSN: 0964-0568
16. Tabi A, Wüstenhagen R (2017) Keep it local and fish-friendly: social acceptance of hydropower projects in Switzerland. Renew Sustain Energy Rev 68:763–773. https://doi.org/10.1016/j.rser.2016.10.006, ISSN: 13640321
17. Jenkins K, Sovacool BK, McCauley D (2018) Humanizing sociotechnical transitions through energy justice: an ethical framework for global transformative change. Energy Policy 117:66–74. https://doi.org/10.1016/j.enpol.2018.02.036, ISSN: 03014215

18. Sovacool BK, Dworkin MH (2015) Energy justice: conceptual insights and practical applications. Appl Energy 142:435–444. https://doi.org/10.1016/j.apenergy.2015.01.002, ISSN: 03062619
19. Mayeda AM, Boyd AD (2020) Factors influencing public perceptions of hydropower projects: a systematic literature review. Renew Sustain Energy Rev 121:109713. https://doi.org/10.1016/j.rser.2020.109713, ISSN: 13640321
20. Venus TE, Hinzmann M, Bakken TH, Gerdes H, Godinho FN, Hansen B, Pinheiro A, Sauer J (2020) The public's perception of run-of-the-river hydropower across Europe. Energy Policy 140:111422. https://doi.org/10.1016/j.enpol.2020.111422, ISSN: 03014215
21. Andersen O, Heidenreich S (2022) Assessment of social acceptance of 30 starts/stops in Ume River
22. Walker G, Devine-Wright P, Hunter S, High H, Evans B (2010) Trust and community: exploring the meanings, contexts and dynamics of community renewable energy. Energy Policy 38(6):2655–2663. https://doi.org/10.1016/j.enpol.2009.05.055, ISSN: 03014215
23. Hayes DS, Bruno MC, Alp M, Boavida I, Batalla RJ, Bejarano MD, Noack M, Vanzo M, Casas-Mulet R, Vericat D, Carolli M, Tonolla D, Halleraker JH, Gosselin M-P, Chiogna G, Zolezzi G, Venus TE (2023) 100 key questions to guide hydropeaking research and policy. Renew Sustain Energy Rev 187:113729. https://doi.org/10.1016/j.rser.2023.113729, ISSN: 13640321
24. Junker B, Buchecker M, Müller-Böker U (2007) Objectives of public participation: Which actors should be involved in the decision making for river restorations?. Water Resour Res 43(10). https://doi.org/10.1029/2006WR005584, ISSN: 0043-1397
25. Cortés-Borda D, Polanco J-A, Escobar-Sierra M (2022) Social perception assessment of hydropower sustainability: a stepwise logistic regression modeling. Environ Sci Policy 134:108–118. https://doi.org/10.1016/j.envsci.2022.03.026, ISSN: 14629011
26. Segreto M, Principe L, Desormeaux A, Torre M, Tomassetti L, Tratzi P, Paolini V, Petracchini F (2020) Trends in social acceptance of renewable energy across Europe–a literature review.Int J Environ Res Public Health 17(24):9161. https://doi.org/10.3390/ijerph17249161, ISSN: 1660-4601
27. Vagnoni E, Gezer D, Anagnostopoulos I, Cavazzini G, Doujak E, Hočevar M, Rudolf P (2024) The new role of sustainable hydropower in flexible energy systems and its technical evolution through innovation and digitalization. Renew Energy 230:120832. https://doi.org/10.1016/j.renene.2024.120832, ISSN: 09601481
28. den Dekker M (2025) PEN@Hydropower STSM WG1: Assessing the impact of flexibility on machine lifetime. https://static1.squarespace.com/static/635d45f550e75155f79c7d18/t/690640dd924cb353be156706/1762017501230/MatthijsDenDekker+STSM+PEN%40hydropower+Report.pdf
29. Čerpinska M, Nurković R, Filannino D, Kuriqi A, Finger DC, Kasiulis E, Doujak E (2025) The role of hydropower on energy market: a comparative study in four EU countries. IOP Conference Series: Earth and Environmental Science, vol 1442, no 1, pp 012001. https://doi.org/10.1088/1755-1315/1442/1/012001, ISSN: 1755-1307
30. Ostraszewski M (2024) PEN@Hydropower STSM WG1: storage needs of future energy system and the role of hydropower. https://static1.squarespace.com/static/635d45f550e75155f79c7d18/t/687a1559f32e90527e4916b2/1752831328336/07_STSM+report+MO+PEN%40hydropower_WG1.pdf
31. Rajić M, Quaranta E, Kuriqi A, Kasiulis E (2025) Renewable energy in the Western Balkans and EU policy: pathway to regional net-zero. IOP Conference Series: Earth and Environmental Science, vol 1442, no. 1, pp 012002. https://doi.org/10.1088/1755-1315/1442/1/012002, ISSN: 1755-1307
32. Seth M, Aparicio MU, Santos CD, Nakomcic-Smargdakis B, Brboric M, Calamita E, Dasic T (2025) Greenhouse gas emissions from hydropower: challenges and opportunities review. https://doi.org/10.5194/egusphere-egu25-6715
33. Lazarevikj M (2024) PEN@Hydropower STSM WG4: the role of hydropower in the EU energy transition. https://static1.squarespace.com/static/635d45f550e75155f79c7d18/t/66ec16acbe49f11ad09256b8/1726748335800/WG4+Updated+STSM+4-1+report+PEN%40hydropower.pdf

34. Akif Bütüner M (2024) PEN@Hydropower STSM WG2: defining digitalization barriers for HPP refurbishment. https://static1.squarespace.com/static/635d45f550e75155f79c7d18/t/687ca0b79e22143862e79514/1752998074572/09_STSM+report+PEN%40hydropower_MAB_WG2_v3.pdf
35. Shanbhag V (2023) PEN@Hydropower STSM WG2: review of trending technologies, and identification of the needs of the hydropower sector from the point of view of digitalization, flexibility, and efficiency. https://static1.squarespace.com/static/635d45f550e75155f79c7d18/t/66ec17d189b8f30606d5cb4f/1726748626522/WG2+STSM_Report_2023_Shanbhag_Vignesh.pdf
36. PEN@Hydropower WG1 (2023) Roundtable event: the role of hydropower in power system flexibility. https://www.pen-hydropower.eu/medianewsletters/v/7tph897wlecpf9ltwxfeyfmgeawal7
37. Telli A, Kırısık A, Quaranta E, Kuriqi A, Kasiulis E, Muntean S (2025) Hydropower's role in enhancing energy justice: preliminary insights from the EU and Turkey. IOP Conference Series: Earth and Environmental Science, vol 1442, no 1, pp 012009. https://doi.org/10.1088/1755-1315/1442/1/012009, ISSN: 1755-1307
38. Chukwuebuka E (2024) PEN@ Hydropower STSM WG5: transboundary dimensions of the hydropower sustainability standard. https://static1.squarespace.com/static/635d45f550e75155f79c7d18/t/670bc0658a37847d0bea65ec/1728823398039/STSM+Scientific+Report+2+%281%29.pdf
39. PEN@Hydropower (2024) Challenges and opportunities in hydropower education and workforce development: insights from the PEN@Hydropower conference. https://static1.squarespace.com/static/635d45f550e75155f79c7d18/t/670bc6dbee4ab439ee215647/1728825051795/Education+and+skills+in+hydropower+-+event+report+PEN%40hydropower+2024.pdf

Part IV
Visions Shaping the Future of Sustainable Hydropower

When discussing sustainable hydropower and future visions for 2030 to 2050, it is important to consider ongoing research projects in this field. This means incorporating today's research topics and their resulting findings into the future research questions in this area. Currently, the focus of ongoing research projects is closely correlated with the need for flexible energy systems and storage capacities. This situation will not change quickly in the short term, given the additional installations of renewable energy sources over the coming decades. To make Europe competitive again, energy must be provided cost-effectively. Moreover, this is only possible, in light of insufficient resources and the set climate goals, through further expansion of renewable energy sources. This energy transition, which began years ago, must reach a meaningful conclusion; otherwise, all previous efforts will have been in vain.

Today's world is more interconnected than ever. New actors have emerged and are contributing to the sustainable development of hydropower. This was already demonstrated in this book during the discussion of the Hydropower Sustainability Standard. In the future, an increasing number of multidisciplinary research projects will be observed, addressing not only technical questions but also environmental, climate change, and biodiversity issues. These issues are closely linked to societal acceptance, as they affect the daily lives of all people. The research community will need to incorporate these concerns into future research topics and projects.

The following chapters of this book summarize some aspects of questions related to stakeholders and the integration of sustainable thinking into the education of young researchers and innovators. However, digitalization will also make a significant contribution. Electronic networking and the use of Artificial Intelligence will revive things in a different form.

The visions presented for the future of the next two decades are based on surveys, special workshops, or the authors' experiences. They are neither comprehensive nor definitive, as the future of hydropower is so complex that, at best, we can only anticipate or speculate about a few aspects.

Stakeholder Perspectives

Iraia Núñez Aguirre

Sustainable hydropower is not only engineered: it is negotiated across interconnected systems, diverse stakeholders, and multiple governance scales.

Abstract This chapter conceptualizes hydropower as a socio-technical system influenced by infrastructure, governance arrangements, environmental constraints, and stakeholder values. It argues that the transition toward sustainable hydropower is not solely a technological challenge, but is fundamentally dependent on coordination, institutional capacity, and societal alignment. The chapter positions stakeholders including policymakers, industry actors, academics, civil society members, and early career professionals, as active agents who influence regulatory frameworks, investment decisions, and knowledge production. This chapter identifies stakeholder engagement as a central enabler of sustainability, linking technical knowledge with policy design and implementation. It emphasizes the need to transition from reactive to anticipatory governance by integrating energy, water, and environmental objectives, while incorporating stakeholder perspectives into decision-making processes from the beginning. In line with international good practice, effective engagement is iterative and inclusive. It is also based on transparent, two-way communication that enhances legitimacy and reduces project risk. Finally, the chapter emphasizes capacity building as a critical condition for sectoral transformation, addressing skills gaps and fostering interdisciplinary competencies. The chapter concludes that sustainable hydropower depends on continuous coordination among stakeholders, supported by shared knowledge, mutual understanding, and coherent governance frameworks.

Keywords Stakeholder · Multidisciplinary · Capacity building · Hydropower

I. Núñez Aguirre (✉)
Zabala Innovation Consulting, Brussels, Belgium
e-mail: inunez@zabala.eu

E. Doujak et al. (eds.), *Rethinking Hydropower: A Holistic Vision for a Sustainable Future*, Springer Tracts in Civil Engineering,
https://doi.org/10.1007/978-3-032-33328-5_12

1 Hydropower as a Socio-Technical System

The transition to a sustainable and climate neutral energy system is not just a technological challenge. It is also a matter of governance, coordination, and societal alignment. Hydropower, one of the most well-established renewable energy technologies, exemplifies this complexity. It operates at the intersection of energy systems, water resource management, environmental protection, and socio-economic development.

Although traditionally considered an engineering discipline, hydropower is increasingly recognized as a socio-technical system, in which infrastructure, institutions, ecosystems, and stakeholder values are tightly interdependent. This perspective aligns with socio-technical transition theory, in which technological systems co-evolve with governance structures and actor networks [1]. In such contexts, technical optimization alone is insufficient without public acceptance, regulatory coherence, and environmental sustainability.

Stakeholders do not merely respond to hydropower development; they actively shape its trajectory. Their decisions influence regulatory frameworks, investment strategies, knowledge production, and societal acceptance. These roles are interconnected and operate across multiple governance levels, affecting both project level outcomes and broader system transformation. This chapter explores how different stakeholder groups interact with, influence, and benefit from the sustainable evolution of hydropower. It positions stakeholder engagement as a core enabler of sustainability and system resilience, not merely a procedural requirement.

2 Stakeholder Value: Translating Knowledge into Action

This volume synthesizes technical knowledge, policy frameworks, and implementation realities, serving as a translation interface between them. While the value differs across stakeholder groups, these benefits are interdependent.

For example policymakers and public authorities benefit from structured, evidence based insights that support integrated and forward-looking policy design. This is particularly relevant in the context of overlapping regulatory frameworks—such as the EU Water Framework Directive, renewable energy policies, and biodiversity strategies,—where coherent management of trade-offs is essential. Industry actors, including developers, operators, and investors, gain a clearer understanding of regulatory trajectories, sustainability standards, and societal expectations. This information supports more robust investment decisions, reduces regulatory uncertainty, and contributes to long-term asset optimization.

Researchers and academic institutions benefit from interdisciplinary synthesis and the identification of knowledge gaps, particularly in areas such as climate resilience, digitalization, and socio-environmental trade-offs. At the same time, civil society organizations gain improved transparency regarding governance processes and decision-making trade-offs, enabling more informed and constructive participation.

Both education providers and early career professionals' benefit from access to applied, interdisciplinary knowledge that reflects real-world challenges. This supports the development of competencies that align with the evolving requirements of the energy transition. By bridging these domains, the volume contributes to a shift from fragmented knowledge production to shared understanding and coordinated action among stakeholder groups.

3 Policymakers: Drivers or Responders?

A recurring question in hydropower governance concerns whether policymakers act primarily as drivers of transformation or as responders to external pressures. In practice, this distinction reflects the interplay between reactive and anticipatory modes of governance.

In the water-energy domain policymaking has often been reactive, with interventions triggered by crises such as energy shortages, extreme hydrological events, or environmental conflicts. This pattern is well documented in the literature, where institutional responses frequently follow emerging risks rather than anticipating them systematically [2, 3]. Theses dynamics result in fragmented policy responses and limited cross-sector integration.

Nevertheless, policymakers remain central structural enablers of the hydropower transition. They define regulatory frameworks, align economic incentives, and coordinate across governance levels. Therefore, their role is not only reactive but also constitutive, shaping the conditions under which hydropower systems evolve.

Strengthening anticipatory governance is essential in this context. This involves integrating long-term planning horizons, incorporating climate scenarios, aligning energy and water policies, and embedding stakeholder participation in the early of decision-making processes. These strategies improve policy coherence and enhance the capacity to manage uncertainty.

Inclusive and forward-looking governance reduces conflict, improves legitimacy, and supports more sustainable outcomes. Hydropower can play a strategic role in climate mitigation and adaptation, not only as a source of renewable energy and as a provider of storage, flexibility, and water management services. Unlocking this potential requires shifting from crisis-driven policymaking to integrated and system-oriented governance.

4 The Book as a Capacity Building Instrument

The transition toward sustainable hydropower is hindered by gaps in education, skills, and institutional capacity. In this context, this volume should be understood as a capacity-building instrument, rather than merely a repository of knowledge.

First, it translates complexity into accessible insights by connecting technical, environmental, and policy domains. This helps overcome the knowledge fragmentation that often limits effective decision making.

Second, it promotes interdisciplinary learning by incorporating diverse stakeholder perspectives and working domains. This systems-level understanding is essential for addressing the interconnected challenges of hydropower development.

Third, it connects theory and practice by provideing applied insights, case studies, and training experiences developed through initiatives such as PEN@Hydropower. This facilitates the transfer of knowledge into operational and policy contexts.

By contributing to knowledge integration and skills development, the book helps reduce the gap between stakeholder needs and the current limitations of education and training systems.

5 Developing Risk and Opportunity Awareness Among Stakeholders

An important dimension of capacity building is the ability of stakeholders to understand not only their own roles, but also the perspectives and constraints of others. Hydropower development involves complex trade-offs between environmental, social, and economic objectives. Stakeholder interactions significantly influence governance outcomes and project risks [4].

Effective stakeholder engagement is therefore essential. International good practice emphasizes that engagement processes should be continuous, inclusive, and based on transparent, two-way communication [5]. Such approaches build trust, improve decision quality, and reduce the likelihood of conflict.

Addressing capacity gaps requires coordinated action across education, industry, and policy domains. Key priorities include:

- developing interdisciplinary educational programs,
- strengthening university–industry collaboration,
- promoting lifelong learning and reskilling,
- enhancing institutional capacity within public authorities,
- and engaging early-career professionals.

The importance of interdisciplinary skills and technical capacity in enabling hydropower innovation and system integration is well established [6]. At the same time, knowledge transfer mechanisms are critical to address workforce aging and preserve operational expertise [7, 8]. Institutional capacity further shapes governance effectiveness and implementation outcomes [9]. Capacity building should therefore be understood not as a supporting element, but as a central condition for sustainable hydropower development.

6 Key Messages for Stakeholders

The analysis yields several stakeholder-specific implications:

For policymakers, there is a need to move from reactive to anticipatory governance. This requires long-term, integrated planning frameworks that align energy, water, and environmental policies. Strengthening institutional capacity and embedding early stakeholder engagement are essential to improving legitimacy and reducing conflict.

For industry actors, sustainability and stakeholder engagement must be integrated into core business strategies. This includes investing in digitalization and modernization, strengthening environmental performance, and engaging proactively with regulators and communities.

For academia and training institutions, the priority lies in developing interdisciplinary and practice-oriented education. Collaboration with industry and public institutions is essential to ensure relevance, alongside the expansion of lifelong learning and professional training.

For civil society organizations, meaningful participation in decision making processes is crucial. Moving toward solution-oriented engagement requires access to information, technical understanding, and recognition of trade-offs.

For early-career professionals, interdisciplinary competencies, including systems thinking, stakeholder engagement, and policy awareness, are increasingly critical. Participation in collaborative learning environments can accelerate knowledge acquisition and professional adaptability.

Across all stakeholder groups, a common principle emerges: sustainable hydropower cannot be achieved through isolated actions, but requires continuous coordination across technical, environmental, and social dimensions. This, in turn, depends on sustained collaboration supported by shared knowledge, mutual understanding, and coherent governance frameworks [10].

Competing Interests The author has no conflicts of interest to declare that are relevant to the content of this chapter.

References

1. Geels FW (2002) Technological transitions as evolutionary reconfiguration processes. Res Policy 31(8–9):1257–1274. https://www.sciencedirect.com/science/article/abs/pii/S0048733302000628
2. OECD (2015) OECD principles on water governance. OECD Publishing. https://www.oecd.org/en/topics/sub-issues/water-governance/the-oecd-principles-onwater-governance-and-implementation-strategy.html
3. Pahl-Wostl C (2017) Water governance in the face of global change. Water Resour Manag 31:2567–2579. https://link.springer.com/book/10.1007/9783-319-21855-7
4. Wang X, et al (2022) Stakeholder behavior risk evaluation of hydropower projects based on social network analysis. Buildings 12(10):1662. https://www.mdpi.com/20755309/12/12/2064

5. Hydropower Sustainability Alliance (2024) How to guide on hydropower communications and consultation. https://www.hs-alliance.org/s/How-to-Guide-on-Communications-and-Consultation-Web-Low-Double-spread.pdf
6. Kougias I, Aggidis G, Avellan F, et al (2019) Analysis of emerging technologies in the hydropower sector. Renew Sustain Energy Rev 113:109257. https://www.sciencedirect.com/science/article/pii/S1364032119304575
7. Zackarias AJ, Bond-Barnard TJ, van Waveren CC (2022) Improving knowledge transfer processes in engineering projects. S Afr J Ind Eng 33(4):147–164. https://sajie.journals.ac.za/pub/article/view/2672
8. PEN@Hydropower (2024) Education and Skills in Hydropower – Event Report. COST Action CA21104–Pan-European Network for Sustainable Hydropower. https://www.pen-hydropower.eu/s/Education-and-skills-in-hydropower-eventreport-PENhydropower-2024.pdf
9. Mills-Novoa M, Borgias S (2025) The new era of contested hydropower development. World Dev 196:107162. https://www.sciencedirect.com/science/article/pii/S0305750X25002487?via
10. PEN@Hydropower (2026) Policy pathways for sustainable hydropower in 2030–2050 scenarios. Workshop during MC meeting in Naples 2026. COST action CA21104–Pan-European network for sustainable hydropower

Hydropower Education and Professional Perspectives

Marko Hočevar and Eduard Doujak

We educate for a better future.

Abstract We discuss strengthening hydropower education by integrating sustainability concepts into existing university courses, based on experiences from training schools on sustainable hydropower organised within the framework of the Hydropower Sustainability Standard. The main idea is not to teach the standard as a certification tool, but to use its performance requirements as a practical framework to connect technical, environmental, social, and economic aspects of hydropower operation within current teaching materials. Hydropower Sustainability Standard performance indicators can be incorporated into courses across different disciplines, including mechanical, civil, environmental, energy, and social sciences. Learning outside the classroom deserves special attention, particularly through direct interaction with hydropower plant operators and field visits. These experiences provide practical professional knowledge that is not available in textbooks and help students understand real operating conditions, maintenance challenges, environmental constraints, and system behaviour. In such model, education and professional perspectives can be well connected and complemented. The chapter also discusses how principles of the Hydropower Sustainability Standard can be applied to other renewable energy technologies, such as wind, solar, and hydrogen, where structured sustainability frameworks are often less developed.

Keywords Hydropower · Education · Hydropower sustainability standard · Assessment · Sustainable energy sources · Capacity building

---

M. Hočevar (✉)
Faculty of Mechanical Engineering, University of Ljubljana, Ljubljana, Slovenia
e-mail: marko.hocevar@fs.uni-lj.si

E. Doujak
Institute for Energy Systems and Thermodynamics, TU Wien, Vienna, Austria
e-mail: eduard.doujak@tuwien.ac.at

E. Doujak et al. (eds.), *Rethinking Hydropower: A Holistic Vision for a Sustainable Future*, Springer Tracts in Civil Engineering,
https://doi.org/10.1007/978-3-032-33328-5_13

1 Introduction

In a report [1] PEN@Hydropower highlighted a growing mismatch between the strategic importance of hydropower in the European green transition and its declining visibility and depth od study within higher education curricula. Training schools can play a key role in addressing this issue by providing an interdisciplinary educational addition to university courses. They can also respond more quickly to the needs of the hydropower sector, particularly with regard to the social acceptance and sustainability of hydropower.

One of the key contributions of training schools lies in reintroducing depth and specialization into hydropower education. As noted in the PEN@Hydropower report [1], university curricula increasingly favour broad overviews of the energy engineering topics rather than hydropower specifically (see also Chapter "Status Quo of European Hydropower Education"). This always comes at the expense of detailed treatment; even courses at mechanical engineering faculties rarely include topics such as hydraulic machinery, detailed turbine or runner blade design, computational fluid dynamics CFD, or experimental methods such as sediment erosion laboratory measurements, and prototype testing. However, topics on sustainable energy may be taught at other universities also, including non technical ones. Training schools can address this lack of detailed treatment by offering focused modules on sustainability related topics. In-depth study is particularly important in hydropower (relative to other engineering disciplines), where hydraulic turbines are prototypes rather than mass-produced machines, and where the underlying knowledge, and skills of engineers are essential.

Of course, a training school focusing on sustainable hydropower is not a direct replacement for traditional in-depth courses on turbine machinery design. The training school provides a framework for integrating sustainability into technical education. Sustainable hydropower encompasses not only energy production, but also sediment management, river ecology, grid flexibility, climate resilience, and long term operation. By combining engineering fundamentals with environmental and systems level perspectives, the training school can help students to understand hydropower as part of a wider water-energy nexus, rather than as purely a civil, mechanical or electrical engineering technology. This directly addresses the perception problem identified in the report [1], whereby hydropower is often considered to be a mature technology with environmental controversies, and limited innovation potential. Including materials on sustainable hydropower could attract students and encourage them to attend more in depth technical and professional courses later on or take them as optional courses.

An especially important aspect of training schools is their potential to support teachers. The report [1] highlights not only a lack of dedicated hydropower courses, but also a shortage of up-to-date textbooks that meaningfully incorporate aspects such as sustainability. The report [1] also mentions uncertainty about how to teach sustainability and digitalization within mechanical and civil engineering programmes.

Providing teacher with high quality and readily available teaching materials such as lecture slides, case study descriptions, datasets, and laboratory exercise materials, can significantly improve the impact of training schools. These materials enable university teachers to improve the content of their hydropower courses without having to wait for slow curriculum reform cycles. They also help to ensure that, hydropower is represented on a balanced way in the curriculum.

In the past, curricula on energy engineering including hydropower have undergone substantial changes. The shift towards wind and solar technologies has served as a testing ground for new educational approaches, including those related to digitalization and artificial intelligence. As discussed in the report [1], AI-driven tools, digital twins, and interactive simulations could transform the teaching of core concepts such as the Euler equation, unsteady flow, and turbine optimization in the future.

However, training school on sustainable hydropower are different in most aspects. They emphasize the importance of methods such as people interaction, discussions, critical thinking and the inclusion of social and environmental aspects. Including them properly in study materials provides an opportunity to train both students and teachers in their use and to translate successful practices back into formal education. Including theses methods in study materials can help improve the future position of hydropower in energy engineering and society.

2 Performance Indicators for Sustainable Hydropower

The Training School on Sustainable Hydropower is organized around the Hydropower Sustainability Standard [2]. The results of the training school could be incorporated into university study materials relevant to a range of subjects, including social, environmental, economic and engineering studies. Ideally the goal would not be to teach the Hydropower Sustainability Standard as a formal compliance or certification tool, but rather to use its performance requirements as a framework through which to explain hydropower sustainability in relation to the course materials that teachers already use for lectures, laboratory exercises, and individual professional projects. In this way, sustainability would be taught as a natural part of hydropower design, operation, and decision-making. Below, we will select individual performance requirements from the Hydropower Sustainability Standard to show how these can be incorporated into relevant study materials.

The requirements of the Hydropower Sustainability Standard relating to *Environmental and social assessment and management (4.1)* could be included in courses focusing on early project planning and system design. This mainly applies to faculties of mechanical engineering, civil engineering, and environmental engineering. Teaching materials could demonstrate the influence of baseline studies, impact assessments, and mitigation measure planning on technical choices such as hydropower plant layout, turbine type, and construction methods. Instead of presenting environmental assessments as separate reports, teachers should ensure that students learn how weak or incomplete assessments can create technical risks at a later stage in the project.

The requirements related to *Labour and working conditions (4.2)* could be incorporated into social science faculty courses. This could cover broader topics such as the drain of young talent from less populated regions and, gender balance. The topics are also relevant to courses related to manufacturing technologies, project execution, construction and operation, which are typically taught at faculties of mechanical and civil engineering. Examples would show how engineering decisions affect site safety and working conditions. This helps students understand that worker safety and long-term plant availability depend partly on design choices made by engineers.

The section on *Water quality and sediments (4.3)* offers links to technical courses. These topics are particularly relevant to programmes in ecology, mechanical engineering and hydraulic engineering. The study materials explain how sediment size and concentration affect turbine geometry, material selection, operational strategies, and efficiency losses. Currently, relatively large volumes of sediment are still prevented from entering turbines by dams. In the future, due to reservoirs filling with sediments and extreme weather events, it is likely that sediments will flow through turbines and cause damage. While the Training School on Sustainable Hydropower cannot provide public available datasets, monitoring results or CFD outputs, these could be used in more general student assignments to analyse trade-offs between sediment management measures and mechanical wear.

The requirements related to *Community impacts and infrastructure safety (4.4)* could be linked to dam safety, hydraulic transients, and operational flexibility. This content is relevant to the faculties of civil engineering, mechanical engineering, and electrical engineering. Teaching examples could discuss how start-stop operations, extreme events, and failure scenarios can pose risks to downstream communities, thereby transforming the concept of safety into a tangible engineering problem or student project. Additionally, the training school could facilitate discussions on relationship between community impacts and the environment in relation to hydropower, which is a relevant topic for social faculties members and courses.

Where relevant, the *Resettlement (4.5)*, *Indigenous peoples (4.7)*, and *Cultural heritage (4.8)* sections of the Hydropower Sustainability Standard performance requirements could be addressed through examples of spatial planning and layout. These topics would be a good fit for courses taught in the fields of social science, civil engineering, environmental engineering, and interdisciplinary sustainability programmes. Students would learn how alternative plant layouts, access roads, or powerhouse locations could reduce social impacts while also affecting costs and technical feasibility. Social science programmes could explore the influence of hydropower on the three aforementioned categories: resettlement, indigenous people and cultural heritage. The latter is particularly important in the majority of European hydropower projects.

The performance requirements relating to *Biodiversity and invasive species (4.6)* could be incorporated into hydraulics and turbine design courses offered by civil engineering and environmental engineering faculties. Teaching materials could demonstrate how intake velocities, fish passage systems, and flow distribution influence ecological outcomes. Students would learn how to translate ecological requirements into hydraulic boundary conditions and design parameters.

Performance indicator *Governance and procurement (4.9)* could be embedded in project management and engineering economics courses, which are often taught in economics, mechanical engineering, electrical engineering, and industrial engineering programmes. Examples would demonstrate the influence on design maturity, risk sharing, and innovation. This could be illustrated using case studies from training schools where technical performance is either limited or supported by opposing requirements.

The requirements related to *Communication and consultation (4.10)* could be introduced through interdisciplinary coursework. These are relevant to all engineering faculties and sustainability-oriented programmes. Students should be aware they will need to explain technical designs and professional decisions to non-technical stakeholders in order to improve the clarity of discussions and justify engineering requirements and assumptions.

Hydrological resources (4.11) requirements fit well into hydraulics and hydrology courses taught at faculties of civil engineering, environmental engineering, and energy engineering. Real inflow data and examples of extreme climate cases from the training schools could be used to show how hydrological assumptions influence turbine sizing, water storage, operational flexibility, and long-term sustainability.

Finally, the *Climate change mitigation and resilience (4.12)* section could be incorporated into various subjects and faculties. Teaching materials would show how design margins, flexible operation, and retrofit options can help hydropower plants to adapt to changing hydrological and market conditions. Students would learn that resilience is not an additional requirement, but rather a design approach affecting efficiency, robustness, and lifecycle performance.

3 Learning Outside Classroom: Individual Discussions with Hydropower Plant Operators

Individual discussions between training school participants and hydropower plant professional operators are one of the most valuable results of the Training Schools on Sustainable Hydropower. These discussions help participants sense and understand how hydropower plant operators work in real life through verbal and nonverbal communication. During these interviews, the operators share their professional expertize, including their daily operational routines, the limits of hydropower equipment, environmental requirements related to flow and head and organizational challenges. This practical knowledge is not available in textbooks or internet forums, but it is highly relevant for university teaching.

Discussions with plant operators often focus on turbine performance, wear, erosion, maintenance and refurbishment for training school participants with a mechanical engineering background. Operators explain why certain design solutions were chosen and how machines behave after many years of operation, and how equipment ages in the face of increasing demands for flexibility. Questions from training school participants focus on acquiring professional information that they have not already learned in lectures, and are thus oriented towards details of maintenance work. For instance, they ask how often hydraulic oil is replaced, how operators decide when to perform selected maintenance work and how long it takes to change a bearing segment. This knowledge is specific to the operator and the plant, so it is difficult to transfer it into course material as it is very professional and less scientific or educational. However, teachers can add small remarks about sustainability to courses on turbomachinery, hydraulics, and machine design, thus making the courses more interesting and appealing. They can use such real-life examples to explain the limits of professional machinery design and maintenance problems.

For training school participants with a background in *civil engineering*, discussions with operators are important for understanding dam safety, hydraulic structures, sediment transport, and aging critical infrastructure. Operators can again provide professional insights into structures respond to floods, sediment loads, and long-term use. Questions from training school participants were focused on how operators maneuver spillway gates, the effect of flexibility requirements on flow downstream of the power plant, how hydropeaking events are handled, whether maneuvering spillway gates allows for sediments transport, how sediment erosion damage is prevented, maintenance intervals and actions taken after high water events. This information can be used in courses on hydrology, water engineering, and hydraulic structures. Case studies from real plants help students understand how design decisions affect safety and long-term operation.

For participants with knowledge of *electrical engineering* and *energy engineering*, discussions with operators provide insight into grid connection, plant control, start-stop operation, and flexibility. Operators can explain how hydropower plants support the electricity system when there are changes to wind and solar power. This knowledge can be incorporated into power systems and energy courses through the analysis of real operating scenarios and control strategies.

In programmes focused on *environmental engineering* and *sustainability* discussions with operators show how environmental measures work in practice. During interviews, operators explain the importance of environmental flows, fish passage operation, sediment flushing, and river water quality monitoring activities. These examples can be used in courses to demonstrate how environmental goals are achieved through technical and operational solutions.

Discussions with operators are also useful for *project management* and *engineering economics* courses, both of which are taught at several faculties. During interviews, operators can talk about costs, risks, maintenance planning, and long-term investment decisions. Teachers can then use these examples to show how technical, financial, and organizational aspects are interconnected throughout the life of a hydropower plant.

These discussions are not expected to result in new standalone courses. Instead, they should be incorporated into existing courses in the form of examples, brief comments, case studies, problem-based homework suitable for individual study, seminar material or invited lectures content, or simply as discussion topics. This approach makes teaching more realistic and application specific without requiring major changes to study programmes.

Other benefits of learning outside the classroom, apart from the direct discussions with plant operators related to sustainable hydropower training school include field visits with well-selected goals, observation of the real operating conditions of the plant, use of real data (if provided by the operator), participants' own assessment of the environmental conditions and their connection with the plant's operation. During *field visits* to hydropower plants, training school participants can get a sense of the size and complexity of the hydraulic power plant installations and equipment. They can link the catchment area size, accumulation size, dam size, water intake diameter, runner diameter, generator size with the variables they encountered during the course, such as the apparent power of the generators. This is something that pictures and videos cannot reveal correctly. *Observing real operating conditions* is another benefit of field visits that must be experienced and cannot be taught solely in the classroom. One example that immediately comes to mind is operating an plant under cavitating conditions and applying the theory of part-load operation under cavitation while standing next to the draft tube of a Francis turbine and experiencing the vibrations, or feeling the massive power of the spillway discharge. When a power plant operator can provide real operating data, students can use it to compare with theoretical calculations and independently verify the validity of equations such as Euler turbine equation or the continuity equation. This strengthens their inherent knowledge, gathered during classroom courses.

The next benefit of field visits is participants' *own assessment of environmental conditions*. Clear naturally colored water usually suggests good river quality, whereas very cloudy or muddy water may indicate high sediment levels or pollution. Unusual colors, such as green, brown, or gray, may indicate algae growth, wastewater, or industrial discharge. The presence of foam, oil films, or strong odors is often a sign of contamination, including from the hydraulic power plant under observation. Healthy rivers usually support visible life, such as fish, insects, and plants along the riverbanks. A lack of living organisms can indicate poor environmental conditions. The state of the river banks is also important. Stable banks with a mix of course, medium and fine sediments and vegetation suggest adequate flow and erosion, while heavily eroded or rocky riverbanks show environmental stress. While visual assessment cannot replace measurements, it provides useful initial information about river water quality.

4 Hydropower Sustainability Standard Principles Transfer to Other Courses in Energy Engineering

New hydropower projects are difficult to implement nowadays [3, 4]. It is important to note that, unlike most other renewable energy technologies, hydropower has a well-developed sustainability assessment standard that is used worldwide. The Hydropower Sustainability Standard is a unique example of a structured framework connecting technical design, environmental aspects, and social responsibility. This makes the standard a valuable reference (see Chapter "Training the Next Generation on Sustainable Hydropower in Europe").

In energy engineering education, sustainability is often considered a less important subject and is generally addressed without reference to the technical details of equipment or environmental characteristics. The Hydropower Sustainability Standard demonstrates how sustainability requirements can be directly linked to engineering constraints, designs and choices. This approach can be meaningfully applied to other energy engineering courses, even when the sustainable electric energy sources and technologies in question do not yet have their own sustainability standards, as we will show below.

The environmental impact of wind turbine operation is relevant for wind energy courses. Wind turbines are known to affect birds and bats [5], and offshore wind farms can impact marine ecosystems. Using the hydropower sustainability standard approach, students could analyse these impacts in a structured way, similar to the way in which fish migration or sediment continuity are treated in hydropower. For example, students could study how wind turbine placement, wind turbine rotor speed, noise or foundation design influence environment, instead of discussing these issues only at a policy level not related to actual turbine type and its installation location.

In solar energy courses, the framework can be used to structure discussions on land use and ecosystem impacts [6]. Large solar plants, for example, are reported to disturb habitats, change the local water balance and affect migratory birds' orientation. By applying the approach of the hydropower sustainability standard, students can learn how site selection and layout can reduce environmental impacts. This helps students understand that even relatively "clean" sustainable energy technologies have environmental effects that must be managed and mitigated through engineering decisions.

In hydrogen-related energy courses, particularly those focusing on hydrogen as an energy storage technology linked to renewable electricity, the framework can serve as a model for life-cycle analysis and thinking. Students can examine the impact of hydrogen production on water resources, energy systems and local communities. Using a structured sustainability framework helps to avoid treating hydrogen as a purely technical solution that is divorced from environment.

An important advantage of using the Hydropower Sustainability Standard in education is that it shows students how sustainability can be assessed and discussed in a clear and transparent way. At present, most other renewable technologies do not have a dedicated sustainability standard or any other evaluation methodology as detailed as hydropower. This can lead to unbalanced comparisons, where hydropower is viewed as more problematic simply because its impacts are analyzed, evaluated and documented more thoroughly. Integrating the Hydropower Sustainability Standard into other energy engineering courses can integrate fairer and more consistent comparisons and discussions. It may also help other electricity generation technologies to address existing gaps and enhance their deployment.

Based on the above brief discussion of the Hydropower Sustainability Standard, sustainable electric energy sources and related education, we can conclude that there are no fundamental differences among the aforementioned electric energy sources or technologies, at least in relation to standard evaluation methods. We therefore assume that the framework of the Hydropower Sustainability Standard can be meaningfully applied—conceptually rather than as a direct one-to-one mapping—to all sustainable electric energy sources and incorporated into university courses at several appropriate levels.

5 Benefits for Teachers and Resulting Pedagogical Innovation

Participation in training schools benefits teachers of all disciplines by strengthening their ability to understand, teach and communicate sustainable hydropower in their daily educational work and social activities. Training schools for teachers work in both directions: during events, participants receive new information and contribute to how hydropower is taught and perceived.

Possibly the most important training school benefit for teachers is access to the most up-to-date interdisciplinary knowledge. Hydropower today encompasses a mix of energy engineering, environmental science, social, climate science, policy and digitalization. Training schools introduce teachers to the latest developments in areas such as sediment management, refurbishment of aging infrastructure, the integration of hydropower with variable renewables, the social and environmental issues of technical solutions, and sustainability assessment. This helps teachers maintain scientific and technical relevance and to avoid an outdated or overly narrow perspective.

Hydropower sustainability training schools also support pedagogical innovation. As highlighted in the PEN@Hydropower report [1], traditional teaching methods focused on static equations and steady-state assumptions will become outdated and insufficient in the era of artificial intelligence and digitalization. Interactive examples, CFD visualization and artificial intelligence assisted learning should integrated into teaching in the future. The PEN@Hydropower report [1] also noted that teachers should gain personal experience with interactive teaching approaches to improve understanding and student engagement, particularly for complex topics such as those observed and studied during the Training Schools on Sustainable Hydropower.

Another important contribution of training schools spezializing in hydropower sustainability is their role in boosting teachers' confidence when it comes to teaching sustainability-related topics. Many engineering teachers are uncertain about how to address sustainability without reducing the technical depth of their teaching and struggle to conned the two concepts. Training schools provide well selected examples of how sustainability can be embedded into technical courses, drawing on their in-depth technical knowledge. This enables teachers to link machinery design choices to environmental performance, lifecycle impacts, and societal issues. It also allows them to teach hydropower sustainability as an engineering challenge rather than as an abstract non-technical concept.

Training schools facilitate knowledge exchange and community building among teachers from different countries, disciplines, and career stages. This is particularly valuable because knowledge and expertise are unevenly distributed across Europe. Through discussion and collaboration, teachers can compare curricula, share teaching materials, and identify common gaps. Training school networks often outlive the training school itself, leading to long-term cooperation in teaching, joint courses, invited lectures, PhD and master's Erasmus exchanges, etc. Therefore, we see a great need to extend the training school on sustainable hydropower beyond the duration of the COST Action PEN@Hydropower.

Finally, training schools on sustainable hydropower help teachers to share clear, science-based information about hydropower. Public opinions on hydropower is often divided and un-scientific, and the available information can be confusing or incomplete. Teachers with access to high-quality teaching materials and a balanced understanding can explain hydropower in a realistic and understandable way to students and to the wider public. In this way, training schools on sustainable hydropower support education, long-term public acceptance, and the well-informed development of sustainable hydropower.

6 Benefits for Professionals

Professionals operating hydropower plants can significantly enhance their professional knowledge and daily work practices in several key areas by participating in the PEN@Hydropower Training School on Sustainable Hydropower [7]. Professionals are bound by the daily routine of their work, whether in the power plant or at the hydropower equipment manufacturer. By participating in the training school, they gain a deeper understanding of a holistic sustainability approach that goes beyond purely professional plant operation. They learn how technical decisions such as operating regimes, water level management, and equipment choices are connected to environmental and social impacts, including those related to fish migration, sediment transport, and the condition of ecosystems. This enables them to make more informed and responsible decisions in everyday plant operation.

Furthermore, professionals can strengthen their competencies in impact assessment and mitigation measures. Using structured approaches such as the Hydropower Sustainability Standard framework, make it possible to identify key issues and implement practical solutions such as adapting operational regimes or supporting river restoration activities. This leads to more effective and environmentally responsible operations.

Another important benefit for power plant operators and other professionals is the improvement of their interdisciplinary and communication skills. By interacting with experts from environmental, social, and technical fields, they gain a better understanding of regulatory requirements and public expectations. This enables professionals to balance different interests more effectively and participate in sustainable development and operation of projects.

Their participation would encourage holistic thinking, enabling them to analyze challenges from the perspective of the entire river basin and energy system. This leads to more sustainable and responsible management practices.

7 Conclusions

Specialized courses in hydropower education are currently in short supply, as many study programmes have shifted towards more general topics instead of a detailed technical approach. In this context, the Hydropower Sustainability Standard provides opportunities to enhance teaching and connect it to real-world challenges. It can be used to improve lectures by incorporating examples from field visits, discussions with plant operators, and interaction between participants from various academic backgrounds.

Specific performance indicators form the Hydropower Sustainability Standard can be incorporated systematically into courses across different disciplines, including mechanical engineering, civil engineering, environmental science, energy science, and social sciences. Rather than treating sustainability as an isolated or purely theoretical subject, this approach links it directly to existing technical and analytical teaching content. Examples from day-to-day operation of hydropower plants show that topics such as environmental assessment, labour conditions, sediment management, biodiversity, hydrology, climate resilience, and governance can be incorporated into lectures, laboratory exercises, and student projects. In this way, students can clearly see how the requirements of the Hydropower Sustainability Standard can influence practical engineering decisions, and how interdisciplinary integration can help to bridge the gap between technical and non-technical perspectives. Engineering students learn to consider environmental and social constraints as design parameters, while students from social and environmental backgrounds gain a better understanding of the technical limitations and trade-offs of hydropower systems. Using real-world examples and performance requirements from the Hydropower Sustainability Standard makes teaching more applied and problem-oriented. Consequently, sustainability is no longer presented as a separate subject, but becomes an integral and essential component of engineering thought processes, decision-making, and professional practice in the hydropower sector.

Attending a Training School on Sustainable Hydropower can help professionals operating hydropower plants to improve their work by gaining a deeper understanding of the connections between technical operation, environmental impacts and sustainability, enabling better daily decision-making. They will also develop broader system knowledge, interdisciplinary skills, and a more holistic approach, helping them operate more responsibly and effectively within modern renewable energy systems.

The Hydropower Sustainability Standard provides a comprehensive evaluation of hydropower projects, highlighting strong aspects as well as weaknesses and potential negative impacts. Hydropower is therefore often discussed more critically than some other renewable energy technologies, which may appear more favorable simply because their impacts are analyzed and debated less frequently. In the future, it will be important to find effective ways to integrate the results and experiences from Hydropower Sustainability Standard assessments and training schools into university courses across different disciplines. This should include engineering, as well as environmental, social, and economic perspectives.

We assume that the framework of the Hydropower Sustainability Standard can be meaningfully extended to all sustainable electricity sources in a conceptual sense, and that it can also be included in university curricula at various levels. When applied to other sustainable electricity sources such as wind, solar, and geothermal energy, the Hydropower Sustainability Standard should preferably be used as a general guiding framework, not as a direct one-to-one model. Its main advantage is that it clearly links sustainability with technical design, environmental impacts, and social aspects. This approach can help other energy technologies to evaluate their impacts in a more structured and transparent way. Rather than treating sustainability as separate from engineering, it encourages its direct inclusion in technical decisions, such as system design, technology choice and site selection. In this way, it can help fill existing gaps, support more balanced comparisons between different energy sources, and contribute to a better and more responsible development of sustainable energy technologies.

Competing Interests The authors have no conflicts of interest to declare that are relevant to the content of this chapter.

References

1. Hočevar M, Stopar RR, Patauner D, Caralis G (2025) Education – knowledge and skills for future hydropower. PEN@Hydropower COST Action CA21104 Event Report, May 22
2. Hydropower Sustainability Alliance (2024) Hydropower sustainability standard: for people, nature and the climate. https://www.hs-alliance.org/hs-standard (accessed 27 May 2026)
3. Quaranta E (2024) Hydropower in the European union: state-of-the-art, future opportunities and main issues. In: Sustainable development of renewable energy, pp 281–302. Elsevier
4. Kasiulis E, Cerpinska M, Pummer E, Finger DC, Malicka E, Fischer-Aupperle B, Doujak E, Anagnostopoulos I, Hočevar M, Cavazzini G (2026) Action towards Increase in sustainable hydropower: a pan-european assessment and strategy. Renew Energy
5. Morant J, Naves-Alegre L, Garcia H, Tena E, Sanchez-Navarro S, Nogueras J, Ibanez C, Sebastian-Gonzalez E, Perez-Garcia J (2025) Mapping bird and bat assemblage vulnerability for predicting wind energy impact. J Environ Manage 380:124961
6. Fleming PA (2025) All that glitters – review of solar facility impacts on fauna. Renew Sustain Energy Rev 224:115995
7. Daw J, Stefek J, DeGeorge E, Philip RT, Cardinal A, Parkhill L (2023) U.S. hydropower workforce: challenges and opportunities. National Renewable Energy Laboratory, Golden, CO

Hydropower Sustainability Challenges

Alain Kilajian and Jian-hua Meng

Sustainability is always a challenge.

Abstract This article examines to what extent the Hydropower Sustainability Standard (HSS) serves to promote improved sustainability practices in hydropower planning, implementation and operation, as a continued governance mechanism following the publication of the World Commissions on Dams (WCD) report. Based on an empirical assessment, the article indicates that the HSS has helped the sector progress significantly on operationalizing the WCD report by providing practical guidance on achieving sustainability outcomes. The article also investigates the remaining gaps in the HSS before providing recommendations to further improve water management practices.

Keywords Affordable and clean energy (SDG 7) · Sustainability · Hydropower · Dams · Hydropower sustainability standard (HSS) · WCD · Hydropower sustainability alliance (HSA)

A. Kilajian (✉)
Hydropower Sustainability Alliance, Lisboa, Portugal
e-mail: alain.kilajian@hs-alliance.org

J. Meng
Department of Civil and Environmental Engineering, Konstanz University of Applied Sciences HTWG, Konstanz, Germany
e-mail: jmeng@htwg-konstanz.de

E. Doujak et al. (eds.), *Rethinking Hydropower: A Holistic Vision for a Sustainable Future*, Springer Tracts in Civil Engineering,
https://doi.org/10.1007/978-3-032-33328-5_14

1 Hydropower's Future: Where Are We and Where Will We Go?

1.1 The Hydropower Sustainability Standard: Can It Deliver on the WCD's Vision

The WCD report set the stage for the International Hydropower Association (IHA), which was tasked with promoting sustainability and developing practical tools to assess hydropower projects.

In 2023, more than two decades of multi-stakeholder dialogue and the evolution of sector guidelines culminated in the launch of the Hydropower Sustainability Standard (HSS), which is now overseen independently by the Hydropower Sustainability Alliance (HSA).

The HSS represents by far the most significant initiative to put the WCD's recommendations into measurable criteria, and has been reinforced by wider industry commitments, such as the San José Declaration on Sustainable Hydropower.

The WCD framework helped kick-start global conversations about the downsides of big dams, but it never offered a clear or practical roadmap for how to make hydropower genuinely sustainable. It faced some challenges in its development, including uneven stakeholder representation, broad and often high-level guidance, and a framework that was difficult for many practitioners to apply in real-world settings.

The HSS, by contrast, feels much more like a real governance system. It brings together industry, governments, financers and civil society; it sets out detailed, transparent criteria; and it offers practical tools that developers can actually use.

But while the HSS represents an attempt to create a practical framework, questions remain. How widely is it applied in practice? Can a standardized tool fully capture the complex social and environmental realities of hydropower projects, especially in regions with weak governance? And does it go far enough in addressing issues like early-stage decision-making, transboundary river management, and gender impacts, which are all areas highlighted by critics of large dams for decades?

Nevertheless, compared with the WCD, the HSS clearly provides a stronger, more grounded model of good governance that feels far better suited to the real-world challenges of hydropower today.

1.2 Testing the Standards: What if Past Projects Were Built Today?

To see how the Hydropower Sustainability Standard (HSS) works in practice, we can look back at three controversial projects highlighted in the WCD report and imagine how they would fare under today's criteria.

Fig. 1 Tucuruí Dam, Brazil. Photo by Repórter do Futuro, distributed under a CC Attribution 2.0 Generic license

1.2.1 Tucuruí Dam, Brazil–Greenhouse Gas Emissions

While large dams provide electricity generation and water management benefits, their reservoirs can release significant greenhouse gases. The HSS requires projects to have a high power density as a proxy for their emissions intensity. Projects that generate relatively little power for their reservoir size tend to have higher emissions.

Tucuruí dam (Fig. 1), one of the world's largest dams, falls into this category. For every square meter of water surface it holds, it generates relatively little electricity. Research shows its annual emissions have consistently exceeded what the HSS would consider sustainable.

Put simply, if Tucuruí dam were being planned today, it would almost certainly fail the climate test. See the climate change mitigation and resilience requirements of the HSS.

1.2.2 Sardar Sarovar Project, India–Resettlement

The Sardar Sarovar Project (Fig. 2) on the Narmada River became infamous for the number of people it displaced. The WCD criticized the authorities for underestimating the number of people affected and for failing to account for communities impacted by canals and wildlife projects [1, 2].

Today, the HSS requires early, rigorous social assessments, public disclosure of resettlement agreements, and ongoing monitoring.

Fig. 2 Sardar Sarovar Dam, India. Photo by Vijayakumarblathur, distributed under a CC 4.0 license

By these standards, Sardar Sarovar would not pass: the baseline work simply was not done properly, and the people affected were not fully considered. See the resettlement requirements of the HSS.

1.2.3 Pak Mun Dam, Thailand–Livelihoods

The Pak Mun Dam (Fig. 3) disrupted local fisheries and livelihoods in Thailand, with many families' means of living deteriorating over time.

The HSS now requires projects to ensure that living standards and livelihoods improve, not decline, and that any economic displacement is fairly compensated.

By these criteria, Pak Mun would also fail. Even when compensation was eventually provided, it came too late to prevent long-term harm. See the community impacts requirements of the HSS.

1.3 Where Does That Leave Us?

Looking at these cases, the verdict is clear: all three projects would fall short of today's minimum HSS requirements. That alone shows how far practice has evolved, and how closely the HSS reflects many of the WCD's original concerns.

Fig. 3 Pak Mun Dam, Thailand. Photo by Laurence McGrath, distributed under a CC 3.0 license

However, meeting the Hydropower Sustainability Standard is not the end of the story. The HSS does not label a project "sustainable"; it shows where good practice is met and where improvements are still needed.

And there is still work to do. Key gaps remain: early-stage decision-making is not fully addressed (though the recent launch of the HydroSelect is a crucial step), transboundary river governance needs strengthening, and gender considerations have still not been incorporated fully.

1.4 How the HSS Is Carving a Different Pathway: Lessons from a Certified Project

1.4.1 The Jirau Hydropower Project in Rondônia, Brazil

The 3750 MW Jirau Hydropower Project (Fig. 4), on the Madeira River in Brazil, shows what is possible when modern sustainability frameworks are applied.

Assessed under a precursor to the HSS during its construction phase back in 2012, Jirau performed strongly across nearly all sustainability criteria, from greenhouse gas management to biodiversity and community engagement. Its power density was high, meaning it generated more electricity per square meter of reservoir than many older projects, resulting in a minimal carbon footprint.

Fig. 4 The Jirau hydropower Project in Rondônia, Brazil

But the original assessment did highlight one area for improvement: resettlement. Even while meeting Brazilian law, Jirau needed to enhance monitoring to ensure that affected communities' livelihoods and living standards improved over time.

Importantly, the project treated this as an opportunity for continuous improvement, developing action plans to address gaps openly and transparently.

More than a decade later, Jirau was assessed under the HSS and awarded Gold certification, reflecting how the framework can guide large-scale projects toward better social and environmental outcomes.

1.5 Hydropower Sector Sustainability: Identifying Gaps and Challenges in Practice

The three case studies analyzed in the previous section demonstrate that there is a clear continuity between the WCD and the HSS. Many of the strategic priorities and recommendations of the WCD reports are embedded into the 12 sections of the HSS and their performance requirements. As the three projects highlighted in the WCD report for their poor ESG performance, projects that are instead able to meet the minimum requirements of the HSS can be considered as generally aligned with the outcomes of the WCD and other global sustainability frameworks, such as the UN SDGs.

Yet, there remain notable differences between the HSS and the WCD. First off, while the HSS covers the main project life cycle stages of hydropower development, it does not have a separate module for early-stage decision making. This was a crucial component of the WCD outcomes. Another focus area of the WCD report was transboundary river sharing. The WCD recommendations have a strong focus on transboundary issues and good governance, while these topics are only slightly addressed in the HSS for example in HSS-9 Governance and Procurement and HSS-11 Hydrological Resource. Finally, gender is not explicitly underlined as a unique guideline or section in either the WCD or the HSS, though it is present throughout both frameworks. Organizations, like Women for Water Partnership, highlight the importance of gender impact assessments and thoughtful consideration of gender-based violence related to dams. This is resonated across the main global water commitments to gender equality [3] and something that can be improved on in further iterations of the HSS.

2 Key Recommendations to Increase Alignment Between the HSS and WCD

Based on lessons learned applying the HSS, three recommendations can be made on two levels. First, the current set of criteria and performances of the HSS can be further expanded to include more robust language and expectations on issues such as gender, transboundary management and human rights issues. Similarly, the HSS would also benefit from an additional module for early-stage decision making and guidance on effective siting and master planning, which would further increase its alignment with the WCD recommendations. Second, more training and capacity building is required for the HSS to be recognized, appreciated and socialized among key decision makers and civil society actors. In particular, more capacity building is needed in developing countries to support local institutions in embedding the key sustainability principles of the HSS and creating an enabling policy environment for HS Certifications. To further facilitate and strengthen the impacts of the HSS, more application is needed. Policy makers and governments play a significant role to drive systemic change through increased HSS application. Policy initiatives that can be considered include but are not limited to:

1. incentivizing the use of voluntary standards in renewable energy certificates (RECs) and carbon markets;
2. embedding the HSS into financial mechanisms and providing incentives for certified projects such as shorter licensing periods, lower interest rates loans and preferential Power Purchase Agreements.

This way, industry players are rewarded for their adoption for meeting strict ESG requirements as exemplified by achieving HS Certification. Only when the HSS is applied and tested, can we further refine and improve it by drawing lessons from its implementation through a process of adaptative management.

3 Future and Outlook

Hydropower is, and will continue to be, an integral part of water and energy systems in many countries and regions across the world. As the global community navigates the multiple challenges of climate change, biodiversity loss, and the transition to clean energy, the role of hydropower -and the standards to which it is held—will only grow in importance.

There is broad consensus today that strong sustainability performance is not optional, but an essential element of good planning, implementation, and operation of hydropower assets. The question is no longer whether sustainability matters, but how it is best pursued, continuously improved, and reliably demonstrated. This shift in perspective marks real progress and it sets the direction of travel for the decades ahead.

3.1 A Journey, Not a Destination

Sustainability in hydropower is, at its core, a journey. It is an evolving and expanding field of knowledge and practice, shaped by scientific advances, changing environmental conditions, and shifting societal expectations. What constitutes good performance today will be refined and elevated tomorrow. As hydropower assets operate over many decades and generations, the frameworks and tools governing them must be designed to allow for adaptation.

For the HSS this means sustaining a culture of integrity, rigor, independence, and openness. It means ensuring that active stakeholder participation remains meaningful and inclusive, continuously bringing in the perspectives, values, and concerns of those affected. Therefore, periodic reviews of criteria, levels of expectation and emerging good and best practices is essential.

3.2 The Vision: Continuous Improvement as Good Practice

The long-term vision is clear: sustainability assessment and continuous improvement should become universally accepted, self-evident practice–as fundamental to hydropower development as for example structural safety is to any infrastructure project. Not a luxury. Not a burden. Simply the way how responsible good practices are done.

Several frontiers will shape the next chapters of this journey. Strengthening system scale and early-stage planning—for instance through multi-criteria and system-scale tools such as HydroSELECT—will help determine where hydropower can genuinely contribute to societal well-being, and equally, where aging dams may create more risk and harm than benefit, making the case for decommissioning and removal.

Making sustainability assessment and management routine across entire portfolios of owners, countries, and river basins will move the field from project-by-project evaluation toward coherent, basin-wide stewardship.

Beyond hydropower itself, the majority of dams and reservoirs serve many other societal purposes: water supply, agriculture, navigation, and flood control among them. The sustainability frameworks and practical experience developed in the hydropower sector may offer useful reference points for these adjacent fields, as they develop and strengthen their own approaches to sustainable infrastructure management. There is potential for meaningful cross-sector exchange, and hydropower has a role to play in that broader conversation–as one contributor among many.

3.3 The Journey Continues

The progress made in hydropower sustainability over the past two decades is real and meaningful. However, this process will not continue on its own. What comes next will depend on the sustained commitment of owners, operators, developers, regulators, financiers, civil society, and the communities living with these projects—and on the willingness to keep questioning, keep improving, and keep updating practice as knowledge, conditions, and expectations evolve.

Competing Interests The author has no conflicts of interest to declare that are relevant to the content of this chapter.

References

1. Brody H (2006) WCD thematic review I.1: contributing paper: assessing the project-social impact and large dams, Working papers id:539, eSocial sciences
2. World Commission on Dams (2000) Dams and development: a new framework for decision-making. Earthscan, London, Sterling, VA
3. Taking stock of progress towards gender equality in the water domain: where do we stand 25 years after the Beijing declaration? Where do we stand 25 years after the Beijing declaration? United nations educational, scientific and cultural organization7, Place de Fontenoy, 75352 Paris 07 SP, France

Perspectives of Hydropower from 2030 to 2050

Eduard Doujak and Behnam Balouchi

The future is digital.

Abstract Hydropower is shifting from a pure baseload provider to a flexible, digitally integrated system component of the power grid. Driven by the energy transition, operators are accelerating the expansion of pumped storage, hybrid solutions, and virtual power plants. However, frequent grid-supporting interventions are accelerating wear. In the short term, what is needed is research into part-load operation. In the long term, multidisciplinary tools will be needed to leverage Earth observation data for runoff and sediment forecasts. Systemically, Europe is prioritizing strategic autonomy, grid resilience, and a socially fair transformation. AI is indispensable for this task: it automates market trading and co-optimizes hybrid battery-hydropower systems. Deep learning models also identify complex hydrological patterns that often escape human analysis. Although 58% of professionals already use AI, an industry survey revealed that a lack of data and transparency is slowing down its adoption. Statistical AI models often lack a fundamental understanding of the physical laws of water. As a result, a healthy mistrust of pure AI persists when it comes to decisions with serious operational consequences. Future-proof architecture, therefore, relies on physics-based digital twins that combine data and natural laws. This hybrid approach ensures that advanced algorithms always operate under essential human oversight.

Keywords Multidisciplinary · Flexible · Multipurpose

E. Doujak (✉)
Institute for Energy Systems and Thermodynamics, TU Wien, Vienna, Austria
e-mail: eduard.doujak@tuwien.ac.at

B. Balouchi
Hydropower group, AFRY Norway AS, Oslo, Norway

E. Doujak et al. (eds.), *Rethinking Hydropower: A Holistic Vision for a Sustainable Future*, Springer Tracts in Civil Engineering,
https://doi.org/10.1007/978-3-032-33328-5_15

1 Future Perspectives for Hydropower

Many factors are shaping the future of hydropower. The kind of isolated analysis and optimization that has taken place so far will no longer be feasible, whether in the near or distant future. The issues at hand will increase significantly in complexity, and additional areas will be involved. We cannot concern ourselves merely with the hydropower plant, we need to adopt a holistic view of the entire energy system in which hydropower will play an essential but no longer dominant role.

Many of these points have already been addressed in earlier chapters of this book, and current trends have been highlighted. These will not be repeated here, but they will be supplemented with additional observations and insights. The following chapters describe various perspectives for hydropower, from different points of view. Both technological and systemic aspects will be taken into account.

1.1 Hydropower Technology Development

Hydropower in Europe is undergoing significant structural change. This shift was initiated by the CET Act and the commitment to renewable energy. The need to reduce greenhouse gas emissions and adopt climate goals are driving a major shift in the operating modes of hydropower plants. While these plants have previously been used for base and peak load production, they are now increasingly utilized for grid support. Consequently, transient load changes and an increase in the number of starts and stops of generating units significantly contribute to equipment degradation. To better address these new challenges, increased research is needed in the following areas:

1. **Expansion and Modernization of Pumped-Storage Power Plants (PSH) Significant Capacity Expansion**:
   There is a noticeable technological trend towards the large-scale construction and expansion of pumped-storage facilities. The industry is set to implement projects with a capacity exceeding 35 GW [17]. Long-term storage is a key technological focus. Pumped-storage power plants are specifically being developed to provide the technological backbone for long-term energy storage, helping to balance the grid over time scales ranging from weeks to months. Doubling the fleet: In the long term, a potential doubling of the total European pumped-storage capacity is anticipated by 2050 [18].

2. **Development of Hybrid Solutions (Hybridization)**:
   Hydropower is transitioning from an isolated technology to an integrated system partner. Hydropower plants are now closely integrated with other systems to offer new flexibility services. They are combined with other renewable energies, including wind farms and floating photovoltaic (PV) systems. Storage integration: Hydropower plants can be coupled with modern battery storage systems.

3. **Digital System Integration and Networking Virtual Power Plants (VPP)**: Hydropower plants are increasingly being integrated into virtual power plants using digital technology. This IT-based networking allows various types of plant (run-of-river, storage and pumped-storage power plants) to be controlled more efficiently as one unified system.

4. **Flexibility Across All Time Scales Expanding Responsiveness**: The technology is optimized and deployed to cover an extremely broad spectrum of system services–from ultra-fast grid interventions in the millisecond range to strategic energy management spanning several months.

These technological topics are closely related to the findings of Working Groups 1 and 2 (see Chapter "Technological Evolution Through Innovation and Digitalization" in Part III of this book) and the recently published position paper on hydropower from EERA [1], in which the COST Action participated during the consultation period. Technological developments for the near future (2030–2040) are more or less apparent now. Topics such as remaining lifetime, influence of start and stops on the degradation of the machine unit or faster transition modes play a key role in current research projects. Additional topics that should be mentioned are the expansion of operational fields towards part load operations and the use of axial machines for grid support.

Development on a longer time horizon will address river discharge forecasts and sediment control. Here, the development of appropriate tools and methodologies, together with validation measurements, will continue over the next decades (2030—onwards). Optimising river restoration alongside hydropower operation is one of the challenging tasks in this case. A combination of different disciplines as well as mutual understanding will be required in order to find common solutions. Sediment particles and machine maintenance time are normally mutually exclusive. The fewer particles in the discharged water, the longer the lifetime of the hydraulic components. As a result, a certain trade-off is required to accommodate both demands.

In the long term, technological developments will become more complex and multidisciplinary. Climate mitigation and adaptation measures will require a database which should be developed in time to be available from 2050 onwards. Climate change is part of a much longer history than any other development so far. In this case, the correlation between operational and climate data is significant. Even today, earth observation techniques provide a longer time history than any other sensor-based solution. The development of synchronising, harmonising and evaluating sequences should be a priority today. Clearly, the challenge here is to predict the data correlation sets needed in the coming decades. This means that creating longer time series of several correlated parameters requires an understanding of how things interact and correlate with each other, and basic investigation is necessary. A holistic view is also important and beneficial in this case.

1.2 Hydropower in the Energy System and Grid Interaction

Hydropower will always play a key role in the European energy system. Often referred to as "the forgotten giant [2]," it nevertheless provides the basis for our electricity consumption in many areas. However, it cannot be viewed in isolation but must always be considered in conjunction with other renewable energy sources. The role of hydropower within this mix ranges from important to secondary, depending on the country. However, this perspective is no longer applicable. In the future, there will be a greater focus on a more global approach to the issue. It will no longer be individual member states that are of interest, but rather Europe as a single economic entity. In order to maintain and expand the competitiveness of industrial locations, a constant and resilient electricity supply will be indispensable. To master the challenges of the future, we will need to move away from old paradigms [3]. Developments in hydropower in the context of the European energy system will require a more complex approach. In this context, European research institutions will be tasked with developing solutions for a sustainable future. Of course, these solutions must take environmental aspects and social acceptance into account. The social willingness to support these solutions will be an essential component of future research efforts. The fundamental general goals already defined in today's European research landscape naturally form a welcome foundation for this. These include goals such as:

1. Improving European strategic independence & resilient supply chains
2. Developing a resilient integrated energy system
3. Enhancing environmental sustainability
4. Deploying an efficient system integration
5. Guaranteeing a cost-effective energy system
6. Ensuring a faire energy transition.

Such targets and goals are based heavily on studies, reports, and policies related to the energy sector and geopolitics. In terms of cross-sectoral policies, these are: the REPowerEU [4], the European Green Deal Industrial Plan [5], the European Green Deal [6], the Net-Zero Industry Act [7]. In terms of energy, the following policies could be mentioned: the EU Strategy for Energy System Integration [8]; the Digitalising the Energy System—EU Action Plan [9]; the Energy Efficiency Directive [10]; the Renewable Energy Directive [11]; the EU Action Plan for Grids [12]; the Electricity Market Design Directive [13]; the Electricity Market Design Regulation [14]; the Energy Performance of Building Directive [15]; the Electrification Action Plan and the European Grids Package [16].

In the context of all European directives and expectations, hydropower plays a complementary role in electricity generation alongside solar and wind energy. Given the expected growth of new renewable energy sources such as solar and wind, hydropower will play an important role, though primarily for flexibility rather than pure baseload power generation. The ageing fleet plays a significant role here, as design specifications from decades ago barely meet today's requirements. The replacement or retrofitting of existing hydropower plants is being considered in many

Table 1 Perspectives of energy systems and grid interaction developments

Goal	Near future	Further ahead
Strategic independence & resilient supply chains	Identification and improvement of own technology leaderships, reduction of dependencies in the supply chain	Regaining technology leadership
Resilient energy system	System improvement for better resilience	Resilient operation of self-adaptive systems under higher stress
Sustainability	Greenhouse gas reduction, circularity, reduction of critical materials	Zero-emission processes
Efficient system integration	Development of digital twins, AI forecasting and united data platforms	Optimisation at system level, AI support
Costs	Handling of flexibility costs, system optimisation to reduce costs	Minimization of life-cycle costs, reduction of long-term prediction uncertainties
Fair energy transition	Creation of green jobs, development of energy communities	Professional development of skills, societal legitimacy, reduction of energy poverty

places. If and when these projects are carried out, it would certainly be worth contemplating a conversion of an existing plant into a storage facility. Fundamental research projects in this area are already underway, although much still remains to be done. Moreover, technological aspects are not the only factors at play–environmental and social issues are often just as important.

Europe's goals for the energy and grid systems of the future are based on key geopolitical objectives that are set to evolve gradually over the coming decades. These objectives aim at strengthening strategic competitiveness, building resilient energy systems, improving environmental sustainability, promoting system integration, adopting an economic efficiency approach, and supporting a fair energy transition for all population groups. Together, they outline a comprehensive path toward a secure, sustainable, and socially inclusive energy future.

The following table provides an overview of the broader topics that may be of interest in the near and long term. It is clear that the further ahead the time horizon, the more uncertain the predictions. Therefore, only tentative areas can currently be identified.

Hydropower can support the topics in Table 1 as follows.

Table 2 Hydropower support to energy systems and grid interaction perspectives

Goal	Near future	Further ahead
Strategic independence & resilience of supply chains	EU is hydropower technology leader, supply chain is manageable due to low raw material dependencies	Remain technology leader in EU through R&D investments
Resilient energy system	Support energy systems with zero carbon emission, increase resilience through flexible generation	Automated operation for a resilient future
Sustainability	Apply sustainability standard, reduce critical materials	Contribute to climate change adaption and mitigation
Efficient system integration	Integrate digital twins into system approach, apply AI forecasting and share data	Be part of energy system & vector optimisation
Costs	Contribute to lower flexibility costs, handle asset degradation costs in the system	Include long-term prediction uncertainties in operation and maintenance
Fair energy transition	Increase employment in this field, become part of energy communities	Increase awareness and strengthen public confidence in the technology

2 Hydropower and Digitalization

In this section, we focus primarily on the future of digitalisation. The most recent advancements are detailed in the book. Over the years, WG2 has discussed various facets of digitalization, ranging from standard practices to further innovations both within and beyond the powerhouse. A significant amount of digitalization occurs at machine level, with an increasing number of sensors and applications being implemented to facilitate predictive maintenance. The process of gathering data from multiple sources and storing it in shared data lakes is beginning. These data lakes will form the basis of standard optimisation processes and interactions between vendors and AI (Table 2).

However, all these areas of digitalization in hydropower will require some general standardization in order to work with the vast amount of data. The following aspects can be helpful in this regard:

- Establish shared data infrastructures to ensure data accessibility
- Facilitate federated energy data spaces
- Create incentives for data holders to share their data
- Develop universal interoperability standards for smart meters
- Pave the way for pan-European smart metering standard initiatives
- Encourage standardized data structures and models within the energy sector
- Expand European data hubs to accommodate energy-related datasets
- Create a cyber resilience action plan.

This tentative list is based on the shared use of energy and operational data. Only by providing appropriate data infrastructures to guarantee data availability at all times will digitalization in the energy sector continue to advance. For this, certain incentives will be needed to enable the respective data owners to make the data publicly accessible. Only through access to effective, real production data, linked to additional optimisation parameters, will a comprehensive European optimisation of the energy system be possible in the first place. Currently, this stage of process optimization is at the level of the Member States. An important aspect in this context is the precise measurement and determination of individual consumption. Smart meters are able to provide measurements at very short intervals, down to the minute or even second. However, to ensure European comparability, a standardization of data structures and models is needed, for only then can energy industry comparisons and energy system optimisation be undertaken. It goes without saying that such a development must be safeguarded by an appropriate cyber security directive. "Data" is far too valuable a commodity to be handled carelessly.

On the basis of these guidelines, it is necessary to define objectives for different time horizons. As in many other cases, these can be classified as short-, medium- and long-term goals. The industry already faces a significant backlog in digitalisation. However, when it comes to the shared use of acquired data, we are still in the early stages of development. Collaborative use of acquired data within one's own assets is far less prevalent than in other sectors. Nevertheless, the critical infrastructure aspect must also be considered, as not all data can be freely accessible. New standards for data structures must therefore be developed to enable exchange and, consequently, the further development of tools and methods based on science. These overarching objectives will shape digitalisation in this area for years and decades to come. Ultimately, they will also lay the groundwork for developing new AI tools and enable a new level of optimisation.

3 Hydropower and Artificial Intelligence (AI)

The use of Artificial Intelligence (AI) in the hydropower domain leads to a profound transformation, evolving from experimental pilot projects into a standardized operational practice. Since the greater weather volatility as a consequence of climate change and the rapid expansion of wind and solar energy both require significant grid flexibility, AI has become indispensable for modern hydropower operators.

A look at the current state of AI adoption in hydropower highlights several key application areas and industry realities:

1. Advanced Inflow and Hydrological Forecasting
2. Automated Production Planning and Market Trading
3. Asset Optimization and "Digital Twins"
4. Co-Optimization of Hybrid Systems.

3.1 Advanced Inflow and Hydrological Forecasting

Traditional statistical methods rely on 'historical normals', but these are becoming increasingly inaccurate due to uncharacteristic droughts, floods and extreme heatwaves. AI is fundamentally changing this:

- **Deep Learning Models**: Modern architectural models such as Long Short-Term Memory (LSTM) and Transformers analyze massive datasets–such as real-time satellite imagery, IoT sensor data, and radar data. By mapping complex, nonlinear relationships between weather conditions, soil moisture, vegetation, and elevation data, these models significantly improve the accuracy of forecasts for inflows into rivers and reservoirs.
- **Hybrid Frameworks**: To ensure forecast reliability, many users still rely on a dual approach. This approach combines data-driven machine learning with physically grounded hydrological laws. The method ensures that forecasts remain grounded in physical reality while also leveraging the benefits of advanced AI-based pattern recognition.

In this area of forecasting, artificial intelligence can reevaluate relationships, as it can analyse and present large amounts of data quickly. In the future, it will become increasingly important to better analyze and understand these complex relationships. Only in this way can we optimize the operation of hydropower plants.

3.2 Automated Production Planning and Market Trading

The role of hydropower has evolved from that of a steady baseline generator to that of a "flexibility champion", balancing the daily and sub-hourly volatility of intermittent renewables such as wind and solar.

- **Closed-Loop Bidding**: Leading operators now feed AI hydrological models directly into energy market trading desks. When a weather forecast changes, AI automatically triggers a recalculation of reservoir dispatch and updates market bids in a rolling, real-time cycle.
- **Predicting Grid Anomalies**: Utility leaders are deploying AI to forecast complex grid loads. Large utilities, for instance, have successfully integrated AI models to accurately anticipate demand anomalies during extreme weather–preventing grid strain where legacy mathematical models failed.

New market models that also incorporate additional parameters affecting generation prices will play a significant role in the future. In particular, due to the volatile generation of other energy sources, hydropower will face ever-increasing demands for flexibility. These new tools can help us better understand the trade-off between profit and loss of asset efficiency. AI tools can also assist with multi-parameter optimization because they reach an optimum much faster than conventional solvers. In the future, these novel tools may be integrated directly into the machine controller, allowing the machine to be operated according to specific objective functions. Such objective functions could be either maximum output or optimal component lifespan. All of this is done to maximise profit through the spot market.

3.3 Asset Optimization and "Digital Twins"

AI is actively being used to extend the lifespan of heavy mechanical infrastructure. It is also being used to optimize physical water flow.

- **Digital Twins**: High-fidelity virtual replicas of entire hydropower networks are used to run simulations for asset refurbishment, environmental flow protection, and greenfield investments.
- **Predictive Maintenance**: Machine learning algorithms continuously screen vibration and temperature data from turbines and generators to flag wear-and-tear before a mechanical failure causes an expensive blackout.

3.4 Co-Optimization of Hybrid Systems

AI-powered control of hybrid hydropower plants offers a decisive advantage: it maximizes market revenue while preserving the physical infrastructure. As the volatile feed-in from wind and solar energy places an increasing strain on power grids, operators must respond flexibly. While the combination of a hydroelectric power plant and a large-scale battery energy storage system (BESS) provides the technological foundation for this, only AI acting as a central control unit can ensure efficient operation. Since heavy water turbines are designed for constant power generation, the constant cycling on and off to capture short-term price spikes leads to massive wear and tear and enormous maintenance costs.

This is where AI acts as a protective shield: It continuously analyzes grid data and market prices in real time and delegates tasks for grid stabilization–such as frequency control–to the responsive battery on a second-by-second basis. This allows the turbines to operate in the highly efficient continuous mode. By linking inflow data and market trends, the AI also automates trading decisions to store or release energy at the economically optimal time. This creates a synchronized ecosystem in which the battery provides speed, hydropower provides endurance, and AI provides maximum efficiency.

3.5 AI Poll of CA21104

A survey about AI and Forecasting tools in the field of Hydropower was conducted in April 2026 in the course of the Naples workshop of the COST Action. The survey was divided into several sections in order to structure the framework and explore the topics of Artificial Intelligence (AI) and Forecasting. The core of this list details a multi-section questionnaire that systematically evaluates the industry's relationship with AI, covering the following sequential topics:

- **Section 1**: Background & Context–Corresponds to the statistical data of the participants
- **Section 2**: Current Tools & Methods–Assesses the existing technologies and practices currently utilized for Forecasting
- **Section 3 & 4**: Awareness, Use, and Perceived Value of AI–Investigates how familiar industry professionals are with AI, how they are currently deploying it, and the specific value they believe it brings to hydropower operations.
- **Sections 5 & 6**: Barriers, Decision-Making, and Trust–A closer look at the primary concerns, challenges, and limitations that prevent wider adoption, combined with an exploration of how trust can be established to enable AI to influence critical decision-making processes.
- **Section 7**: Future Outlook–Gathers perspectives on the long-term time frame and evolving role of AI in the sector.
- **Environmental Impact and Prioritization**: The final segment of the text addresses a critical sustainability paradox. This specific inquiry focuses on the future outlook for AI expansion, specifically highlighting the significant irony that greater reliance on AI will simultaneously trigger a massive surge in electricity and water consumption.

The following paragraphs synthesize the survey findings. A cohesive overview of the following issues is provided: who is steering this technology, how does current infrastructure perform, what limitations persist, and how does the relationship between human decision-makers and automated algorithms evolve.

3.5.1 Background and Context

The survey data reveal that the persons engaging with AI in this field are highly specialised, balancing academic research with direct industrial application.

Industry and Domain Distribution

The respondents primarily operate at the intersection of energy production and advanced engineering:

- **Academia & Research**: Strong presence from universities and dedicated R&D departments in electrical engineering and hydropower.

Table 3 Question: how often do you make forecasts or predictions as part of your job?

Frequency of forecasts	Percentage (%)
Daily	9
Weekly	27
Monthly	23
Rarely	36
Never	5

- **Industry & Engineering**: Professionals spanning hydropower operation, power utilities, power transmission, small hydropower, mechanical engineering, and hydraulic structure design.
- **Emerging Tech**: Cross-disciplinary presence from Earth Observation and Climate Technology sectors.

Key Roles

The survey gathered insights from both senior decision-makers and primary researchers. This cohort includes Professors, Associate Professors, and Research Scientists, as well as corporate leaders such as CEOs, Heads of R&D Departments, Heads of Divisions, Project Managers, and technical experts. Predictive modelling is an integral part of their daily routines, albeit with varying frequency, and is used by senior electrical engineers and young researchers and innovators.

Forecasting Frequency and Type

Predictive modelling is an integrated component of their daily routines, though the frequency varies (Table 3).

Interestingly, when asked about the types of forecasting they conduct, standard corporate categories such as Demand (14%), Financial (9%), Sales (5%), and Risk (5%) scored relatively low. A striking 68% categorised their predictive tasks as 'other', which suggests the use of domain-specific physical modelling for tasks such as hydrological runoff, reservoir inflows, and turbine performance matrix calculations.

3.5.2 Current Tools and Methods

Current Tool Usage

In the evaluation of baseline infrastructure, specialized and dedicated tools demonstrate a slight advantage over manual spreadsheets (Table 4).

When asked about their satisfaction with the predictive tools they use, respondents stated the following: The overall user satisfaction score is a modest 3.5 out of 5. The largest group of users (44%) rated their satisfaction as a '4', while 45% rated it as a '3' or lower.

Table 4 Question: what tools do you currently use for forecasting?

Tools	Percentage (%)
In-house/custom tools	30
Dedicated forecasting software	25
AI/ML-based tools	25
Spreadsheets (e.g., Excel)	20

Table 5 Question: are you currently using AI-powered forecasting tools?

Answer	Percentage (%)
YES	58
NO	37
Not sure	5

System Limitations
The technical friction points that prevent satisfaction from reaching higher levels are heavily influenced by data constraints and physical gaps:

- **Data Deficits**: A universal bottleneck exists due to a *lack of sufficient data* and *in-situ data availability for training ML models*, plus the general *unreliability of input databases*.
- **The "Physics" Gap**: Open-text feedback strongly highlighted a *lack of insight into physics* within standard models. Traditional statistical engines often fail to accurately model the sheer complexity of highly interconnected, non-linear hydraulic systems.
- **Boundary Conditions**: Current systems suffer from reduced accuracy when forced to project under *'exceptional conditions'* or extreme weather anomalies.

3.5.3 Awareness, Use, and Perceived Value of AI

Despite data constraints, AI has moved past theoretical exploration and is actively deployed in the sector (Table 5).

Of those utilising AI, 50% use *'Generative AI'* (e.g. for scenario simulation), followed by 33% who use *'Predictive analytics platforms'* and 17% who use *'Automated time-series forecasting tools'*. Crucially, users admit that their familiarity with these systems is still evolving.

Over half of the respondents (56%) self-identify as *'Beginners'*, 28% as *'Intermediate'*, and only 17% claim *'Advanced understanding'*. 0% checked the *'Expert'* box. This gap underscores a significant issue: practitioners are implementing advanced AI

Table 6 Question: Would you rely on AI forecasts for high-stakes decisions?

Answer	Percentage (%)
YES	10
NO	70
DEPENDS	20

tools although they are still only in the process of acquiring a better understanding of their functionality.

Perceived Value vs. Practical Friction

What is driving this adoption? AI is perceived to offer strong structural benefits across three primary vectors: identification of obscured patterns, time savings, and enhanced raw numerical precision. User Agreement Scores (Out of 5).

- **Identifying Hidden Patterns**: 4.3/5 score. A staggering 58% of respondents awarded a score of 4 out of 5, agreeing that AI is excellent at identifying complex systemic relationships that human observation misses.
- **Time-Saving Capabilities**: 3.9/5 score. Users enjoy faster data processing, automated screening, and streamlined reporting structures.
- **Improving Accuracy**: 3.8/5 score. AI enables a holistic assessment of complex variables to achieve better overall optimization.

3.5.4 Barriers, Decision-Making, and Trust

Barriers to Full Adoption

Conversely, significant structural concerns prevent complete integration. The primary concerns are split fairly evenly between *'Over-reliance on automation'* (35%) and *'Data privacy'* (35%), alongside *'Lack of transparency'* (15%), *Cost* (10%), and *'System integration hurdles'* (5%).

When asked what specifically prevents them from adopting these systems, professionals pointed directly to AI hallucinations, data manipulation/corruption, a lack of clarity regarding the source of data, and the fundamental question of accountability ("who is responsible" when an automated forecast fails).

The Trust Dilemma in High-Stakes Environments

When it comes to high-stakes decision-making–where a bad forecast can compromise grid stability or damage multimillion-dollar civil infrastructure–the industry exercises extreme caution. When asked how much they trust AI compared to human intuition, 55% say it is *'About the same'*. The remaining responses are divided: 15% trust AI *'More'*, while 30% trust it *'Less'* or *'Much less'* (Table 6).

The open-text explanations for the *'Depends'* demographic further clarify this defensive attitude. Industry professionals view AI strictly as an auxiliary asset, not as an autonomous pilot. They state that AI will only be trusted if it is used to complement human decision-making processes, if it occurs only after expert validation or if it is vetted via extensive testing across different simulated scenarios. Human interaction and legacy operational experience remain non-negotiable insurance policies.

3.5.5 Future Outlook

Looking to the future, the momentum behind AI integration shows no signs of stopping. Respondents indicated a very high likelihood of increasing their use of AI tools over the next two years, giving it an impressive score of 4.1 out of 5.

Desired Future Features
To unlock maximum value, professionals are asking for targeted capabilities:

- Deep integration with physics-based tools and models.
- The structural ability to test alternative policy interventions, regulatory adjustments, and shifting economic outcomes.
- Higher fidelity data pipelines and easier onboarding systems for existing workflows.

The consensus regarding AI's future role is massive. Respondents continuously used terms such as *'a necessary pillar combined with digitalization'*, *'will dominate'*, *'a principal role'*, and a tool that will eventually be utilized across all stages of forecasting.

The Resource & Geopolitical Trade-off
The survey concluded with a thought-provoking question: while AI has the potential for significant growth, this would come at the cost of a substantial ecological footprint, specifically a huge increase in electricity and water consumption. Respondents were asked to rank a series of geopolitical and operational statements by importance (on a scale of 1 to 4) to express their views.

1. **Strategic Autonomy (Score: 3.93)**:
   'This is important for ensuring Europe's independence in this area. I support this even if I have no advantage out of it'.
2. **Operational Reality (Score: 3.60)**:
   'I know that this is important and it will have a significant impact on my daily work.'
3. **Hardware Sovereignty (Score: 3.53)**:
   'Geopolitically, Europe plays a minor role in the development of AI hardware and tools. This should change.'
4. **Infrastructure Expansion (Score: 3.13)**:
   'European policies aim to increase the number of data centres threefold. This is important, and I support it.'

3.5.6 Environmental Impact and Conclusion

The data paint a clear picture: In hydropower and energy supply, the future of resource forecasting is digital. Even in the face of inherent risks–such as the massive ecological footprint of AI systems and their theoretical decoupling from real physical processes–the European professional community stands united behind the expansion of its own data infrastructure. It views the use of local resources as a necessary sacrifice for the continent's strategic autonomy. However, the future-proof system architecture will not be purely AI; rather, it will be based on a deeply integrated hybrid model that respects both the laws of nature and the irreplaceable control function of humans.

4 Hydropower in the Context of Long-Term Flow and Sediment Modelling Under Climate Change

The future of hydropower toward 2050 depends strongly on our ability to understand and predict long-term changes in river systems. Climate change is expected to significantly alter hydrological regimes through shifts in precipitation patterns, increased frequency of extreme events, glacier retreat, and changes in sediment supply. These processes introduce substantial uncertainty into both flow and sediment dynamics, making long-term modelling an essential tool for sustainable planning and operation of hydropower systems.

Conventional approaches, such as hydrological models and one-dimensional (1D) hydraulic models, remain useful for large-scale assessments of discharge and water levels. However, they are often insufficient for capturing the spatial variability and complexity of sediment transport and river morphology. In contrast, two-dimensional (2D) hydraulic and morphodynamic models provide a more detailed and physically representative description of flow patterns, sediment transport processes, and bed evolution. This makes them particularly valuable for long-term analysis under changing climate conditions.

Despite these advantages, the application of 2D sediment modelling over long time horizons–such as 50 years or more–remains challenging. One of the primary limitations is the high computational demand. To make such simulations feasible, modelers often rely on simplifications and acceleration techniques. For example, discharges below the critical threshold for sediment transport can be excluded from the input hydrograph, reducing simulation time without significantly affecting overall sediment dynamics [8]. In addition, the use of morphological acceleration factors enables scaling of sediment processes, effectively shortening simulation periods while preserving long-term trends. However, these approaches require careful calibration and a clear understanding of their limitations to avoid introducing bias.

Model setup also involves important trade-offs. The selection of mesh size, for instance, directly influences both accuracy and computational efficiency. Similarly, the choice of key parameters–such as the Manning roughness coefficient, sediment transport functions, hiding and exposure corrections, and bed roughness predictors–can significantly affect model outcomes [19]. Achieving a balanced and reliable model therefore requires both technical expertise and physical insight.

A major constraint in long-term sediment modelling is the limited availability of high-quality calibration data. In many regions, sediment measurements are sparse, discontinuous, or entirely lacking. This creates uncertainty in model predictions and limits confidence in future projections. Addressing this issue requires stronger collaboration between researchers, industry stakeholders, and authorities to improve monitoring strategies and expand data collection efforts.

At the same time, new technologies provide opportunities to overcome some of these limitations. Remote sensing tools, including drones and satellite imagery, allow for improved spatial and temporal observation of river systems. When combined with numerical modelling, these data sources can enhance model calibration and validation, leading to more reliable long-term simulations under different climate scenarios.

From an educational perspective, these challenges emphasize the importance of training future engineers with a strong and integrated understanding of hydrology, hydraulics, and sediment dynamics. Beyond technical skills, there is a need to develop critical thinking regarding model assumptions, uncertainties, and limitations. Future professionals must be equipped not only to use advanced modelling tools but also to interpret their results responsibly in the context of long-term decision-making.

In the vision of hydropower in 2050, improving long-term flow and sediment modelling is essential for designing resilient and adaptive systems. Equally important is ensuring that the next generation of engineers is prepared to address these challenges, bridging the gap between advanced modelling techniques and practical, sustainable solutions.

5 Future Perspectives and Visions

During the International Conference of COST Action CA21104 in Naples from 15–17 April 2026, a round table on the topic "Vision on Hydropower's Future" was held with representatives from industry and various associations. Together, a vision for the future use of hydropower in the context of other stakeholders within the Nexus framework was outlined. This vision is complex, multidisciplinary, and multifunctional. The following passage provides a summary of this event.

The discussion was framed as a forward-looking exchange on the future of hydropower, focusing on long-term developments over a time horizon of 20 to 50 years. The focus is on exploring visions, expectations, and potential roles of hydropower within a changing energy system. From the outset, hydropower is considered not as an isolated technology, but as part of a broader, interconnected system that includes water management, energy production, policy frameworks, and societal factors.

A central theme emerging early in the discussion is the multidisciplinary nature of hydropower. It becomes evident that the sector increasingly requires collaboration across engineering, environmental sciences, policy, and social sciences. This multidisciplinary character is described as both a strength and a challenge. Future developments are expected to depend strongly on closer cooperation between research and industry, particularly in areas such as operation, monitoring, and maintenance of hydraulic machinery. At the same time, technological advancements alone are not sufficient. Environmental and societal aspects must be integrated into future solutions. Additionally, the geographically dependent nature of hydropower is emphasized, meaning that solutions must be adapted to specific local conditions rather than simply replicated.

The discussion broadened to include the perspective of water as a resource within the water-energy nexus. Water is increasingly recognized not only as a resource for energy production but also as a structural factor for economic stability, social cohesion, ecosystem integrity, and geopolitical resilience. In this context, hydropower occupies a unique position at the intersection of water and energy systems. However, it is also noted that the importance of water for energy systems is not yet fully acknowledged in policy-making, highlighting the need for stronger integration. Regulatory frameworks, particularly environmental regulations, are identified as both constraints and opportunities, requiring balanced approaches that ensure the continued role of hydropower while addressing ecological concerns.

Economic considerations played a significant role in the discussion. Hydropower, especially new large-scale projects, is often perceived as capital-intensive. Pumped storage is highlighted as a key future application, particularly in the context of increasing renewable energy penetration. However, significant challenges remain, including environmental constraints, limited suitable locations, and high investment costs. In addition, the applicability of pumped storage is considered to be limited in marine environments due to saltwater-related issues such as increased corrosion and biological growth (e.g. algae), which can negatively affect infrastructure durability and maintenance requirements. Innovative concepts, such as underground storage systems using tunnels, are introduced as potential alternatives, although uncertainties remain regarding their practical feasibility, efficiency and associated costs.

A central debate developed around the comparison between pumped storage hydropower and battery storage technologies. Batteries are described as rapidly advancing, with increasing efficiency, decreasing costs, and technological innovations such as sodium-ion batteries. These developments raise concerns about the long-term competitiveness of hydropower-based storage. At the same time, hydropower offers distinct advantages, including long lifetimes, large-scale storage capabilities, and the provision of essential system services such as grid stability and inertia. The discussion converged on the view that both technologies are likely to coexist, as they fulfill different roles within the energy system and rely on different resources and value chains.

From a system-level perspective, the role of hydropower is closely linked to the broader transformation of the energy sector. A key assumption discussed is the increasing electrification of energy demand, potentially leading to a significant rise in the share of electricity in total energy consumption by 2050. Under such scenarios, the need for flexibility in the energy system is expected to increase substantially. Hydropower is identified as a critical contributor to this flexibility across multiple time scales, although it cannot meet these demands alone. Instead, it must be integrated with other technologies, including battery storage.

At the same time, the expansion potential for new hydropower plants in Europe is considered to be limited. The focus is therefore expected to shift toward modernization and refurbishment of existing facilities. However, electricity production remains a key concern in light of increasing electrification. Additional potential may arise from powering existing dams that are not yet equipped for energy generation. This would require new technologies to enable their electrification and highlights the role of continued research. Hybrid solutions combining hydropower and battery systems are also highlighted as a promising approach, allowing the strengths of each technology to complement the limitations of the other. Changes in the structure of the energy system are also discussed. Increasing decentralisation, driven by the rise of prosumers (i.e. consumers who also produce energy), local energy production, and decentralized storage solutions such as residential battery systems, are expected to reshape electricity markets. However, this does not eliminate the need for centralized infrastructure. Instead, a more complex system emerges in which local and large-scale solutions must coexist. In this context, hydropower is positioned as a key provider of flexibility at regional and national levels, contributing to grid stability in systems with high shares of variable renewable energy. In addition, technologies such as synchronous compensators can support decentralized grid stability at critical points within the grid.

Climate change introduces additional challenges. Hydrological patterns are already shifting, with increasing frequency of extreme events such as droughts and intense precipitation. While the total volume of water may remain relatively constant, its temporal distribution is becoming more uneven. This leads to operational challenges, including increased flooding risks and periods of reduced water availability. Adaptation strategies are therefore required, including improved forecasting methods, optimised operation, and potentially new infrastructure concepts.

Beyond energy production, the multifunctional role of hydropower is emphasized. Hydropower systems contribute to water management tasks such as flood control, water supply, and environmental services. This broader contribution is often underrepresented in public discourse, where hydropower is sometimes primarily associated with environmental impacts. A more balanced perspective is therefore suggested, recognizing both the benefits and challenges associated with hydropower.

Water scarcity is identified as a critical issue for the future. Changes in precipitation patterns and the reduction of glacier volumes are expected to affect water availability, with direct implications for hydropower generation. In some cases, this may require technical adaptations, such as resizing turbines or optimising operational strategies. At the same time, new opportunities may arise from improved utilization of peak water flows, which are currently not fully exploited.

The importance of digitalization was also highlighted. Tools such as digital twins of river catchments are presented as promising approaches for optimising water use across sectors. These models can integrate hydrological data, energy market conditions, and competing water demands, enabling more efficient and coordinated decision-making. Such approaches illustrate the increasing integration of hydropower into broader water resource management systems. From a research perspective, two complementary approaches are identified. On the one hand, research should address immediate industry needs and support practical innovation. On the other hand, exploratory research is necessary to develop new concepts and technologies, including unconventional methods of utilizing hydropower potential. Both approaches are considered essential for the long-term development of the sector.

Finally, broader societal and political factors were addressed. Investment decisions are often influenced by short-term economic and political cycles, which can conflict with the long lifetimes of hydropower infrastructure. This creates challenges for long-term planning and development. At the same time, policy frameworks, funding mechanisms, and international cooperation are identified as key factors influencing the future of hydropower.

Overall, hydropower is expected to remain an important component of the future energy system, although its role will evolve. It will function as part of an increasingly complex and interconnected system rather than as a standalone solution. Its main contributions lie in providing flexibility, supporting system stability, and enabling integrated water and energy management. Achieving this potential will require close collaboration across disciplines, sectors, and regions.

Acknowledgements The authors would like to thank all contributors who participated to the poll either being participants to the workshop or remotely present.

Competing Interests The author has no conflicts of interest to declare that are relevant to the content of this chapter.

References

1. European Energy Research Alliance (EERA) (2026) Joint program—hydropower: hydropower for a resilient, secure, and competitive Europe. Maximizing impact through FP10 and a dedicated SET plan implementation working group. https://eeraset.sharepoint.com/:b:/s/JP-HP/IQBwDd4WMUv7SJl-EXcPI9i7ASvn88MV9h0yfLKCTnLOWgE?e=o6qeXl
2. World Economic Forum (2021) Hydropower energy is the 'Forgotten Giant' of clean electricity. IEA says. https://www.weforum.org/stories/2021/07/hydropower-energy-low-carbon-clean-electricity-iea/
3. Grigoraş G, Scarlatache F, Comanescu D, Neagu BC (2017) Expert system for optimal power allocation in hydropower dispatchable units. https://doi.org/10.1109/optim.2017.7975035
4. REPowerEU (2022) Affordable, secure and sustainable energy for Europe. https://commission.europa.eu/topics/energy/repowereu_en
5. Communication: a green deal industrial plan for the net-zero age (2023). https://commission.europa.eu/document/41514677-9598-4d89-a572-abe21cb037f4_en
6. The European Green Deal (2020) Striving to be the first climate-neutral continent. https://commission.europa.eu/strategy-and-policy/priorities-2019-2024/european-green-deal_en

7. The Net-Zero Industry Act (2024) Making the EU the home of clean technologies manufacturing and green jobs. https://single-market-economy.ec.europa.eu/industry/sustainability/net-zero-industry-act_en
8. Powering a climate-neutral economy: an EU strategy for energy system integration (2020). https://eur-lex.europa.eu/legal-content/EN/ALL/?uri=COM:2020:299:FIN
9. Digitalisation of the energy system (2022) The EU is promoting the availability of safe, secure, and sustainable digital energy services. https://energy.ec.europa.eu/topics/eus-energy-system/digitalisation-energy-system_en
10. Energy Efficiency Directive (2023) Using energy more efficiently will contribute to reducing the EU's overall energy consumption. The Energy Efficiency Directive is therefore a key driver of Europe's energy transition. https://energy.ec.europa.eu/topics/energy-efficiency/energy-efficiency-targets-directive-and-rules/energy-efficiency-directive_en
11. Renewable Energy Directive (2024) The renewable energy directive sets targets to increase the EU's renewable energy use. It supports cooperation between EU countries towards this goal. https://energy.ec.europa.eu/topics/renewable-energy/renewable-energy-directive-targets-and-rules/renewable-energy-directive_en
12. Grids, the missing link—an EU action plan for grids (2023). https://eur-lex.europa.eu/legal-content/EN/TXT/?uri=COM:2023:757:FIN
13. Directive (EU) 2024/1711 of the European parliament and of the council of 13 June 2024 amending directives (EU) 2018/2001 and (EU) 2019/944 as regards improving the Union's electricity market design (2024). https://eur-lex.europa.eu/legal-content/EN/TXT/?uri=OJ:L_202401711
14. Regulation (EU) 2024/1747 of the European parliament and of the council of 13 June 2024 amending regulations (EU) 2019/942 and (EU) 2019/943 as regards improving the Union's electricity market design (2024). https://eur-lex.europa.eu/legal-content/EN/TXT/?uri=OJ:L_202401747
15. Energy Performance of Buildings Directive (2024) Aiming to achieve a fully decarbonised building stock by 2050, the revised energy performance of buildings directive contributes directly to the EU's energy and climate goals. https://energy.ec.europa.eu/topics/energy-efficiency/energy-performance-buildings/energy-performance-buildings-directive_en
16. A competitiveness compass for the EU (2025). https://commission.europa.eu/document/download/10017eb1-4722-4333-add2-e0ed18105a34_en
17. International Hydropower Association (IHA) & Eurelectric (2025) The Paris pledge: committing to pumped storage to secure Europe's clean energy future. Joint commitment by industry to double pumped storage hydropower (PSH) capacity by 2050. https://indd.adobe.com/view/385c4753-2b01-4a7f-8ab7-e2ae22ac0a99
18. Quaranta E, Boes R, Hunt JD, Szabo S, Tattini J, Pistocchi A (2024) Considerations on the existing capacity and future potential for energy storage in the European Union's hydropower reservoirs and pumped-storage hydropower. J Energy Storage **104**, 114431. https://doi.org/10.1016/j.est.2024.114431 (JRC140280)
19. Balouchi B, Rüther N, Schwarzwälder K (2024) Temporal variation of braided intensity and morphodynamic changes in a regulated braided river using 2D modeling and satellite images. River Res Appl 40(11). https://doi.org/10.1002/rra.4268
20. Balouchi B, Mirzaahmadi A (2025) Assessing the effect of a morphological acceleration factor on 2D morphodynamic modeling in a braided river. In: Proceedings of the 41st IAHR world congress–innovative water engineering for sustainable development, Singapore. https://doi.org/10.64697/978-90-835589-7-4_41WC-P2046-cd